# Question Bank in Food Science

## ABOUT THE AUTHORS

**Er. A. Poshadri** is currently working as Assistant Professor (Food Technology) at Professor Jayashankar Telangana State Agricultural University, Hyderabad. Some of the milestones in his career are listed below:

- Established College of Food Science & Technology, Rudrur, one of its constituted college of Professor Jayashankar Telangana State Agricultural University, Hyderabad
- Founder member of Food Research Laboratory (NutriPlus Knowledge Programme) in ICRISAT, India
- Established Spices and Seasonings Division R&D in Bambino Agro Industries Ltd, Hyderabad
- Secured All India First Rank in ICAR-PG (JRF) examination 2007 (Roll No. 100258)
- Cleared Agricultural Research Service (ARS)-2012 examination of Agricultural Scientist Recruitment Board (ASRB) of ICAR-New Delhi (Roll No.13550034)
- Secured All India Third Rank in ICAR-PhD (SRF) examination 2016 (Roll No. 159144)
- Secured All India Second Rank in ICAR-PhD (SRF) examination 2017 (Roll No. 159169)
- Cleared IARI, New Delhi, Ph.D entrance examination-2017 (Roll No. 2210031)
- Cleared Osmania University, Hyderabad, PhD entrance examination-2017 (Roll No. 633120016)

**Dr. T. V. N. Padmavathi** Ph.D (Food Science & Nutrition) is working as Assistant Professor (Food Safety and Quality Assurance) at Professor Jayashankar Telangana State Agricultural University, Hyderabad. She is also empanelled member for FSSAI standards and regulations for cereals, pulses and oil seeds.

**Dr. Aparna Kuna,** Ph.D is working as Sr. Scientist in MFPI – QualityControl Laboratory, Professor Jayashankar Telangana State Agricultural University, Hyderabad, India. Her research centers around functional foods, value addition to underutilized foods, extrusion cooking and food quality control.

**Dr. P. Ramesh** Ph.D (Mathematics) is working as Assistant Professor (Engineering Mathematics and statistics) at Professor Jayashankar Telangana State Agricultural University, Hyderabad. He is passionate about helping students and young researchers on statistics and data analysis to get good interpretations in agriculture and allied sciences in the university.

# Question Bank in Food Science

A. Poshadri
T.V.N. Padmavathi
Aparna Kuna
P. Ramesh

nipa® books for everyone
PAPERBACKS

nipa books for everyone
PAPERBACKS

*A Paperback Division of*
**NEW INDIA PUBLISHING AGENCY**
101, Vikas Surya Plaza, CU Block, LSC Market
Pitam Pura, New Delhi 110 034, India
Phone: + 91 (11)27 34 17 17 Fax: + 91(11) 27 34 16 16
Email: info@nipabooks.com
Web: www.nipabooks.com

Feedback at feedbacks@nipabooks.com



ISBN 978-93-87973-33-6



Composed, Designed & Printed in India

# Preface and Acknowledgements

Food processing is one of the emerging sectors in India with immense potential to contribute to achieve the laudable objective of Government of India to “Double Famer’s Incomes by 2022". The consumption of processed food in India is gaining greater momentum and the demand for healthy, safe, hygienic and convenience driven food is slated to increase at an even more significant pace in the future. The combination of traditional Indian food, with modern technology, processing and packaging, can help the world rediscover the health benefits, and refreshing taste of Indian food ingredients. In the years to come, it is expected that Indian food sector has the potential to create four million additional jobs and entrepreneurial opportunities by 2025, to youth and women. In view of the above advantages, competent dynamic human resources are required in the field of Food Science and Technology in the country. This book “Question Bank in Food Science” is an important readymade tool covering frequently asked multiple choice questions from topics like Cereals, Pulses, Oil seeds, Fruits and Vegetable Products, Milk and Milk Products, Poultry, Meat and Fish Products, Food Microbiology, FSSAI rules and regulations, Food Chemistry and Nutrition, Food Engineering, Sensory evaluation and Product development, Food Safety and Quality Assurance.

This book is a compilation of more than 6500 Multiple Choice Questions (MCQ’s) which can serve as a readymade referral material for the students preparing for competitive exams like GATE, ICAR – PG(JRF), ICAR-Ph.D (SRF), ICAR-NET Examination, ICAR-ARS Examination, P.G entrance examinations of CFTRI, SAUs,SLIT, IICPT, NIFETM, UICT. Italso serves as an extremely important resource material for those preparing for other competitive examinations like Food Safety Officer, Food Analyst, Food Inspector and other related State/Central service examinations.

We would like to express our sincere thanks to our teachers, faculty and students who helped us in all possible ways to bring this book into useful form.

Special appreciation is extended to the team of New India Publishing Agency, New Delhi for accepting to publish this book.

# Contents

# 1
# Cereals, Pulses and Oil Seeds

1. Rice is a

   a) Edible fruit
   b) Edible Vegetable
   c) Edible flower
   d) All

2. Major components in cereal seeds are

   a) Endosperm
   b) Germ or embryo
   c) a & b
   d) None

3. Bran is composed of

   a) Pericarp
   b) Testa
   c) aleuron layer
   d) All

4. Oats contains ______________ percent of hull

   a) 50
   b) 40
   c) 30
   d) 20

5. Cereals are the main source of

   a) Energy
   b) Protein
   c) Carbohydrates
   d) a & c

6. ______________ percent of dry matter of cereals is carbohydrate.

   a) 80
   b) 70
   c) 60
   d) 50

7. Cereals contain ____________ percent of protein

   a) 2-20
   b) 6-12
   c) 5-40
   d) 1-40

8. Cereals are deficient in ______________but rich in methionine

   a) Glycine
   b) Lysine
   c) Proline
   d) All

9. Pulses are rich in lysine but deficient in _________
   a) Glycine b) Lysine
   c) Proline d) Methionine
10. Wheat germ contains _______________ percent lipids
   a) 6-11% b) 20-25%
   c) 1-15% d) 4-25%
11. Lipids in cereals are mostly the tiglycerides of
   a) Palmitic b) Oleic
   c) linoleic acid d) All
12. Phosphorus in cereals is present in the form of ______
   a) Phytin b) Tannin
   c) Phosphate d) Saponin
13. Presence of Phytates in cereals decrease the absorption of
   a) Iron b) Calcium
   c) Magnesium d) All
14. Ragi is a rich source
   a) Iron b) Calcium
   c) a & b d) Magnesium
15. Cereal grains oils are rich in
   a) Vitamin-C b) Vitamin-E
   c) a & b d) None
16. ________ is a phenomenon of equilibrium moisture Characteristics of cereal grains, reached by adsorption/sorption or desorption of the moisture
   a) Hysteresis b) Isotherm
   c) Relative humidity d) Water activity
17. Drying of the wet solids in a suspended or fluidized state is called as
   a) Tray drying b) Fluidized bed drying
   c) Pneumatic drying d) Both b & c
18. The first cleaning operation of paddy after threshing is called
   a) Scalping b) Grading
   c) Sorting d) Winnowing

19. In general freshly harvested paddy has a moisture content of
    a) 18-25% b) 12-14%
    c) 20-30% d) 30-40%
20. For longer storage, the grains should be dried to a moisture level of ________ percent
    a) 14 b) 13
    c) 15 d) 18
21. The huller mill combines ______________ and ___________ unit operations
    a) Dehusking b) Polishing
    c) Destining d) a & b
22. The average milled rice recovery from raw rice in huller mill is
    a) 85% b) 75%
    c) 65% d) 70%
23. Destoner is working based on
    a) Surface tension b) Denisty
    c) Viscosity d) Length
24. The rubber –roll sheller consists of two rubber rollers rotating
    a) Same direction b) Opposite direction
    c) a & b d) None
25. In rubber –roll sheller, shelling (dehusking) efficiency is
    a) 65 b) 55
    c) 75 d) 85
26. The average husk content in paddy is
    a) 20-22% b) 20-30%
    c) 15-30% d) 25-40%
27. More the husk is
    a) Higher the rice b) Lower the rice
    c) No change in yield d) Both a & b
28. By law in India the degree of milling of rice is fixed at
    a) 4-5% b) 1-2 %
    c) 3-8% d) 4-10%

29. In practice rice is milled at as low as
    a) 4-5%  b) 1-2 %
    c) 3-8%  d) 4-10%
30. Raw rice bran contains _________ percent of oil
    a) 14-20  b) 18-25
    c) 15-30  d) 15-35
31. Parboiled rice bran contains _________ percent of oil
    a) 14-20  b) 18-25
    c) 15-30  d) 15-35
32. The free fatty acids content in crude brain oil should not be more than _________ percent
    a) 5  b) 8
    c) 10  d) 15
33. _______________ is a hydrothermal treatment followed by drying before milling
    a) Steaming  b) Instantization
    c) Parboiling  d) Wet cooking
34. Parboiling imparts
    a) Gelatinization  b) Disintegration of protein
    c) a & b  d) None
35. Parboiled rice takes a _______ time to cook to the same degree of softness than raw rice of the same variety
    a) Lesser  b) Longer
    c) Equal time  d) All
36. Parboiling involves
    a) Soaking  b) Steaming
    c) Drying  d) All
37. The gelatinization temperatures of rice is
    a) 65-75°C  b) 75-85°C
    c) 55-75°C  d) 65-85°C
38. Desirable final moisture content in parboiled paddy is _______ percent
    a) 15-18  b) 14-16
    c) 15-20  d) 14-20

39. ___________ is prepared by throwing rice into sand heated to a high temperature in an iron or Earthen pan.
   a) Parched rice  b) Puffed rice
   c) Flaked rice  d) Poha
40. _________________ is preferred in making of parched rice.
   a) Raw rice  b) Parboiled rice
   c) Both a & b  d) None
41. Old rice has
   a) Increased levels of amylase  b) Increased levels of amylopectin
   c) Both a &b  d) None
42. Puffed rice from paddy is produced by heat treatment at _____________ temperature for 30 -40 sec
   a) 200-270°C  b) 250-270°C
   c) 250-350°C  d) 200-350°C
43. Flaked rice is popularly known as
   a) Parched rice  b) Puffed rice
   c) Parboiled rice  d) Poha
44. Instant rice processing involves
   a) Pre-cooking  b) Soaking
   c) Dehydration  d) a&c
45. The flour from the weak wheat is ideal for
   a) Pasta  b) Biscuits and cakes
   c) Macaroni  d) Bread
46. As per FSSAI regulations, the moisture content in Rice is not more than _____________ percent by weight
   a) 13  b) 14
   c) 16  d) 15
47. Whisky is made by fermentation and distillation of fermented
   a) Grapes  b) Potato
   c) Cereal grains  d) Molasses
48. Average yield of oat meal is
   a) 50 %  b) 57%
   c) 60  d) 67%

49. Bread flour should contain in excess of _____________ percent of protein

a) 10.5 b) 8.6

c) 9.5 d) 8.8

50. The marketing guide recommends minium falling number index for bread making is

a) Above 150 b) Above 250

c) Below 250 d) Below 100

51. The marketing guide recommends minium falling number index for biscuit making is

a) Above 140 b) Above 100

c) Below 140 d) Below 100

52. The hard spring and hard winter wheat's are the types most desirable for production.

a) Pasta b) Biscuits and cakes

c) Macaroni d) Bread

53. Durum wheat is suitable for the production of

a) Pasta b) Biscuits and cakes

c) Cakes d) Bread

54. Endosperm content in wheat is

a) 72.5% b) 82.5%

c) 92.5% d) 62.5%

55. From 100kg of wheat, 72% extraction is known as _______ straight flour consisting of all the streams.

a) 100% b) 72%

c) 80% d) 90%

56. First Patent constitutes __________ separation from 72% extraction

a) 80% b) 70%

c) 100% d) 90%

57. Short Patent constitutes __________ separation from 72% extraction

a) 100% b) 72%

c) 80% d) 90%

58. Medium Patent constitutes __________ separation from 72% extraction
   a) 100%   b) 72%
   c) 80%   d) 90%

59. Long Patent constitutes __________ separation from 72% extraction
   a) 100%   b) 95%
   c) 80%   d) 90%

60. First Patent is used as __________ and is obtained from soft wheat's
   a) Bread flour   b) Cake flour
   c) Pasta flour   d) Vermicelli

61. Short Patent flour is used for
   a) Premium brands of breads   b) Featured brands of breads
   c) Competitive brands of breads   d) All

62. Medium Patent flour is used for
   a) Premium brands of breads   b) Featured brands of breads
   c) Competitive brands of breads   d) All

63. Long Patent flour is used for
   a) Premium brands of breads   b) Featured brands of breads
   c) Competitive brands of breads   d) All

64. What are the enzymes present in wheat flour and are essential in bread production
   a) Beta-amylase   b) Alpha amylase
   c) Both a & b   d) None

65. Wheat gelatinizes at temperature in the range of
   a) 57°C to 70°C   b) 47°C to 70°C
   c) 67°C to 70°C   d) 70°C to 75°C

66. Amylograph measures
   a) Alpha-amylase activity   b) Strength of flour
   c) Loaf volume potentialities   d) All

67. Farinograph measures
   a) Alpha-amylase activity
   b) Strength of flour for commercial mixing operation
   c) Loaf volume
   d) All

68. Extensograph indicates
    a) Alpha-amylase activity b) Strength of flour
    c) Loaf volume potentialities d) All
69. MacMicheal Viscosimeter indicates
    a) Alpha-amylase activity b) Strength of flour
    c) Bleaching of flour d) All
70. ________ yeast perishes more rapidly under normal conditions.
    a) Compressed bakers b) Dried
    c) Both a &b d) None
71. Active dried yeast is approximately ____________ percent solids and _______ percent moisture
    a) 30 & 70 b) 92 & 8
    c) 70 & 30 d) 60 & 20
72. Compressed yeast is approximately ____________ percent solids and _______ percent moisture
    a) 30 & 70 b) 92 & 8
    c) 70 & 30 d) 60 & 20
73. The ideal temperature for storage of dried bakers yeast over prolonged periods is
    a) 15°C b) 7°C
    c) 20°C d) 27°C
74. Baking Soda is known chemically as
    a) $NaHCO3$ b) $Na_2CO3$
    c) $Na_2NH_4CO3$ d) $NH_4CO3$
75. Double acting Baking powder consists of two acid ingredients
    a) One fast acting b) Slow acting
    c) a & b d) None
76. The fast acting part of double acting Baking powder usually consists of
    a) Calcium acid phosphate b) Sodium pyrophosphate
    c) Sodium aluminiumsulphate d) All
77. The slow acting part of double acting Baking powder usually consists of
    a) Sodium pyrophosphate b) Sodium aluminiumsulphate
    c) Calcium acid phosphate d) a & b

78. Ammonium bicarbonate is used in cookies as

a) Bleaching agent b) Flour improver

c) Leavening agent d) All

79. Amber durum wheat's are used for the production of

a) Noodles b) Pasta

c) Macaroni d) All

80. __________ is used to examine vitreousity of wheat

a) MacMicheal Viscosimeter b) Pohlfarinator

c) Pcyanometer d) None

81. Optimum storage moisture for wheat is less than

a) 12.5% b) 13.5%

c) 14.5% d) 15.1%

82. Hag berg Falling number test is used for

a) α–amylase activity b) Protein content

c) Gluten content d) None

83. Sedimentation value is used for

a) α–amylase activity b) Protein content

c) Gluten quality d) None

84. As per FSSAI regulations, the moisture content in Atta (wheat flour) is not more than __________ percent by weight

a) 13 b) 14

c) 16 d) 15

85. As per FSSAI regulations, the moisture content in Maida is not more than __________ percent by weight

a) 14 b) 13

c) 16 d) 15

86. As per FSSAI regulations, the alcoholic acidity of Atta is is not more than __________ percent by weight

a) 0.18 b) 0.12

c) 0.19 d) 0.11

87. As per FSSAI regulations, the moisture content in Semolina (Suji or rawa) is not more than _____________ percent by weight

a) 13.4 b) 14.5

c) 15.5 d) 12.5

88. As per Codex standards, the moisture content in wheat is not more than _____________ percent by weight

a) 14.5 b) 13.5

c) 16.5 d) 15.5

89. _________________ is a saccharifying enzyme

a) α–amylase b) β–amylase

c) Amylopectinase d) All

90. _________________ is a liquefying enzyme

a) α–amylase b) β–amylase

c) Amylopectinase d) All

91. Optimum pH for wheat flour is usually between

a) 5.5 and 6.5 b) 4.5 and 6.5

c) 5.5 and 7.5 d) 3.5 and 6.5

92. Pekar color test indicates

a) Alpha-amylase activity b) Strength of flour

c) Separation d) All

93. ______________ is a biological leavening agent

a) Bacteria b) Yeast

c) Fungi d) $NaHCO_3$

94. Commercially Pectin is prepared from

a) Citrus peel b) Citrus rags

c) a & b d) None

95. Dutch-process in cocoa processing involves

a) Acid treatment b) Alkali treatment

c) Sodium bicarbonate d) b & c

96. Dough mixing stages are

a) Pick up b) Clean up

c) Final development d) Letdown

e) Breakdown f) All

97. In which bread making process all the ingredients in the recipe are put into the mixing bowl and followed by a sequence of steps for bread making

a) Straight-Dough Method b) Sponge dough method

c) Commercial method d) a & c

98. __________ the stage where the production of gas in the dough is at the final stage givingvolume to the bread.

a) Proofing b) Mixing

c) Dough formation d) All

99. Under-proofing will produce bread of

a) Small volume b) Large volume

c) Medium volume d) None

100. α–amylase hydrolysis occurs during __________ phase of baking

a) Final stage b) Initial stage

c) Both a & b d) None

## Answers

1. a 2. c 3. d 4. c 5. d 6. a 7. b 8. b 9. d 10. a
11. d 12. a 13. a 14. c 15. b 16. a 17. d 18. a 19. a 20. b
21. d 22. c 23. b 24. c 25. d 26. a 27. b 28. a 29. b 30. a
31. b 32. b 33. c 34. c 35. b 36. d 37. a 38. b 39. a 40. b
41. a 42. b 43. d 44. d 45. b 46. c 47. c 48. b 49. a 50. b
51. a 52. d 53. a 54. b 55. a 56. b 57. c 58. d 59. b 60. b
61. a 62. b 63. c 64. c 65. a 66. a 67. b 68. c 69. c 70. a
71. b 72. a 73. b 74. a 75. c 76. a 77. d 78. c 79. d 80. b
81. a 82. a 83. c 84. b 85. a 86. a 87. b 88. a 89. a 90. b
91. a 92. c 93. b 94. c 95. d 96. f 97. d 98. a 99. a 100. b

1. Bread yeast is
   a) *S. cerevisiae*  b) *S. uvarum*
   c) *S. carlbergens*  d) All
2. The tenderizing action of sugars in baked products results in
   a) Texture  b) Volume
   c) Symmetry  d) All
3. ____________are machines that separate particles on the basis of differences in size, air resistance,and particle specific gravity, simultaneously
   a) Planetary sifter  b) Purifiers
   c) Vibrosifter  d) All
4. The main ingredient of premium quality pasta is 100 percent
   a) Hard wheat semolina  b) Soft wheat semolina
   c) Durum wheat semolina  d) None
5. ________the only cereal grain other than wheat capable of forming gluten
   a) Oats  b) Millet
   c) Rye  d) Barley
6. To claim as a healthy food a whole-oat-containing food must provide at least _______ soluble fiber per serving.
   a) 0.55g  b) 0.75g
   c) 0.65g  d) 0.45g
7. _____________ amount of soluble fiber is needed for a lowering effect on cholesterol levels
   a) 3g/day  b) 5g/day
   c) 2g/day  d) 6g/day
8. Yeast in dough is active between
   a) 0 and 55°C  b) 0 and 65°C
   c) 10 and 45°C  d) 50 and 55°C
9. The optimum temperature for fermentation in bread making process is
   a) 17 and 38°C  b) 27 and 38°C
   c) 7 and 38°C  d) 17 and 28°C

10. The yield of endosperm products and hominy feed of Maize are about ________ percent and ____ percent respectively.
    a) 60 & 40 b) 65 & 45
    c) 70 & 30 d) 80 & 20
11. The optimum moisture content for degerming of maize in the Beall Degermer is
    a) 21-25 % b) 14-20%
    c) 15-25% d) 16-20%
12. The germ constitutes in corn is ________ percent
    a) 11 b) 12
    c) 8 d) 10
13. Kilning is a process involved in
    a) Parboiling b) Mashing
    c) Malting d) All
14. Malt is dried to a moisture content of ________ percent
    a) 10 b) 13
    c) 15 d) 8
15. Principal storage proteins in wheat
    a) Gliadin b) Gluten
    c) Glutenin d) a&c
16. Pelshenke value of biscuit flour is
    a) Less than 40 min b) Less than 30 min
    c) Less than 20 min d) More than 30 min
17. ______________ is responsible for the characteristic sticky property of gluten
    a) Gliadin b) Proline
    c) Glutenin d) a&c
18. Rum is an alcoholic beverage produced from
    a) Sugarcane molasses b) Cereal grains
    c) Potato d) Grapes
19. Alcohol content in Rum is
    a) 50% b) 60%
    c) 40% d) 30%

20. Alcohol content in Gin is

a) 50% b) 60 %

c) 40% d) 30%

21. Vodka is an alcoholic beverage produced from

a) Sugarcane molasses b) Cereal grains

c) Potato d) b&c

22. Naked Caryopsis grains are

a) Wheat b) Rye

c) Maize d) All

23. __________ percent of aleurone layer presents in wheat

a) 5 b) 7

c) 10 d) 12

24. ________ is an additional layer to the caryopsis, also referred to as the husk.

a) Glume b) Bran

c) Germ d) Endosperm

25. __________________ is the incremental area under the blood glucose curve after consumption of 50g carbohydrate from a test food divided by the area under the curve after eating a similar amount of control food

a) Glycaemic index b) Glycaemic load

c) a & b d) None

26. ______________________ are the anti nutritional factors present in food causes goitre

a) Goitrogens b) Phytins

c) Tannins d) Saponins

27. ______________ is the blending of grains prior to milling to produce a flour of the required quality

a) Mixing b) Gristing

c) Conditioning d) Tempering

28. ______________ is a polishing process that removes the outer husk and part of the grain, leading to rounding of a cereal grain.

a) Pearling b) Dehusking

c) Debraning d) All

29. ______________ is that resists digestion and is only partially digested in the small intestine.

a) Modified starch
b) Resistant starch
c) Soluble starch
d) All

30. ____________ is the addition of water to cereal grain prior to milling.

a) Tempering
b) Conditioning
c) Both a & b
d) None

31. Commercial importance Maize varieties are

a) Dent maize
b) Flint maize
c) Sweet corn
d) Popcorn
e) All

32. _____________ the first cereal produced by man by crossing wheat and rye

a) Triticale
b) *Triticosecale*
c) *T. vavilovii*
d) a & b

33. As per the Codex standards maximum moisture content in Rice is _____ percent

a) 14.5
b) 15.5
c) 13.5
d) 12.5

34. As per the Codex standards maximum moisture content in Maize is _____ percent

a) 14.5
b) 13.5
c) 15.5
d) 12.5

35. As per the Codex standards maximum moisture content in Durum Wheat is _____ percent

a) 14.5
b) 13.5
c) 15.5
d) 12.5

36. The first grinding stage in wheat milling process is referred as

a) Reduction
b) Breaking
c) a & b
d) None

37. Milling and decortication of cereal grains reduces the level of

a) Lysine
b) Methionine
c) Cysteine
d) Glycine

38. The most important mycotoxin with respect to cereals is

a) *Aspergillus flavus* b) *Aspergillus niger*

c) *Aspergillus oryze* d) All

39. Mycotoxin produced by *Aspergillus flavus* is

a) Aflatoxins B1 b) Patulin

c) Aflatoxins B2 d) Both a &c

40. Ochratoxin is produced by

a) *Penicillium verrucosum* b) *Penicillium expansum*

c) *Penicillium patulin* d) All

41. *Aspergillus ochraceus produces* __________ mycotoxin

a) Aflatoxins b) Ochratoxin

c) Botulin d) Patulin

42. *Fusarium species* produces _____________mycotoxin

a) Fumonisins b) Deoxynivalenol

c) Zearlenone d) All

43. Ergotism is caused by

a) *Penicillium patulin* b) *Penicillium expansum*

c) *Claviceps purpurea* d) All

44. The Mycotoxin Ergot is produced in

a) Rye b) Sorghum

c) a & b d) None

45. ______________ Resistant starch is chemically modified starch

a) RS1 b) RS2

c) RS3 d) RS4

46. ______________ Resistant starch is retrograded starch

a) RS1 b) RS2

c) RS3 d) RS4

47. ______________ Resistant starch is native resistance starch granules

a) RS1 b) RS2

c) RS3 d) RS4

48. ______________ Resistant starch is Trapped and inaccessible for digestion

a) RS1 b) RS2

c) RS3 d) RS4

49. The storage protein in Rice is

a) Gliadens b) Glutenins

c) Glutelin (oryzenin) d) Zein

50. The storage protein in Maize (corn) is

a) Prolamin (zein) b) Glutenins

c) Glutelin (oryzenin) d) Gliadens

51. The storage protein in barley is

a) Gliadens b) Glutelins

c) Hordeins d) Both b & c

52. ____________________ are the type of phytoestrogen found in cereals

a) Phytin b) Lignans

c) Tannins d) Saponins

53. Phytate can bind ________________ minerals

a) Iron b) Calcium

c) Zinc d) All

54. Goitrogen (thioamide) is found in

a) Pearl millet b) Barley

c) Sorghum d) a & b

55. Trypsin inhibitors are found in

a) Cereals b) Legumes

c) Fruits & vegetables d) a & b

56. Dhurrin is a cyanogenetic glucoside found in seedlings of

a) Sorghum b) Finger millet

c) Pearl millet d) Maize

57. Maize is deficient in

a) Vitamin B1 b) Vitamin B2

c) Niacin d) All

58. Dry maize milling products are
    - a) Grits (40%)
    - b) Coarse meal (20%)
    - c) Fine meal (10%)
    - d) Flour (5%)
    - e) All
59. Tannins are present in
    - a) Rice
    - b) Wheat
    - c) Sorghum
    - d) None
60. The organism involved in production of Ang-kak (Red rice)
    - a) *Monascus purpureus*
    - b) *Pediococcus*
    - c) *Saccharomyces cerevisiae,*
    - d) None
61. The organism involved in production of Dhokla
    - a) *Monascus purpureus*
    - b) *Pediococcus*
    - c) *Leuconostoc mesenteroides*
    - d) None
62. The organism involved in production of Dosa or Idli
    - a) *Saccharomyces sake*
    - b) *Pediococcus*
    - c) *Leuconostoc mesenteroides*
    - d) None
63. The organism involved in production of Miso
    - a) *Aspergillus oryzae*
    - b) *Pediococcus*
    - c) *Leuconostoc mesenteroides*
    - d) None
64. The organism involved in production of Sake
    - a) *Aspergillus oryzae*
    - b) *Saccharomyces sake*
    - c) *Leuconostoc mesenteroides*
    - d) None
65. The organism involved in production of Koji
    - a) *Aspergillus oryzae*
    - b) *A. soyae*
    - c) *Leuconostoc mesenteroides*
    - d) a & b
66. Polished rice is poor in
    - a) Calcium
    - b) Iron
    - c) Thiamine
    - d) All
67. Celiac disease in human caused by consumption
    - a) Wheat
    - b) Rye
    - c) Barley
    - d) All

68. Celiac disease is caused by _______ protein
    a) Gluten b) Gliadin
    c) Tuberin d) All
69. The aroma of Basmati rice is due to
    a) 2-acetyl- 1-proline b) 2-acetyl- 1-pyrrolane
    c) 2-acetyl- 1-pyrroline d) All
70. Swelling power of Gluten is assed by
    a) Sedimentation value b) Saponification value
    c) Iodine value d) None
71. The organism involved in production of Natto
    a) *Bacillus nato* b) *Bacillus subtilis*
    c) a & b d) None
72. Protease inhibitors are present in
    a) Beans and cereals b) Chick-pea
    c) Peas d) Potatoes
    e) All
73. Hemagglutinins are present in
    a) Soya b) Kidney bean
    c) Lentils and peas d) All
74. Saponins are glycosides present in
    a) Soya bean b) Pea nuts
    c) Spinach d) All
75. Glycosinolates are thioglycosides present in
    a) Cabbage b) Radish
    c) Mustard and rape seed d) All
76. Glutenin is responsible for ___________ nature of wheat gluten
    a) Elastic and rubbary b) Sticky
    C) Viscous and extensible d) None
77. Cyanogens are Cyanogenic glucosides present in
    a) Peas b) Linseed
    c) Cassava d) All

78. Gossypol is an anti nutritional factor present in

a) Cotton seed b) Lin seed

c) Peas d) None

79. Lathyrogens are β-aminopropionitrile and derivatives present in

a) Chick-pea b) Linseed

c) Cassava d) All

80. Cycasin is a Methylazoxy-methanol present in

a) Peas b) Nuts of *Cycas*genus

c) Rice d) None

81. Favism is a toxin present in

a) Fava beans b) Wheat

c) Linseed d) None

82. Phytoalexins are present in

a) Fava beans b) Wheat

c) Sweet potatoes d) Linseed

83. α-Amantin is a poisonous toxic substance present in ______________ mushroom

a) *Amanita phalloide* b) *Ganoderma lucidum*

c) *Agaricus bisporus* d) None

84. Pulse proteins are predominantly

a) Albumins b) Globulins

c) Glutelins d) All

85. ____________ oligosaccharides are known for the flatulence production in man and animals

a) Raffinose b) Stachyose

c) Verbiscose d) All

86. Dough development is measured by

a) Brabender farinograph b) Extensio graph

c) Amylograph d) Mixograph

87. The higher the α-amylase activity________ the Falling Number.

a) Higher b) Lower

d) No change d) None

88. Bread dough is prepared by mixing water and flour at ________ ratio

a) 60:40 b) 70:30

c) 80:20 d) 50:50

89. Legumin from soybeans is called

a) Gluten b) Arachin.

c) Glycinin d) None

90. Legumin from Pea nut is called

a) Gluten b) Arachin.

c) Glycinin d) None

91. Gluten comprises ________ percent of total proteins of wheat flour

a) 60 b) 70

c) 50 d) 80

92. ________________ is the most common bleaching agent for flour treatment

a) Potassium bromide b) Potassium Iodide

c) Benzyl peroxide d) All

93. ____________ phenomenon is partly responsible for staling of bread.

a) Gelatinization b) Swelling of gluten

c) Conditioning d) Retrogradation

94. _____________ are acts as anti staling agents

a) Glycerin Monosteareate b) MSG

b) KMS d) All

95. The most widely used acidulants in biscuit industry are

a) Acid calcium phosphate b) Sodium acid pyrophosphate

c) a & b d) None

96. The Protein content in soya protein concentrate

a) 72 b) 62

c) 42 d) 52

97. The Protein content in soya Isolates

a) 85 b) 75

c) 65 d) 95

98. ______________ rolls are used to get maximum purifiable semolina and least amount of flour

a) Scratch b) Breaking

c) Reduction d) None

99. In wheat milling process, the reduction rolls have a roll differential of

a) 1:35 b) 1:1.25

c) 1:45 d) 1:65

100. Sound wheat has FFA value of

a) 30 mg b) 40 mg

c) 20 mg d) 50 mg

## Answers

| | | | | | | | | | |
|---|---|---|---|---|---|---|---|---|---|
| 1. a | 2. d | 3. b | 4. c | 5. c | 6. b | 7. a | 8. a | 9. b | 10. c |
| 11. a | 12. b | 13. c | 14. b | 15. d | 16. b | 17. a | 18. a | 19. c | 20. b |
| 21. d | 22. d | 23. b | 24. a | 25. a | 26. a | 27. b | 28. a | 29. b | 30. c |
| 31. e | 32. d | 33. b | 34. c | 35. a | 36. b | 37. a | 38. a | 39. d | 40. a |
| 41. b | 42. d | 43. c | 44. c | 45. d | 46. c | 47. b | 48. a | 49. c | 50. a |
| 51. d | 52. b | 53. d | 54. d | 55. d | 56. a | 57. c | 58. e | 59. c | 60. a |
| 61. c | 62. c | 63. a | 64. b | 65. d | 66. d | 67. d | 68. a | 69. c | 70. a |
| 71. c | 72. e | 73. d | 74. d | 75. d | 76. a | 77. d | 78. a | 79. a | 80. a |
| 81. a | 82. c | 83. a | 84. b | 85. d | 86. a | 87. b | 88. b | 89. c | 90. b |
| 91. d | 92. c | 93. d | 94. a | 95. c | 96. a | 97. d | 98. a | 99. b | 100. c |

1. The angle between the base and slope of the cone formed on a free vertical fall of the grain mass to a horizontal

   a) Angle of repose  b) Angle of degree
   c) Angle of curve  d) None

2. Angle of repose of wheat is

   a) 20-28  b) 23-28
   c) 23-25  d) 20-23

3. Angle of repose of Paddy is

   a) 20-30  b) 23-28
   c) 30-40  d) 30-45

4. Angle of repose of Millets is

   a) 20-25  b) 23-28
   c) 30-40  d) 30-45

5. Commercial flow dryers are designed based on

   a) Thick layer drying  b) Thin layer drying
   c) Both a & b  d) None

6. What is the thickness of grain bed in thin layer drying

   a) 20 cm  b) 30 cm
   c) 20 mm  d) 30 mm

7. Debranned (polished) and cracked parboiled wheat is known as

   a) Bulgur  b) Semolina
   c) Pasta  d) Vermicelli

8. Starch is insoluble in

   a) Hot water  b) Cold water
   c) Both a & b  d) Nonc

9. The angle of repose is depends on

   a) Size  b) Shape
   c) Moisture content  d) Orientation of the grains
   c) All

10. ______ is the angle of friction taken up by granular material to first slide upon itself.

   a) Static angle of repose  b) Dynamic angle of repose
   c) a & b  d) None

11. Natural air drying cannot be used if the ambient relative humidity exceeds________ percent

    a) 80 b) 90

    c) 70 d) 75

12. Natural drying of grains cannot achieved, if the moisture content of the grains more than ___ %

    a) 10 b) 15

    c) 17 d) 20

13. __________ is a continuous flow non mixing type of grain dryer

    a) LSU dryer b) PHTC type dryer

    c) a & b d) None

14. __________ is a continuous flow mixing type of grain dryer

    a) LSU dryer b) PHTC type dryer

    c) a & b d) None

15. __________is the temperature of steam used for parboiling of paddy in CFTRI method

    a) 85°C b) 75°C

    c) 65°C d) 80°C

16. Hydrothermal treatment of food grains before milling is

    a) Conditioning b) Soaking

    c) Steeping d) All

17. Pneumatic separation is based

    a) Thermal properties b) Aerodynamic properties

    c) Magnet properties d) All

18. Concave type husker removes

    a) Bran b) Husk

    c) a & b d) None

19. Whitening in milling is also referred as

    a) Polishing b) Pearling

    c) Scouring d) All

20. The abrasive roll is made of

    a) $SiO_2$ b) SiC (Carborundum)

    c) $SiCl_4$ d) All

21. The ratio of horny to floury endosperm in corn

a) 2:1 b) 1:2

c) 1.5:2 d) 2:1.5

22. Milling of wheat is carried out by

a) Ball mills b) Huller mills

c) Roller mills d) None

23. In Wheat milling process, the surface of reduction rolls is

a) Smooth b) Corrugated

c) a & b d) None

24. In Wheat milling process, the surface of break rolls is

a) Smooth b) Corrugated

c) a & b d) None

25. In Wheat milling process, break rolls mainly used for production of

a) Flour b) Middlings

c) both a & b d) None

26. In Wheat milling process, reduction rolls mainly used for production of

a) Flour b) Middlings

c) Both a & b d) None

27. The yield of spilt pulses in traditional mills is ________ percent

a) 65-70 b) 85-90

c) 75-85 d) 70-85

28. The Oil requirement for pretreatment of 1 tonne of pulses before milling is

a) 1.0 to 2.0 kg b) 1.5 to 2.0 kg

c) 2.0 to 5.0 kg d) 1.5 to 2.5 kg

29. Pulses are dried to ________ percent moisture before milling

a) 10-12 b) 10-18

c) 10-20 d) None

30. Pitting or scratching of pulses is done in a

a) Break mill b) Roller mill

c) Hammer mill d) Ball mill

31. Emery-coated roller machine in pulse milling is called as
    a) Sheller b) Chakki
    c) Gota machine d) All

32. __________ percent of coarse stone powder is mixed with the grains as abrasive material
    a) 0.3 to 0.75 b) 0.5 to 0.75
    c) 0.2 to 0.55 d) 0.3 to 0.55

33. The yield of spilt pulses commercial dhal mills is _________ percent
    a) 65-70 b) 68-75
    c) 75-85 d) 70-85

34. In dry milling of black gram, the scratched grains are coated with _____ percent of oil
    a) 2-5 b) 5-9
    c) 1-2 d) 3-9

35. The physiological factors deteriorating the food grains during storage are
    a) Respiration b) Heating
    c) Moisture d) a & b

36. The Chemical factors deteriorating the food grains during storage are
    a) Oxygen b) Carbon dioxide
    c) Moisture d) All

37. The Physical factors deteriorating the food grains during storage are
    a) Temperature b) Humidity
    c) both a & b d) None

38. A grain sprouts only when its moisture content is
    a) 20-25% b) 30-35%
    c) 40-45% d) 10-15%

39 Gaseous fumigant is
    a) Methyl bromide b) ALP
    c) Ethylene bromide d) None

40 _____ is solid fumigant
    a) Methyl bromide b) Aluminium phosphide
    c) Ethylene bromide d) None

41. Generally FCI uses ___________quantity of Aluminium phosphide for tone of bulk grains for period of 5 to 7 days

a) Full tablet  b) Half tablet
c) 2 tablets  d) 5 tablets

42. Disk separator separates the material on the basis of

a) Size  b) Shape
c) Length  d) All

43. Indented cylinder separator separates the material on the basis of

a) Size  b) Length
c) Shape  d) All

44. Spiral separator separates the grains on the basis of

a) Roundness  b) Shape
c) Length  d) All

45. Specific gravity separator separates the materials on the basis of

a) Density  b) Specific gravity
c) Length  d) a & b

46. Inclined belt draper separates the materials on the basis of

a) Surface texture  b) Shape
c) Length  d) a & b

47. Velvet roll separator separates the materials on the basis of

a) Surface texture  b) Shape
c) Length  d) a & b

48. Fluidized bed cleaner makes the separation of grains based on

a) Density  b) Specific gravity
c) Size  d) a & c

49. Magnetic separator performs separation on the basis of

a) Surface texture  b) Stickiness properties
c) a & b  d) None

50. Cyclone separator works on the principle of

a) Centrifugal force  b) Weight of the particle
c) a & b  d) None

51. The moistur content in wet basis is

a) *Wm/Wm+Wd*×100
b) *Wd/Wm+Wd*×100
c) *Wm/Wd+Wd*×100
d) None

52. The moisture content in dry basis is

a) M = *Wm/Wd*×100
b) *m*/100-*m*×100
c) Both a & b
d) *Wm/Wm+Wd*×100

53. Two tones of paddy with 22 percent moisture content on wet basis (wb) are to be dried to 13 percent moisture content on dry basis. Calculate the weight of bone dry products.

a) 1756.8 kg
b) 1762.8 kg
c) 1760.2 kg
d) 1761.8 kg

54. __________ desiccant is used to determine moisture content in grains

a) $CaCl_2$
b) $CaCO_3$
c) $CaO_2$
d) All

55. __________ is a rice weevil

a) *Rhizopertha dominica*
b) *Sitophilus oryzae*
c) *Trogoderma granariam*
d) All

56. __________ Lesser grain borer

a) *Rhizopertha dominica*
b) *Sitophilus oryzae*
c) *Trogoderma granariam*
d) All

57. ______________ is a long head flour beetles

a) *Latheticus oryzae*
b) *Sitophilus oryzae*
c) *Trogoderma granariam*
d) All

58. __________ is a pulse beetles

a) *Latheticus oryzae*
b) *Sitophilus oryzae*
c) *Bruchus* sp.
d) All

59. The opening size of the sieves is based on the ______________ mesh sieve

a) 150
b) 180
c) 200
d) 210

60. Albumins are soluble in

a) Water
b) Dilute salt
c) 70 % Alcohol
d) Dilute acids/base

61. Globulins are soluble in
    a) Water b) Dilute salt
    c) 70 % Alcohol d) Dilute acids/base

62. Gliadins are soluble in
    a) Water b) Dilute salt
    c) 70 % Alcohol d) Dilute acids/base

63. Glutenin are soluble in
    a) Water b) Dilute salt
    c) 70 % Alcohol d) Dilute acids/base

64. Loss of moisture in bread is
    a) Steeping b) Staling
    c) Soaking d) Drying

65. Anti staling agent is
    a) Glycerin Monosteareate b) Acid calcium phosphate
    c) a & b d) None

66. Legumin from soybeans is called
    a) Arachin b) Glycinin
    c) Oryzein d) All

67. Storage proteins are
    a) Albumins b) Globulins
    c) Glutenin d) All

68. Fats and oils are building blocks
    a) Triglycerides b) Diglycerides
    c) Mono glycerides d) None

69. Free fatty acid content (FFA) in refined fats and oils is
    a) Less than 3% b) Less than 0.1%
    c) Less than 2.5% d) Less than 1.3%

70. Oil content in Canola
    a) 40-45 b) 18-20
    c) 45-50 d) 65-68

71. Oil content in Cotton seed
    a) 40-45 b) 18-20
    c) 45-50 d) 65-68
72. Oil content in Soya bean
    a) 40-45 b) 18-20
    c) 45-50 d) 65-68
73. Oil content in Ground nut
    a) 40-45 b) 18-20
    c) 45-50 d) 65-68
74. Oil content in dry copra (Coconut)
    a) 40-45 b) 18-20
    c) 45-50 d) 65-68
75. Oil content in Sesame seeds
    a) 40 b) 50
    c) 30 d) 60
76. Oil content in Palm kernel
    a) 40-45 b) 18-20
    c) 44-53 d) 65-68
77. The most predominant saturated fatty acids
    a) Lauric C-12:0 b) Myristic C-14:0
    c) Palmitic C-16:0 d) Stearic C-18:0
    e) All
78. The most important monounsaturated fatty acids
    a) Oleic (C-18:1) b) Erucic (C-22:1)
    c) Both a & b d) None
79. Omega 3/n-3 fatty acids
    a) Eicosapentaenoic acid (C-20:5) b) Docosahexaenoic acid C-22:6
    c) a & b d) None
80. Mono- and diglycerides are
    a) Emulsifiers b) Antioxidants
    c) Acidulants d) Stabilizers

81. Two common phosphatides in vegetable oils are
    a) Lecithins b) Cephalins
    c) a & b d) None
82. Hydratable phosphatides are
    a) Phosphatidylcholine b) Phosphatidylinositol
    c) a & b d) None
83. Non Hydratable phosphatides are
    a) Phosphatidic acid b) Lysophosphatidic acid
    c) a & b d) None
84. Lecithin is present in
    a) Egg b) Soya
    c) a & b d) None
85. Hydratable and non hydratable phosphatides are removed by adding
    a) Phosphoric acid b) Citric acid
    c) NaOH d) a & b
86. Alpha-tocopherol has
    a) Highest vitamin E activity b) Lowest antioxidant activity
    c) a & b d) None
87. Gossypol is a pigment found in ______ oil
    a) Soya b) Cotton seed
    c) Ground nut d) Canola
88. Squalene is present in
    a) Ricc bran oil b) Shark liver oil
    c) Olive oil and amaranth seed oil d) All
89. Vegetable oil sterols are
    a) Campesterol b) Stigmasterol
    c) Sitosterol d) All
90. Animal fat sterol is
    a) Campesterol b) Cholesterol
    c) Stigmasterol d) All

91. Commercially RBD soya oil means

a) Refined b) Bleached

c) Deodorized d) All

92. Characteristic yellow color of cottonseed oil is due to

a) Gossypol b) Chlorophyalls

c) Xanthophills d) None

93. ________ is the major saturated fatty acid in cottonseed oil

a) Lauric C-12:0 b) Myristic C-14:0

c) Palmitic C-16:0 d) Stearic C-18:0

94. Cotton seed oil is indentified by __________ test

a) Bauden test b) Halphen test

c) Hold test d) Baume test

95. _____________ are blamed for a "light-struck" off-flavor in potato chips

a) Peroxides b) Oxidized Melnoidin

c) Oxidized cyclopropenoid d) None

96. Characteristic color of pea nut oil is due to

a) Beta-carotene b) Lutelin

c) Gossypol d) Both a &b

97. The principal fatty acid in peanut oil is

a) Lauric C-12:0 b) Oleic (C-18:1)

c) Palmitic C-16:0 d) Stearic C-18:0

98. Naturally occurring fatty acids are in

a) Cis form b) Trans form

c) L-form d) D-form

99. Canola is the registered trademark of the Canola Council of Canada for the genetically modified crop of

a) *Brassica napus* b) *B. campestris*

c) a & b d) None

100. _________oil contains highest levels of chlorophyll among common vegetable oils

a) Soya b) Sesame

c) Cotton d) Canola

## Answers

| | | | | | | | | | |
|---|---|---|---|---|---|---|---|---|---|
| 1. a | 2. b | 3. d | 4. a | 5. b | 6. a | 7. a | 8. b | 9. e | 10. a |
| 11. c | 12. d | 13. b | 14. a | 15. a | 16. a | 17. b | 18. c | 19. d | 20. b |
| 21. a | 22. c | 23. a | 24. b | 25. b | 26. a | 27. a | 28. d | 29. a | 30. b |
| 31. c | 32. b | 33. b | 34. c | 35. d | 36. d | 37. c | 38. b | 39. a | 40. b |
| 41. b | 42. c | 43. b | 44. a | 45. d | 46. d | 47. d | 48. d | 49. c | 50. c |
| 51. a | 52. c | 53. b | 54. a | 55. b | 56. a | 57. a | 58. c | 59. c | 60. a |
| 61. b | 62. c | 63. d | 64. b | 65. a | 66. b | 67. b | 68. a | 69. b | 70. a |
| 71. b | 72. b | 73. c | 74. d | 75. b | 76. c | 77. e | 78. c | 79. c | 80. a |
| 81. c | 82. c | 83. c | 84. c | 85. d | 86. c | 87. b | 88. d | 89. d | 90. b |
| 91. d | 92. a | 93. c | 94. b | 95. c | 96. d | 97. b | 98. a | 99. c | 100. d |

1. ___________ Olive Oil has a flavor rating of 6.5 or better and a 1.0 max free acidity.
   a) Extra virgin b) Fine virgin
   c) Semifine virgin d) None
2. ___________ Olive Oil has a flavor rating of 5.5 or better and a 1.5 max free acidity.
   a) Extra virgin b) Fine virgin
   c) Semifine virgin d) None
3. ___________ Olive Oil has a flavor rating of 3.5 or better and a 3.3 max free acidity.
   a) Extra virgin b) Fine virgin
   c) Semifine virgin d) None
4. Virgin olive oil with ___________ and ______________ is considered not fit for human consumption
   a) Flavor rating of less than 3.5 b) FFA of 3.3 max
   c) a & b d) None
5. The color of virgin olive oil is mainly due to the presence of
   a) Chlorophyll b) Pheophytin
   c) Xanthophills d) a & b
6. ______________ is the trade name for the dried coconut meat or kernel
   a) Dessicated coconut c) Grated coconut
   c) Copra d) None
7. ______________ oil is an important group in the lauric acid
   a) Coconut oil b) Soya oil
   c) Sesame Oil d) All
8. ______________ is the hard fat of ruminants.
   a) Butter b) Cream
   c) Shortening d) Tallow
9. ________________ has been associated with the beneficial health effects, including cancer protection, heart disease defense, reduction in body fat
   a) Conjugated linoleic fatty acid b) Palmitic acid
   c) Eruic acid d) None

10. Delinting process is involved in ____________ processing

a) Soya b) Sesame

c) Cotton d) Peanut

11. Lipoxygenase causes in oils

a) Higher peroxide b) Secondary oxidation products

c) Rancidity d) All

12. Commercially used solvent for extraction of oils is

a) Alcohol b) Ether

c) Hexane d) None

13. Oil-bearing fruits of commercial importance are

a) Olive b) Palm

c) Sesame d) a & b

14. General oil extraction methods are

a) Hydraulic presses b) Continuous centrifuges

c) Adhesion filtering d) All

15. Palm oil is expressed by

a) Pressing b) Cooking

c) a & b d) None

16. Recovery of animal fat form fat tissue by cooking process is called as

a) Ageing b) Rendering

c) Pressing d) Solvent extraction

17. _________ chemical is used to stabilize the fresh rice bran

a) Sodium Metabisulphite b) NaoH

c) $Na_2CO_3$ d) All

18. ____________ process is used to remove the waxes in oil refining

a) Degumming b) Winterization

c) Bleaching d) None

19. Winterization is carried out at _________________ temperature

a) 8-10°C b) 20-40°C

c) 10-40°C d) 9-40°C

20. The strength of the caustic solution is measured in terms of specific gravity, expressed in
    a) Brix b) Baume
    c) Beta d) All
21. Winterized oil is also called as
    a) Refined oil b) Vegetable oil
    c) Double refined oil d) Salad oil
22. Winterization is used to remove
    a) Triglycerides of fats b) Waxes
    c) Short chain fatty acids d) All
23. What are the predominant triglyceride forms Palmitic (P), Stearic (S), and Oleic (O) acids in cocoa butter
    a) POS b) SOS
    c) PSS d) a & b
24. The most common oil to be degummed is
    a) Sesame oil b) Pea nut oil
    c) Soya d) Mustard
25. Rearranging of the fatty acids in triglyceride molecules is
    a) Hydrogenation b) Interesterification
    c) Dewaxing d) Esterification
26. Interesterification is used principally in
    a) Confectionery fats b) Table spreads
    c) Shortenings d) All
27. Summer oils, or oils that had not been subjected to winterization, became known as
    a) Cooking oil b) Salad Oil
    c) Esterified oil d) Winterized oil
28. ________ are those naturally occurring in food fats and oils
    a) Cis isomers b) Trans Isomers
    c) Positional Isomerism d) None
29. Important dietary sources of vitamin K are
    a) Soybean oil b) Canola oil
    c) Soya Oil d) a & b

30. Elevated levels of __________ are associated with increased risk of coronary heart disease

a) LDL cholesterol
b) HDL cholesterol
c) Both a & b
d) None

31. __________ is the temperature at which smoke is first detected in a laboratory apparatus

a) Smoke point
b) Flash point
c) Fire point
d) None

32. __________ is the temperature at which the volatile products are evolved at such a rate that they are capable of being ignited but not capable of supporting combustion

a) Smoke point
b) Flash point
c) Fire point
d) None

33. _______ is the temperature at which the volatile products will support continued combustion.

a) Smoke point
b) Flash point
c) Fire point
d) None

34. ________-is a naturally occurring phosphatides which has emulsifying, wetting and antioxidant properties

a) Lecithin
b) BHA
c) TBHQ
d) BHT

35. ______ is the process of creating a solid crystal structure in a fat or oil product resulting in a smooth appearance and firm consistency

a) Hydrogenation
b) Interesterification
c) Plasticize
d) Esterification

36. _______ is a aqueous byproduct from the chemical refining process that is comprised of soap, hydrated gums, water, oil and other impurities

a) Soap stock
b) Soap solution
c) Alkali stock
d) Alkali solution

37. Examples of fats having a higher percentage of C18 saturated fatty acids

a) Large
b) Tallow
c) a & b
d) None

38. Smoke Point is effected by

a) Free fatty acids b) Surface area of oil exposed
c) Presence of suspended matter d) All

39. If free fatty acids are present in large amounts, the smoking point is

a) Lowered b) Higher
c) No change d) None

40. When oleic acid is added to cotton seed oil the smoking point is

a) Lowered b) Higher
c) No change d) None

41. Hydrogenation of refined deodorized oils is carried out in presence of _______ catalysts

a) Ni b) Pt
c) Pd d) All

42. ____________ is the hydrogenated ground nut oil and is used widely in India

a) Vanaspathi b) Dalda
c) Butter d) None

43. __________ is a butter substitute manufactured in large quantities in U.S.A. the ingredients are hydrogenated fat, soup, salt and vitamin-A.

a) Vanaspathi b) Dalda
c) Margarine d) All

44. ___________ are do not become cloudy when kept in refrigerators due to separation of saturated triglycerides

a) Cooking oils b) Salad Oils
c) Esterified oils d) Winterized oils

45. Natural salad oils are

a) Corn oil and Soya oil b) Sesame oil
c) Sunflower seed oil d) All

46. The conversion of cotton seed oil to a salad oil is done by a process called

a) Hydrogenation b) Interesterification
c) Winterization d) Esterification

47. __________ is a semisolid emulsion of vegetable oils, egg yolk or whole egg, vinegar and Seasonings.

a) Vanaspathi b) Mayonnaise

c) Margarine d) All

48. __________ is a 20 times as active as Iron (Fe) pro oxidant

a) Copper b) Sodium

c) Potassium d) All

49. According to F.D.A mayonnaise should contain not less than __________ % of oil

a) 50 b) 60

c) 65 d) 55

50. Lipoxidases are present in

a) Cereals b) Oil seeds

c) Legumes d) a & b

51. The steps involved in the auto oxidation are

a) Formation of free radical

b) Reaction of free radical with $O_2$ to form peroxide

c) Formation and breakdown of hydro peroxides

d) All

52. A compound which in habits or delays the auto oxidation of fats is known as

a) Emulsifier b) Antioxidant

c) Pro-oxidant d) None

53. Compounds which delay auto oxidation by reinforcing the action of antioxidants are called as

a) Emulsifier b) Antioxidant

c) Pro-oxidant d) Synergists

54. Examples of synergists are

a) Citric acid b) Ascorbic acid

c) EDTA d) All

55. What are the changes occur in fats and oils during frying of foods

a) Increase in FFA b) Decrease in Iodine Value

c) Increase in peroxide value d) All

56. Vitamin-A occurs only in
    a) Vegetable foods b) Animal foods
    c) Both a & b d) None
57. _______ is heat stable and is not destroyed when its solutions are exposed to light
    a) Vitamin A b) Vitamin C
    c) Vitamin D d) All
58. _________ are fats with chemical compositions similar to cocoa butter and capable of replacing it in any proportion in chocolates.
    a) Cocoa butter equivalents b) Lauric cocoa butter replacers
    c) Nonlauric CBRs d) All
59. In general, deodorization temperatures will vary from
    a) 204 to 246°C b) 104 to 146°C
    c) 180- 200°C d) 190-210°C
60. Mayonnaise is an ____________ emulsion
    a) Oil in water b) Water in oil
    c) Both a & b d) None
61. Melted fat is sprayed under high pressure into a cooled ("35°C) air blast crystallization chamber, where the fat particles solidify to _________
    a) Mayonnaise b) Fat Powder
    c) Vanaspathi d) Shortenings
62. ________ is the weight of KOH (in mg) needed to hydrolyze 1 g of fat or oil under standardized conditions
    a) Saponification Value b) Acid Value
    c) Iodine Value d) Hydroxyl Value
63. ______is the number of milligrams of KOH needed to neutralize the organic acids present in 1 g of fat.
    a) Saponification Value b) Acid Value
    c) Iodine Value d) Hydroxyl Value
64. _________ is the number of grams of halogen, calculated as iodine, which bind to 100 g fat.
    a) Saponification Value b) Acid Value
    c) Iodine Value d) Hydroxyl Value

65. ______ reflects the content of hydroxy fatty acids, fatty alcohols, mono- and diacylglycerols and free glycerol.

a) Saponification Value b) Acid Value

c) Iodine Value d) Hydroxyl Value

66. What test is used for identification of Sesame oil

a) *Baudouin* b) *Halphen*

c) *Fitelson* d) None

67. What test is used for identification of Cottonseed oil

a) *Baudouin* b) *Halphen*

c) *Fitelson* d) None

68. What test is used for identification of Teaseed oil

a) *Baudouin* b) *Halphen*

c) *Fitelson* d) None

69. The FFA content in refined oil or fat

a) 0.1% b) 1%

c) 3% d) 0.5%

70. ________ is a preferred method for detecting lipid peroxidation in biological systems.

a) Peroxide value b) Iodine value

c) Thiobarbituric acid test d) None

71. _________ is the most widely accepted method for solids fat measurement in the United States.

a) Solids fat index b) Baume

c) Dilatometry d) None

72. _________ is determined by cooling a melted sample while stirring until the fat becomes cloudy

a) Congeal point b) Baume

c) Dilatometry d) None

73. The Mesocarp in palm oil is pulped and separated from the nut by a process known as

a) Soaking b) Cooking

c) Maceration d) None

74. Immediately after milling the Free fatty acid content of rice bran is normally less than ______ percent

a) 1 b) 2

c) 3 d) 5

75. _________ of fats and oils increases with the increase in USFAs content and with molecular weight.

a) Refractive index b) Baume

c) Dilatometry d) None

76. What test is used for identification of Linseed oil

a) *Baudouin* b) *Halphen*

c) *Fitelson* d) Hexa bromide test

77. What test is used for identification of Mineral oil

a) *Baudouin* b) *Halphen*

c) *Fitelson* d) Holde Test

78. ________ is designed to detect oxidative spoilage of fats, due to oxidation, epihydrin aldehyde is formed.

a) *Baudouin* b) Kries test

c) *Fitelson* d) Holde Test

79. The Kries test is between epihydrin aldehyde formed by oxidation and Phloroglucinol in HCl forms pink color indicates

a) Extensive oxidation b) Slight oxidation

c) No change d) None

80. The Kries test is between epihydrin aldehyde formed by oxidation and Phloroglucinol in HCl forms Red color indicates

a) Extensive oxidation b) Slight oxidation

c) No change d) None

81. Protein rich (paushtik) atta means the product obtained by mixing wheat atta with groundnut flour "or soya flour", or a combination of both". flour up to an extent of _______ percent

a) 12.0 b) 10.0

c) 15.0 d) 20.0

82. As per FSSAI regulations, Alcoholic acidity (with 90 per cent alcohol) of Protein rich (paushtik) atta is not more than ____________ percent by weight

a) 0.28 b) 0.15

c) 0.12 d) 0.31

83. As per FSSAI regulations, Moisture content of corn flour is not more than ____________ percent by weight

a) 10.5 b) 12.5

c) 13.5 d) 14.5

84. As per FSSAI regulations, Moisture content of Pasta (Macaroni, spaghetti, Vermicelli) products is not more than ____________percent by weight

a) 10.5 b) 12.5

c) 13.5 d) 14.5

85. As per FSSAI regulations, Nitrogen content of Pasta (Macaroni, spaghetti, Vermicelli) products is not less than ____________ percent by weight

a) 1.2 b) 1.5

c) 1.7 d) 1.9

86. As per FSSAI regulations, Bacterial count in Malted milk food without cocoa powder products is not more than ____________ per gram

a) 10,000 b) 50,000

c) 40,000 d) 25,000

87. As per FSSAI regulations, Protein content in Malt Based Foods is not less than ___________ per cent by weight.

a) 5 b) 6

c) 7 d) 8

88. As per FSSAI regulations, calcium carbonate fortification in 100 parts of atta should not be less than ______ parts

a) 0.30 b) 0.35

c) 0.45 d) 0.50

89. If the Maida product is to be used for bakery purpose, the Benzoyl peroxide treatment should not be more than ________ ppm

a) 20 b) 40

c) 60 d) 80

90. If the Maida product is to be used for bakery purpose, the Ascorbic acid treatment should not be more than _________ ppm

a) 200 b) 300

c) 400 d) 500

91. As per FSSAI regulations, the moisture content in rice is not more than _________ percent by weight

a) 16.0 b) 18

c) 13 d) 14

92. As per FSSAI regulations, the moisture content in Black gram dhal and green gram dhal is not more than _________ percent by weight

a) 16.0 b) 18

c) 13 d) 14

93. As per FSSAI regulations, the moisture content in Channa dhal is not more than _________ percent by weight

a) 16.0 b) 18

c) 13 d) 14

94. As per FSSAI regulations, the moisture content in Corn flakes is not more than _________ percent by weight

a) 8.5 b) 7.5

c) 9.5 d) 10.5

95. As per FSSAI regulations, the moisture content in Custard powder is not more than _________ percent by weight

a) 8.5 b) 7.5

c) 12.5 d) 10.5

96. As per FSSAI regulations, the moisture content in rolled oats (quick cooking oats) is not more than _________ percent by weight

a) 8 b) 7

c) 9 d) 10

97. As per FSSAI regulations, the Protein content in defatted soya flour (Solvent Extract Soya Flour) is not less than _________ percent by weight

a) 38 b) 40

c) 42 d) 48

98. As per FSSAI regulations, the Acid Insoluble Ash (AIA) in Biscuits is not more than __________ percent by weight

a) 0.5 b) 0.1

c) 0.3 d) 0.4

99. As per FSSAI regulations, the acidity of extracted fat (as oleic acid) in Biscuits is not exceeding __________ percent by weight

a) 1.5 b) 0.1

c) 1.35 d) 1.4

100. As per FSSAI regulations, Alcoholic acidity (with 90 per cent alcohol) in Bread is not exceeding Shall be not more than equivalent __________ ml. N NaOH per 100 g of dried substances

a) 7.5 b) 8.5

c) 9.5 d) 10.5

## Answers

| | | | | | | | | | |
|---|---|---|---|---|---|---|---|---|---|
| 1. a | 2. b | 3. c | 4. c | 5. d | 6. c | 7. a | 8. d | 9. a | 10. c |
| 11. d | 12. c | 13. d | 14. d | 15. c | 16. b | 17. a | 18. b | 19. a | 20. b |
| 21. d | 22. d | 23. d | 24. c | 25. b | 26. d | 27. a | 28. a | 29. d | 30. a |
| 31. a | 32. b | 33. c | 34. a | 35. c | 36. a | 37. c | 38. d | 39. a | 40. a |
| 41. d | 42. a | 43. c | 44. b | 45. d | 46. c | 47. b | 48. a | 49. c | 50. d |
| 51. d | 52. b | 53. d | 54. d | 55. d | 56. b | 57. c | 58. d | 59. a | 60. a |
| 61. b | 62. a | 63. b | 64. c | 65. d | 66. a | 67. b | 68. c | 69. a | 70. c |
| 71. c | 72. a | 73. c | 74. c | 75. a | 76. d | 77. d | 78. b | 79. b | 80. a |
| 81. b | 82. c | 83. b | 84. b | 85. c | 86. b | 87. c | 88. a | 89. b | 90. c |
| 91. a | 92. d | 93. a | 94. b | 95. c | 96. d | 97. d | 98. b | 99. a | 100. a |

1. As per FSSAI regulations, Butyro-refractometer reading at 40°C for Coconut oil is

   a) 55.6 to 60.2 b) 34.0 to 35.5

   c) 54.0 to 57.1 d) 51.0 to 55.6

2. As per FSSAI regulations, acid value for Coconut oil is not more than

   a) 6 b) 3

   c) 1.5 d) 3.5

3. As per FSSAI regulations, Iodine value for Cotton seed oil is

   a) 98 to 112 b) 85 to 99

   c) 75 to 94 d) 75 to 92

4. As per FSSAI regulations, Test for argemone for edible vegetable oil shall be

   a) Positive b) Negative

   c) Both a & b d) None

5. As per FSSAI regulations, The refined oil shall not contain Hexane more than __________ ppm

   a) 2 b) 3

   c) 4 d) 5

6. As per FSSAI regulations, acid value for Groundnut oil, Safflower seed oil, Til Oil , Palmolein, Sun flower seed oil and Palm kernel oil is not more than

   a) 6 b) 3

   c) 1.5 d) 3.5

7. As per FSSAI regulations acid value for Palm oil is not more than

   a) 6 b) 3

   c) 10 d) 9

8. As per FSSAI regulations, Oryzanol Content in rice bran oil is not less than

   a) 2 b) 1

   c) 3 d) 2.5

9. As per FSSAI regulations, acid value for rice bran oil is not more than

   a) 0.1 b) 0.5

   c) 3 d) 6

10. As per FSSAI regulations, Percentage of free fatty acids (calculated as oleic acid) in Cocoa butter is not more than

a) 0.1  b) 0.5
c) 1.5  d) 6

11. As per FSSAI regulations, Vitamin A content in Table margarine is not less than ____ I.U. per gram of the product at the time of sale.

a) 20  b) 30
c) 40  d) 50

12. As per FSSAI regulations, Melting point of extracted fat (Capillary Slip Method) of Table margarine is

a) 31°C to 35°C  b) 31°C to 47°C
c) 31°C to 57°C  d) 31°C to 37°C

13. As per FSSAI regulations, Table margarine prepared with colour and flavors shall only be sold in sealed packages weighing not more than

a) 200g  b) 250g
c) 100g  d) 500g

14. As per FSSAI regulations, Melting point of extracted fat (Capillary Slip Method) of Bakery and Industrial Margarine is

a) 31°C to 35°C  b) 31°C to 41°C
c) 31°C to 57°C  d) 31°C to 37°C

15. As per FSSAI regulations, Melting point of extracted fat (Capillary Slip Method) of Fat spread is

a) 35°C  b) 47°C
c) 57°C  d) 37°C

16. As per FSSAI regulations, Fat content in Bakery and Industrial Margarine is not less than ____ per cent mass/mass

a) 60  b) 80
c) 70  d) 100

17. As per FSSAI regulations, Addition of Diacetyl to the Vanaspatı shall not exceed __________ ppm

a) 4.0  b) 5.0
c) 3.0  d) 10.00

18. As per FSSAI regulations, Vanaspati shall always positive to __________ test

    a) *Baudouin*　　b) *Halphen*

    c) *Fitelson*　　d) None

19. As per FSSAI regulations, Melting point of extracted fat (Capillary Slip Method) of Vanaspati is

    a) 31°C to 35°C　　b) 31°C to 41°C

    c) 31°C to 57°C　　d) 31°C to 37°C

20. As per FSSAI regulations, Synthetic Vitamin A content in Vanaspati is not less than ____ I.U. per gram at the time of packing and shall test positive when tested with Antimony Trichloride (carr-Price Reagent)

    a) 35　　b) 45

    c) 25　　d) 50

21. As per FSSAI regulations, Free fatty acid (as oleic acid) content in Vanaspathi is not more than

    a) 0.15　　b) 0.25

    c) 0.35　　d) 0.55

22. _________ enzyme is used for the production of sugars from starch

    a) Amylase　　b) Protease

    c) Diastase　　d) Lyase

23. The Enzyme used in crackers dough is _______

    a) Amylase　　b) Protease

    c) Diastase　　d) Lyase

24. Lecithin is present in

    a) Egg yolk　　b) Butter

    c) Soya　　d) All

25. Short dough lack in

    a) Elasticity　　b) Extensibility

    c) Crumbly nature　　d) All

26. Research water absorption meter is used to determine

    a) Flavour　　b) Elasticity

    c) Consistency　　d) Texture

27. __________ mixer is normally used for bread dough

a) Reciprocating arm mixer b) Horizontal mixer
c) Vertical mixer d) None

28. What are the changes takes place during baking

a) Development of rigid porous structure
b) Reduction in moisture
c) Surface coloration
d) All

29. Short dough biscuits are

a) High in fat b) High in sugar
c) both a & b d) None

30. Hard dough biscuits are

a) Low fat b) Low sugar
c) Both a & b d) None

31. Salted or Savoury biscuits are

a) Fermented b) Non fermented
c) Either a or b c) None

32. Examples of Short dough biscuits are

a) Glucose b) Milk Biscuits
c) Cookies d) All

33. Examples of Hard dough biscuits are

a) Marie b) Thin arrow root
c) Tea time biscuits d) All

34. __________ is the preferred packaging material for biscuits

a) Pearlized polypropylene b) HDPE
c) LDPE d) None

35. Baking powder consists of 3 parts namely

a) $NaHCO_3$ b) An acid salt
c) Filler-usually starch d) All

36. Chocolate liquor contains approximately __________ percent of cocoa butter

a) 44 b) 54
c) 64 d) 74

37. _________ flour will give best results when it is used for the sponge dough method

a) Strong flours b) Weak flours

c) Either a or b d) None

38. Strong flours requires

a) Longer mixing time b) Longer fermentation time

c) Short mixing time d) a & b

39. Over mixing causes

a) Breaks down of gluten structure

b) Slows down the fermentation

c) Low volume

d) All

40. Under mixing causes

a) Less elastic b) Low volume

c) Bread collapse in proofing d) All

41. Staling of bread is greatly a matter of

a) Proofing b) Moisture loss

c) Low volume d) None

42. The most favorable temp for the action of yeast in bread dough is between

a) 20 to 30°C b) 24 to 30°C

c) 26 to 30°C d) 24 to 34°C

43. Over Fermented dough causes

a) Stales quickly b) Soft and sticky

c) Low yield d) All

44. Under fermented dough causes

a) Improper baking b) Crumbly heavily

c) Tendency to flatten out d) All

45. Under proofing will produce bread of

a) Greater volume b) Low volume

c) No change d) None

46. Over proofing will cause
    a) Staling  b) Collapse in the oven
    c) Greater shrinkage  d) All
47. Common results of over fermentation
    a) Pale crust color  b) Poor symmetry
    c) Sticky dough  d) All
48. Lack of volume of bread is due to
    a) Use of weak flour  b) Under/over mixing
    c) Too much salt  d) All
49. Too much volume of bread is due to
    a) Over proofing  b) Cool oven
    c) Not enough salt  d) All
50. Poor texture of bread is due to
    a) Weak flour  b) Improper mixing
    c) Lack of shortening  d) All
51. Which ingredient provides structure to the cake
    a) Flour  b) Egg
    c) Milk  d) All
52. What are the ingredients which makes the cake tender
    a) Shortening  b) Sugar
    c) Baking powder  d) All
53. Cookies are also known as
    a) Small sweet cakes  b) Biscuits
    c) Small breads  d) Muffins
54. Example for fast acting part of baking powder____________
    a) Calcium acid phosphate  b) Sodium acid sulphate
    c) Sodium pyrophosphate  d) Marsh mellow
55. An enzyme possessing the power to convert maltose sugar into two units of dextrose
    a) Invertase  b) Maltase
    c) Zymase  d) All

56. _______ Is the enzyme that finally turns the trick in dough fermentation of sugars by

a) Invertase  b) Maltase
c) Zymase  d) All

57. An enzyme possessing the power to convert sucrose into glucose and fructose

a) Invertase  b) Maltase
c) Zymase  d) All

58. The most important ingredients in marsh mellow are

a) Air & moisture  b) Sugar and moisture
c) Water and moisture  d) All

59. By the process of compressing of powder in die the following are prepared

a) lozenges  b) Tablets
c) Hand boiled candy  d) Caramel

60. The traditional whipping agent used in nougat preparation is

a) Egg albumi  b) Milk
c) Air  d) None

61. An Example for hydrocolloid sweets are

a) Jellies  b) Toffee
c) Gums  d) b&c

62. As per FSSAI regulations, maximum permitted level of artificial sweetener called Aspartame (methylester) in Biscuits, Bread, Cakes and Pasteries is

a) 2000 ppm  b) 3000 ppm
c) 2200 ppm  d) 700 ppm

63. As per FSSAI regulations, maximum permitted level of artificial sweetener called Sucralose in Biscuits, Bread, Cakes and Pasteries

a) 2000 ppm  b) 3000 ppm
c) 2200 ppm  d) 750 ppm

64. As per FSSAI regulations, maximum permitted level of Sorbic acid and its salts as preservative in Flour confectionery is

a) 2000 ppm  b) 1500 ppm
c) 2200 ppm  d) 750 ppm

65. As per FSSAI regulations, permitted preservative in chapattis is__________ and allowed to add maximum of 1500 ppm

a) Sorbic acid b) Sulphur dioxide

c) Propionic acid d) Citric acid

66. As per FSSAI regulations, Butylated Hydroxanisole (BHA) is allowed to add to ready-to-eat dry breakfast cereals at ________ ppm

a) 100 b) 50

c) 40 d) 25

67. Antifoaming agents in edible oils and fats.

a) Try imethyl Polysiloxane b) Dimethyl Polysiloxane

c) a & b d) None

68. As per FSSAI regulations, Dimethyl Polysiloxane is added as an antifoaming agents in edible oils and fats for deep fat frying up to a maximum limit of______ ppm

a) 10 b) 50

c) 40 d) 25

69. __________ used, as release agent in confectionery

a) Try imethyl Polysiloxane b) Dimethyl Polysiloxane

c) Backing powder d) NaOH

70. Types of Caramel are

a) Type-I- Plain Caramel

b) Type-II-Caustic sulphite caramel

c) Type - III - Ammonia Process Caramel

d) Type-IV- Ammonia Sulphite Caramel

e) All

71. Examples of wheat flour Improvers used for bread and biscuit production

a) Fungal alpha amylase b) Bacterial amylase

c) Amylases and other enzymes d) All

72. As per FSSAI regulations, ____________ ppm of calcium carbonate is used as an improver for bread production

a) 5000 b) 3000

c) 2000 d) 6000

73. As per FSSAI regulations, ____________ ppm (maximum) of Calcium or sodium propionate or Acid sodium pyrophosphate is used as preservative for production of bread

a) 3000 b) 3000

c) 5000 d) 6000

74. As per FSSAI regulations, ____________ ppm (maximum) of Sorbic acid or its Sodium, Potassium or Calcium salts or Acid calcium phosphate are used as preservative for production of bread.

a) 3000 b) 1000

c) 5000 d) 6000

75. As per FSSAI regulations, ____________ ppm (maximum) of Ammonium carbonate is used as leavening agent in production of bread and biscuits

a) 3000 b) 1000

c) 5000 d) 6000

76. India's total wheat production (2011-12) is ______ million tons from an area of more than 28.0 million hectares.

a) 85.1 b) 88.31

c) 86.3 d) 90.4

77. India's total Rice production (2011-12) is ______ million tons from an area of more than 44.41 million hectares.

a) 102.74 b) 88.31

c) 110.3 d) 100.1

78. India's total coarse cereal grains production is _______ million tones in the year of 2011-2012

a) 32.08 b) 42.08

c) 52.08 d) 62.08

79. India's total cereal grains production is _______ million tones in the year of 2011-2012

a) 132.08 b) 242.08

c) 233.14 d) 262.08

80. India's total Pulses production is _______ million tones in the year of 2011-2012

a) 17.28 b) 12.08

c) 22.08 d) 32.08

81. India's total food grains production is _______million tones in the year of2011-2012

a) 232.08 b) 250.42

c) 233.14 d) 262.08

82. India's total Oil seeds production is _______ million tones in the year of 2011-2012

a) 17.28 b) 28.04

c) 22.08 d) 32.08

83. India's total fruit production is _________ million tones in the year of 2011-2012

a) 77.58 b) 70.34

c) .71.02 d) 70.56

84. India's total Vegetable production is _______ million tones in the year of 2011-2012

a) 147.01 b) 149.61

c) 140.14 d) 170.56

85. India's total sugar cane production is _______million tones in the year of 2011-2012

a) 387.86 b) 327.86

c) 347.86 d) 337.86

86. India's total spices production is _______million tones in the year of 2011-2012

a) 3.74 b) 5.73

c) 8.74 d) 6.78

87. The percapita Net availability of rice is _________g/day in India in the year of 2011-2012

a) 184.8 b) 170.48

c) 210.1 d) 250.1

88. The percapita Net availability of wheat is _________g/day in India in the year of 2011-2012

a) 184.8 b) 167.9

c) 210.1 d) 250.1

89. The percapita Net availability of cereals is _________ g/day in India in the year of 2011-2012

a) 484.8 b) 367.9

c) 407.0 d) 450.1

90. The percapita Net availability of pulses is _________g/day in India in the year of 2011-2012

a) 18.8 b) 17.48

c) 21.1 d) 31.6

91. The percapita Net availability of Food grains is _________g/day in India in the year of 2011-2012

a) 484.8 b) 438.6

c) 407.0 d) 450.1

92. __________ is a good source of omega 3 fatty acids

a) Cow milk b) Buffalo milk

c) Human milk d) Sheep milk

93. In omega 3 poly unsaturated fatty acids, the double bond lies between the third and fourth carbon atom counting form the ________ end

a) Carboxyl b) Methyl

c) COOH d) None

94. EPA and DHA are found in

a) Fish oil b) Algal Oil

c) Human milk d) All

95. ______________is a fat substitute

a) Olestra b) Salatrim

c) Caprenin d) Sorbesein

e) All

96. ________ is a group of structured triacylglycerols that provides the physical properties of fat but only 5 calories per gram

a) Olestra b) Salatrim

c) Caprenin d) Sorbesein

97. ______ is a low enrgy fat compound and comprises caprylic, capric and behenic acid

a) Olestra b) Salatrim

c) Caprenin d) Sorbesein

98. Fat mimetics are short chain carbohydrates or proteins used to replace fat and provides _______ calories per gram

a) 9 b) 4

c) 5 d) 7

99. Oatrim is a

a) Fat substitute

b) Carbohydrate based fat substitute

c) Oat Maltodextrin

d) All

100. Zein is a protein present in

a) Rice b) Maize

c) Wheat d) Potato

## Answers

| | | | | | | | | | |
|---|---|---|---|---|---|---|---|---|---|
| 1. b | 2. a | 3. a | 4. b | 5. d | 6. a | 7. c | 8. a | 9. b | 10. c |
| 11. b | 12. d | 13. d | 14. b | 15. d | 16. b | 17. a | 18. a | 19. b | 20. c |
| 21. b | 22. c | 23. b | 24. d | 25. d | 26. c | 27. a | 28. d | 29. c | 30. c |
| 31. c | 32. d | 33. d | 34. a | 35. d | 36. b | 37. a | 38. d | 39. d | 40. d |
| 41. b | 42. b | 43. d | 44. d | 45. b | 46. d | 47. d | 48. d | 49. d | 50. d |
| 51. d | 52. d | 53. a | 54. a | 55. b | 56. c | 57. a | 58. b | 59. b | 60. a |
| 61. d | 62. c | 63. d | 64. b | 65. a | 66. b | 67. b | 68. a | 69. b | 70. e |
| 71. d | 72. a | 73. c | 74. b | 75. c | 76. b | 77. a | 78. b | 79. c | 80. a |
| 81. b | 82. b | 83. a | 84. b | 85. c | 86. b | 87. a | 88. b | 89. c | 90. d |
| 91. b | 92. c | 93. b | 94. d | 95. e | 96. b | 97. c | 98. b | 99. d | 100. b |

# 2

# Fruits and Vegetables

1. Asepsis is a process of
   a) Removal of microorganisms
   b) Keeping out microorganisms
   c) Hindering the growth & activity of microorganisms
   d) Killing the microorganisms
2. Self-decomposition of the food is prevented by
   a) Inactivation of enzymes  b) Delay of chemical reactions
   c) a & b  d) None
3. Commercially available fruit beverages are generally pasteurized at
   a) 85°C for 25 to 30 min  b) 62.5°C for 25 to 30 min
   c) 71.1°C for 15 to 30 sec  d) 92.1°C for 15 to 30 min
4. Fruit and tomato products are generally processed at
   a) 70°C for 30 min  b) 100°C for 30 min
   c) 80°C for 30 min  d) None
5 The temperature employed for aseptic packaging food is
   a) 149°C/1 sec  b) 62.8°C/1 sec
   c) 100°C/1 sec  d) 119°C/1 sec
6. Cellar storage temperature for fruits and vegetables is
   a) 15°C  b) 0-5°C
   c) -18 to 40°C  d) 0°C
7. Refrigeration or chilling temperature for storage of fruits and vegetable is
   a) 15°C  b) 0-5°C
   c) -18 to 40°C  d) 0°C

8. Freezing temperature for storage of fruits and vegetable is
   a) 15°C b) 0-5°C
   c) -18 to -40°C d) 0°C
9. In which freezing method small ice crystals are formed
   a) Slow freezing b) Quick freezing
   c) Both a & b d) None
10. What is the best preserving method for preservation of pure juice
   a) Refrigeration b) Chilling
   c) Freezing d) Spray drying
11. What is the frozen temperature for keeping the juice in best condition for long time
   a) -12 to -17°C b) 0-5°C
   c) -18 to -40°C d) 0°C
12. Sulphur dioxide acts effectively against
   a) Bacteria b) Moulds
   c) a & b d) Yeast
13. ____________________ is the preservative, which acts as an antioxidant and bleachingagent.
   a) Sulphur dioxide b) Benzoic acid
   c) Propionic acid d) Sorbic acid
14. Naturally colored juice are preserved with
   a) Sulphur dioxide b) Benzoic acid
   c) Propionic acid d) Sorbic acid
15. Hydrogen sulphide has formed in fruit juices packed and preserved with ______________ preservative
   a) Sulphur dioxide b) Benzoic acid
   c) Propionic acid d) Sorbic acid
16. The antibacterial action of benzoic acid is increased in presence of
   a) Carbon dioxide b) Acid
   c) a & b d) None
17. Benzoic acid is more effective against
   a) Yeast b) Moulds
   c) Bacteria d) Viruses

18. Carbonation of water or beverage is

a) Dissolving of $CO_2$ b) Passing through $CO_2$

c) Exposing to $CO_2$ d) None

19. Minimal microbial inhibitory concentration of $CO_2$ in preservation of water or beverage is

a) 13.6g/Litre b) 14.6g/Litre

c) 10.6g/Litre d) 12.6g/Litre

20. Carbonation of water or beverage helps in

a) Creation of anaerobic conditions

b) Reduces oxidation of ascorbic acid

c) Prevents browning

d) All

21. At what concentration sugar acts as a preservative

a) 50% b) 66%

c) 15% d) 16%

22. Decomposition of carbohydrates by microorganisms or enzymes is called as

a) Putrefaction b) Fermentation

c) Proteolysis d) Lipolysis

23. At what concentration alcohol acts as a preservative in wines

a) 12% b) 13%

c) 14% d) 7%

24. At what concentration acetic acid acts as a preservative in fruits and vegetable products

a) 2% b) 4%

c) 10% d) 3.6%

25. An example antibiotic used for food preservation

a) Nisin b) Subtilin

c) Pimaracin d) All

26. Nisin is produced by

a) *Streptococcus lactis* b) *Bacillus subtilis*

c) Both a & b d) None

27. Subtilin is produced by

a) *Streptococcus lactis*  b) *Bacillus subtilis*
c) Both a & b  d) None

28. As per FSSAI regulations 2011, maximum permitted dose of Irradiation (KGY) for Onion is

a) 0.09  b) 0.03
c) 0.06  d) 0.02

29. As per FSSAI regulations 2011, maximum permitted dose of Irradiation (KGY) for Mango is

a) 0.19  b) 0.75
c) 0.26  d) 0.12

30. As per FSSAI regulations 2011, maximum permitted dose of Irradiation (KGY) for Potatoes is

a) 0.15  b) 0.75
c) 0.26  d) 0.12

31. As per FSSAI regulations 2011, maximum permitted dose of Irradiation (KGY) for Raisins, Figs and Dried Dates is

a) 0.15  b) 0.75
c) 0.26  d) 0.12

32. Cold sterilization is also called as

a) Microwave cooking  b) High pressure processing
c) Irradiation  d) Ohmic heating

33. Weight of the dried sample by drained weight of the rehydrated sample is

a) Rehydration ratio
b) Rehydration Coefficient
c) both a & b
d) Per cent water in the rehydrated material

34. In which freezing method large ice crystals will be formed

a) Sharp freezing  b) Slow freezing
c) Quick freezing  d) a & b

35. What are the refrigerants used in cryogenic freezing

a) Liquid nitrogen  b) Liquid carbon dioxide
c) a & b  d) CFC

36. In which freezing method partial dehydration of foods is involved
    a) Dehydro-freezing b) Freeze drying
    c) Cryogenic freezing d) IQF
37. The physical effects of freezing of foods are
    a) Expansion of volume b) Ice crystals formation
    c) Size of crystals d) All
38. What is the expansion volume percentage of water changes state from liquid to solid
    a) 5% b) 9%
    c) 50% d) 90%
39. After thawing extreme undesirable texture changes are observed in
    a) Potatoes b) Tomatoes
    c) a & b d) Brinjal
40. Hermetically sealing and sterilizing of food stuffs by heat in containers is called as
    a) Canning b) Pouching
    c) Packaging d) All
41. Father of Canning
    a) Appert b) Saddington
    c) Peter Durand d) William Underwood
42. Canning is also known as
    a) Sterilization b) Cold sterilization
    c) Appertization d) None
43. Who obtained the first British Patent on canning of foods in tin containers
    a) Appert b) Saddington
    c) Peter Durand d) William Underwood
44. The process of removal of air from cans is known as
    a) Evacuation
    b) Exhausting
    c) Vaccum
    d) All

45. Exhausting of cans helps in
    a) Minimizes discoloration
    b) Avoids corrosion of the tinplate
    c) Reduces chemical reaction
    d) All
46. What is the temperature of water used in exhaust box during exhausting
    a) 82 to 87°C
    b) 82 to 100°C
    c) 90 to 100°C
    d) 62 to 90°C
47. What is the temperature for sealing of cans in canning process
    a) Not less than 74°C
    b) Not less than 64°C
    c) Not less than 71.1°C
    d) None
48. An example of non acid (low acid) foods
    a) Peas & Corn
    b) Lima bean
    c) Potato & Spinach
    d) All
49. Lye peeling is done for
    a) Peaches & apricots
    b) Carrots and sweet potatoes
    c) Sweet oranges, Mandarin oranges
    d) All
50. What is the concentration of caustic soda (lye) solution is used in lye peeling
    a) 10-20%
    b) 5-10%
    c) 1-2%
    d) 0.1- 0.2%
51. Flame peeling is done for
    a) Onion
    b) Ginger
    c) Garlic
    d) a & c
52. Blanching of fruits and vegetables helps in
    a) Inactivates enzymes
    b) Reduces area to surface volume
    c) Enhances the green colour of vegetables
    d) All
53. What is the pH range of non acid (low acid) foods
    a) > 5.0
    b) 4.5 to 5.0
    c) 3.7 to 4.5
    d) <3.7

54. What is the pH range of Medium acid foods

    a) > 5.0 b) 4.5 to 5.0

    c) 3.7 to 4.5 d) <3.7

55. What is the pH range of acid foods

    a) > 5.0 b) 4.5 to 5.0

    c) 3.7 to 4.5 d) <3.7

56. What is the pH range of high acid foods

    a) > 5.0 b) 4.5 to 5.0

    c) 3.7 to 4.5 d) <3.7

57. An example of medium acid foods

    a) Turnip & carrot b) Okra & cabbage

    c) Pumpkin& green bean d) All

58. An example of acid foods

    a) Tomato & Pear b) Banana & jackfruit

    c) Mango &Pineapple d) All

59. An example of acid foods

    a) Apple b) Tomato

    c) Peach d) All

60. An example of high acid foods

    a) Rhubarb b) Sauerkraut

    c) Pickle d) All

61. Sauerkraut is a

    a) Acid food b) Medium acid food

    c) High acid food d) Non acid food

62. Tomato is a

    a) Acid food b) Medium acid food

    c) High acid food d) Non acid food

63. Apple is a

    a) Acid food b) Medium acid food

    c) High acid food d) Non acid food

64. An example of high acid foods
   a) Pickle  b) Potatoes
   c) Tomato  d) Banana
65. Mango is a
   a) Acid food  b) Medium acid food
   c) High acid food  d) Non acid food
66. Soups and sauces are examples of
   a) Acid food  b) Medium acid food
   c) High acid food  d) Non acid food
67. Corn is an example of
   a) Acid food  b) Medium acid food
   c) High acid food  d) Non acid food
68. A canned product having pH less than 4.5 can be processed at
   a) 121°C  b) 100°C
   c) 115°C  d) 150°C
69. A canned product having pH more than 4.5 can be processed at
   a) 121°C  b) 100°C
   c) 105°C  d) 150°C
70. Thermally processed cans are cooled at
   a) 39°C  b) 50°C
   c) 15°C  d) 10°C
71. Tin cans are made of thin steel plate of
   a) Low carbon content  b) Medium carbon content
   c) High carbon content  d) All
72. Coating of cans with a material, which can prevents discoloration of foods is called as
   a) Lacquering  b) Layering
   c) Deposition  d) None
73. Cans coated with acid-resistant lacquer are called
   a) C-enamel  b) R-enamel
   c) A.R cans  d) b & c

74. R-enamel coated cans are used for packing of acid fruits

a) Raspberry
b) Strawberry & Red plum
c) Colored grapes
d) All

75. Can coated with Sulphur-resistant lacquer is called

a) C-enamel
b) A.R-enamel
c) S.R can
d) a & c

76. Non-acid foods are generally packed in

a) C-enamel cans
b) A.R-enamel cans
c) R-enamel cans
d) a & c

77. C-enamel cans are used for

a) Pea
b) Corn
c) Lima bean & kidney bean
d) All

78. Corn is packed in

a) C-enamel
b) A.R-enamel
c) S.R can
d) a & c

79. A perfect can with good vacuum become bulged is termed as

a) Swell
b) Blower
c) a & b
d) None

80. Bulging of can is due to the hydrogen gas produced by the action of food acids on the metal of the can is called as

a) Swell
b) Hard swell
c) Hydrogen swell
d) Springer

81. Can appears normal, but when struck against a tabletop onc or both ends become convex and springs or flips is termed as

a) Swell
b) Flipper
c) Hydrogen swell
d) Springer

82. Flipper in canned products is due to

a) Overfilling
b) Under-exhausting
c) Gas pressure due to spoilage
d) All

83. A mild swell at one or both ends of a can is called as

a) Swell
b) Flipper
c) Hydrogen swell
d) Springer

84. _____ is the final stage of swell. The bulged ends cannot be pressed back to normal position and the cans ultimately burst

a) Swell
b) Flipper
c) Hard swell
d) Springer

85. Which organic acid check the formation of hydrogen swell in canned low acid fruits

a) Citric acid
b) Phosphoric acid
c) Tartaric acid
d) Succinic acid

86. What is the percentage of citric acid is required to check swell in canned low acid fruits

a) 1%
b) 0.5%
c) 5%
d) 10%

87. What is the minimum head space to be maintained in can before exhausting

a) 0.5 to 1.5 cm
b) 0.6 to 0.9 cm
c) 0.1 to 0.5cm
d) 2 to 5 cm

88. A leaking can is known as a

a) Springer
b) Leakage
c) Breathing
d) Flipper

89. Leakage of can is due to

a) Defective seaming
b) Nail holes
c) Excessive internal pressure
d) All

90. A very tiny leak in can is called as

a) Springer
b) Leakage
c) Breathing
d) Flipper

91. Improper cooling and distortion of the can which resembles a swell is called as

a) Buckling
b) Leakage
c) Breathing
d) Flipper

92. Formation of peak or small ridge on the spoiled can is known as

a) Buckling
b) Leakage
c) Peaking
d) Flipper

93. Enzymatic browning is caused by an enzyme

a) Polyphenol oxidase b) Phosphatase

c) Lipase d) Protease

94. Non-enzymatic browning (Maillard reactions) is caused by reactions between

a) Nitrogenous compounds and sugars

b) Nitrogenous compounds and organic acids

c) Sugars and organic acids

d) All

95. Metallic contamination in canned foods is mainly due

a) Copper b) Iron

c) Magnesium d) a & b

96. Improper cooling of processed cans is causes

a) Buckling b) Stack burning

c) Peaking d) Flipper

97. What is the temperature of superheated steam in flash evaporators for concentration of fruit juices

a) 120°C b) 150°C

c) 115°C d) 250°C

98. ________________ is a polymer used in reverse osmosis membrane

a) Cellulosicesters

b) Mixed esters of cellulose with acetate/propionate/butyrate

c) Polyacrylonitrile and its copolymers

d) All

99. What is the moisture content in intermediate moisture foods

a) 10-25% b) 20-90%

c) 20-50% d) 10-15%

100. The following one is a right statement

a) Intermediate moisture foods do not requires refrigeration

b) Intermediate moisture foods requires refrigeration

c) Both a & b

d) None

## Answers

| | | | | | | | | | |
|---|---|---|---|---|---|---|---|---|---|
| 1. b | 2. c | 3. a | 4. b | 5. a | 6. a | 7. b | 8. c | 9. b | 10. c |
| 11. a | 12. c | 13. a | 14. b | 15. a | 16. c | 17. a | 18. a | 19. b | 20. d |
| 21. b | 22. b | 23. c | 24. a | 25. d | 26. a | 27. b | 28. a | 29. b | 30. a |
| 31. b | 32. c | 33. a | 34. d | 35. c | 36. a | 37. d | 38. b | 39. c | 40. a |
| 41. a | 42. c | 43. c | 44. b | 45. d | 46. b | 47. a | 48. d | 49. d | 50. c |
| 51. d | 52. d | 53. a | 54. b | 55. c | 56. d | 57. d | 58. d | 59. d | 60. d |
| 61. c | 62. a | 63. a | 64. a | 65. a | 66. b | 67. d | 68. b | 69. a | 70. a |
| 71. a | 72. a | 73. d | 74. d | 75. d | 76. a | 77. d | 78. d | 79. c | 80. c |
| 81. b | 82. d | 83. d | 84. c | 85. a | 86. b | 87. b | 88. b | 89. d | 90. c |
| 91. a | 92. c | 93. a | 94. d | 95. d | 96. b | 97. b | 98. d | 99. c | 100. a |

1. An example of natural intermediate moisture food
   a) Honey  b) Raisins
   c) Jelly  d) All
2. An example of intermediate moisture food
   a) Jam & Jellies  b) Fruit cake
   c) Sweetened condensed milk  d) All
3. The following one is a partially dried and comes under intermediate moisture food
   a) Figs  b) Dates
   c) Pemmican  d) All
4. When a food is in moisture equilibrium with its environment then the water activity of food will be quantitatively equal to the
   a) RH×100  b) RH/100
   c) 100/RH  d) None
5. The minimum water activity below which most important food bacteria will not grow is
   a) 0.99  b) 0.95
   c) 0.90  d) 0.93
6. The water activity of semi moist textured products is in between
   a) 0.70 & 0.85  b) 0.91 &0.99
   c) 0.85 & 0.90  d) None
7. Hurdle of reduced water activity can be done by
   a) Drying  b) Curing
   c) Conserving  d) All
8. What is the traditionally processed and preserved sea food of Japan by application of hurdle concept
   a) Surmi  b) Sushi
   c) Suko  d) Sikoo
9. __________a process where food is exposed to ionizing energy, utilizing gamma photons emitted by radioisotopes
   a) Irradiation  b) Microwave cooking
   c) Dielectric heating  d) None

10. What are the source of irradiation
    a) Radioisotopes of $^{60}Co$ or $^{137}Cs$
    b) Machine-generated X-rays ("Bremsstrahlung") of max. 5 MeV,
    c) Accelerated electronsof max. 10 MeV kinetic energy
    d) All
11. What is the unit of irradiation
    a) kilo gray (kGy) b) kilo gram (kGy)
    c) kilo c-ray (kGy) d) None
12. Gy is equal to _____J/kg absorbed energy
    a) 1 J/kg b) 100 J/kg
    c) 10 J/kg d) 1000 J/kg
13. What is the general food irradiation dose requirement for Insect disinfestation
    a) 0.03 to 0.12 kGy b) 0.2 to 0.8 kGy
    c) 0.1 to 3 kGy d) 0.5 to 3 kGy
14. What is the general food irradiation dose requirement for Parasite disinfestation
    a) 0.03 to 0.12 kGy b) 0.2 to 0.8 kGy
    c) 0.1 to 3 kGy d) 0.5 to 3 kGy
15. What is the food irradiation dose requirement for Shelf-life extension ("radurization")
    a) 0.03 to 0.12 kGy b) 0.2 to 0.8 kGy
    c) 0.1 to 3 kGy d) 0.5 to 3 kGy
16. What is the food irradiation dose requirement for Production of meat, poultry and fishery products shelf-stable at ambient temperature ("radappertization")
    a) 0.03 to 0.12 kGy b) 25 to 60 kGy
    c) 0.1 to 3 kGy d) 15 to 30 kGy
17. What is the food irradiation dose requirement for Elimination of non-spore forming pathogenic bacteria ("radicidation")
    a) 1.5 to 7 kGy b) 25 to 60 kGy
    c) 0.1 to 30 kGy d) 15 to 30 kGy

18. What are the right foods for radiation decontamination

    a) Poultry meat/ carcasses b) Egg products & Fish products

    c) Spices & dry ingredients d) All

19. X-ray machines convert electron energy to electromagnetic X-rays called

    a) Baramsstrahlung b) Bremsstrahlung

    c) Bourdenstrahgling d) None

20. Gamma rays and X-rays have ________ penetrating characteristics than that of accelerated electrons

    a) Low b) High

    c) No change d) None

21. Microbial inactivation by ionizing radiation is mainly due to

    a) Damage of nucleic acids b) Radiolysis of water

    c) Both a & b d) None

22. ________ is a is a product made by boiling fruit pulp with sufficient sugar to a reasonably thick consistency, firm enough to hold the fruit tissues in position

    a) Jam b) Jelly

    c) Fruit bar d) Fruit concentrate

23. Tutti-frutti is prepared from

    a) Pieces of fruit b) Fruit scraping

    c) Pulp adhering to cores d) All

24. What is the ideal content of Invert sugar in jam

    a) 30 to 50 % b) 20 to 30%

    c) 10 to 20% d) 5 to 10%

25. What will happen to jam, if invert sugar is more than 50%

    a) Become thick viscous honey like mass

    b) Forms small crystals of glucose

    c) a & b

    d) None

26. Stickiness or gumminess of jam is due to

    a) High TSS b) Low TSS

    c) High pectin d) All

27. Stickiness or gumminess of jam can be solved by addition of
    a) Pectin b) Citric acid
    c) both a & b d) Sugar
28. Premature setting of jam is due to
    a) Low TSS b) High pectin
    c) a & b d) High sugar
29. Premature setting of jam can be prevented by
    a) Addition of more sugar b) Addition of high pectin
    c) Addition of excess acid d) All
30. Surface graining and shrinkage of jam is caused by
    a) Loss of moisture b) Microbial action
    c) Low acid d) All
31. As per FSSAI regulations-2011, maximum permitted level of Sorbic acid & its salts in Jams is
    a) 100 ppm b) 500 ppm
    c) 750 ppm d) 1500 ppm
32. As per FSSAI regulations-2011, maximum permitted level of Sulphur dioxide in Jams is
    a) 40 ppm b) 20 ppm
    c) 50 ppm d) 65 ppm
33. As per FSSAI regulations-2011, maximum permitted level of Benzoic acid in Jams is
    a) 400 ppm b) 200 ppm
    c) 500 ppm d) 650 ppm
34. As per FSSAI regulations-2011, Jam manufacture from not less than ______ percent of original prepared fruit
    a) 65 b) 45
    c) 25 d) 68
35. As per FSSAI regulations-2011, TSS (m/m) of Jam not less than
    a) 65% b) 45%
    c) 25% d) 68%

36. As per FSSAI regulations-2011, Jam of strawberry or raspberry may manufacture from not less than ______ percent of original prepared fruit

a) 65 b) 45

c) 25 d) 68

37. Commercially available pectin is in _______ color

a) White b) Brown

c) a & b d) Orange

38. Pectin is building blocks of

a) Galacturonic acid. b) Gluconic acid

c) Fructonic acid d) xylose acids

39. Pectin is commercially extracted from

a) Mango b) Citrus

c) Pineapple d) Guava

40. Pectin is used as a

a) Jelling agent b) Stabilizing agent

c) Thickening agent d) All

41. What is the typical dose of pectin as a food additive

a) 0.5-1.0% b) 1.5-5.0%

c) 2.5-10.0% d) 3.5-8.0%

42. The speed of gelatin of pectin is depends on

a) Degree of methoxylation b) Degree of methyalation

c) Degree of sugars d) All

43. Pectin having degree of methoxylation of more than 50% is called

a) Low methoxyl [LM] pectin b) High methoxyl [HM] pectin

c) Medium methoxyl [MM] pectin d) None

44. Pectin having degree of methoxylation of less than 50% is called

a) Low methoxyl [LM] pectin b) High methoxyl [HM] pectin

c) Medium methoxyl [MM] pectin d) None

45. Strength of pectin jellies depends on

a) Quantity of pectin b) Quantity of acid

c) Quantity of sugar d) All

46. ___________ is a semi-solid product prepared by boiling a clear, strained solution of pectin containing fruit extract, free from pulp, after the addition of sugar and acid

a) Jam b) Jelly

c) Preserve d) Candy

47. As per FSSAI regulations-2011, Jelly manufacture from not less than ______percent of original prepared fruit

a) 65 b) 45

c) 25 d) 68

48. As per FSSAI regulations-2011, TSS (m/m) of Jelly not less than

a) 65% b) 45%

c) 25% d) 68%

49. An example of fruit rich in pectin and acid

a) Sour and crab apple b) Sour guavas

c) Lemon and oranges (sour) d) All

50. An example of fruit rich in pectin but low in acid

a) Apple (low acid varieties) b) Unripe banana

c) Peel of orange and grapefruit d) All

51. An example of fruit low in pectin but rich in acid

a) Apricot (sour) b) Pineapple

c) Strawberry d) All

52. An example of fruit low in pectin and acid

a) Ripe apricot b) Peach (ripe)

c) Pomegranate d) All

53. What are the essential ingredients in jelly production

a) Pectin b) Sugar

c) Acid & water d) All

54. Firmness of fruits is due to

a) Calcium pectate b) Fruit sugars

c) Organic acids d) All

55. Which enzyme is responsible for production of pectin from protopectin

a) Pectinase b) Protopectinase

c) Calcium pectinase d) Pectic methyl esterase

56. Why immature and overripe fruits are not suitable for production of jelly
    a) Presence of Pectinase
    b) Presence of Protopectinase
    c) Presence of Calcium pectinase
    d) Presence of Pectic methyl esterase

57. Pectin content is determined by
    a) Alcohol test
    b) Jelmeter test
    c) a & b
    d) None

58. In alcohol test for pectin extract, a single, transparent lump or clot indicates in jelly production is
    a) An equal amount of sugar is to be added to the extract
    b) ¾ th amount of sugar is to be added to the extract
    c) ½ amount of sugar is to be added to the extract
    d) None

59. In alcohol test for pectin extract , a clot will be less firm and fragmented indicates in jelly production is
    a) An equalamount of sugar is to be added to the extract
    b) ¾ th amount of sugar is to be added to the extract
    c) ½ amount of sugar is to be added to the extract
    d) None

60. In alcohol test for pectin extract , a numerous small granular clots indicates in jelly production is
    a) An equalamount of sugar is to be added to the extract
    b) ¾ th amount of sugar is to be added to the extract
    c) ½ amount of sugar is to be added to the extract
    d) None

61. Which acid is produces better results in jelly production
    a) Citric acid
    b) Malic acid
    c) Tartaric acid
    d) Phosphoric acid

62. The optimum pH for a jelly containing 1 per cent pectin is approximately 3.0, 3.2 and 3.4 for __________ per cent TSS, respectively.
    a) 60, 65 and 70
    b) 65, 60 and 70
    d) 70, 65 and 60
    d) 70,60 and 65

63. The optimum pH value for jelly is
    a) 3.0    b) 3.2
    c) 3.4    d) 3.6
64. End point of jelly is judged by
    a) Sheet or flake test    b) Drop test
    c) Temperature test    d) All
65. A TSS of 65 % in Jelly obtained at temperature
    a) 125°C    b) 105°C
    c) 100°C    d) 98°C
66. The phenomenon of spontaneous exudation of fluid from a gel is called
    a) Syneresis    b) Weeping
    c) a & b    d) None
67. Syneresis or weeping of jelly is caused by
    a) Excess acid    b) Insufficient pectin
    c) Premature gelation    d) All
68. Suspension of shredded peels of orange, lemon in product is called as
    a) Murrabba    b) Marmalade
    c) Preserve    d) Candy
69. As per FSSAI regulations-2011, Marmalades manufacture from not less than ______percent of original prepared fruit except fruit peel
    a) 65    b) 45
    c) 25    d) 68
70. As per FSSAI regulations-2011, TSS (m/m) of Marmalade not less than
    a) 65%    b) 45%
    c) 25%    d) 68%
71. As per FSSAI regulations-2011, peel in suspension of Marmalade not less than
    a) 10%    b) 5%
    c) 15%    d) 20%
72. As per FSSAI regulations-2011, maximum permitted level of Sorbic acid & its salts in Marmalade is
    a) 100 ppm    b) 500 ppm
    c) 750 ppm    d) 1500 ppm

73. As per FSSAI regulations-2011, maximum permitted level of Sulphur dioxide in Marmalade is

a) 40 ppm b) 20 ppm

c) 50 ppm d) 65 ppm

74. As per FSSAI regulations-2011, maximum permitted level of Benzoic acid in Marmalade is

a) 400 ppm b) 200 ppm

c) 500 ppm d) 650 ppm

75. A mature fruit/vegetable or its pieces impregnated with heavy sugar syrup till it becomes tender and transparent is known as

a) Jam b) Jelly

c) Preserve d) Marmalade

76. A fruit/vegetable impregnated with cane sugar or glucose syrup, and subsequently drained free of syrup and dried, is known as

a) Jam b) Candy

c) Preserve d) Marmalade

77. Covering of candied fruits/vegetables with a thin transparent coating of sugar, which imparts them a glossy appearance, is known as

a) Glossiness b) Glazing

c) Translucent d) All

78. Covering or coating of candied fruits/vegetables with crystals of sugar and allowing to deposit on it is called as

a) Crystallized fruits b) Candied fruit

c) Fruit Preserve d) Marmalade

79. The floating of fruit preserve or candied fruit in jar is due to

a) Filling the preserve without cooling

b) Fermentation

c) Inadequate blanching

d) Cooking of fruits in heavy syrup

80. As per FSSAI regulations-2011, maximum permitted level of Sulphur dioxide in Crystallized glace or Cured fruit (including candid peel) is

a) 150 ppm b) 200 ppm

c) 250 ppm d) 650 ppm

81. As per FSSAI regulations-2011, maximum permitted level of Sulphur dioxide in Preserve canned cherry and fruit jelly is

a) 40 ppm b) 20 ppm
c) 50 ppm d) 65 ppm

82. As per FSSAI regulations-2011, maximum permitted level of Benzoic acid in Preserve canned cherry and fruit jelly is

a) 400 ppm b) 200 ppm
c) 500 ppm d) 650 ppm

83. As per FSSAI regulations-2011, maximum permitted level of Sorbic acid & its salts in preserve, crystallized glazed or candid fruits including candid peels fruit bars is

a) 100 ppm b) 500 ppm
c) 750 ppm d) 1500 ppm

84. A good quality chutney should be

a) Palatable b) Appetizing
c) a & b d) None

85. Preservation of food in common salt or in vinegar is

a) Chutney b) Pickling
c) Natural storage d) All

86. Pickling is the result of

a) Salting b) Acidification
c) Fermentation d) Addition of Spices

87. Which group of bacteria is responsible for fermentation of pickle

a) Lactic acid bacteria b) Acetic acid bacteria
c) both a & b d) None

88. Most lactic acid bacteria are active at

a) 50°C b) 40°C
c) 30°C d) 60°C

89. What is the optimum concentration of salt to inhibit most of the bacteria

a) 5% b) 10%
c) 15% d) 25%

90. What should be the ideal concentration of acetic acid in finished pickle
    a) 10% b) 20%
    c) 2% d) 8%
91. Sauerkraut is
    a) Acid cauliflower b) Acid cabbage
    c) Acid lime d) Acid Potato
92. Lactic acid concentration in Sauerkraut is not less than
    a) 1.5% b) 5%
    c) 2.5% d) 10%
93. Stimulation of peptic glands and mild laxative property of Sauerkraut is due to
    a) Acetylcholine b) Lactylcholine
    c) a & b d) None
94. Ideal cabbage for production of Sauerkraut is
    a) Cabbage with low polyphenols b) White cabbage
    c) a & b d) None
95. Shriveling is a problem in pickle, due to
    a) Strong solution of salt b) Strong solution of sugar
    c) Strong solution of Vinegar d) All
96. Softness and slipperiness of pickle is due to
    a) Weak brine b) Strong brine
    c) Strong acid d) All
97. Blackening of pickle is due to
    a) Iron in the brine b) Metallic contamination
    c) a & b d) None
98. As per FSSAI regulations-2011, maximum permitted level of Sulphur dioxide in Pickles and chutneys made from fruits or vegetables is
    a) 100 ppm b) 200 ppm
    c) 500 ppm d) 650 ppm
99. As per FSSAI regulations-2011, maximum permitted level of Benzoic acid in Pickles and chutneys made from fruits or vegetables is
    a) 400 ppm b) 250 ppm
    c) 500 ppm d) 650 ppm

100. As per FSSAI regulations-2011, Mould count in 25 gm / ml of Pickle is
   a) Absent b) 10
   c) 40 d) 100

## Answers

| | | | | | | | | | |
|---|---|---|---|---|---|---|---|---|---|
| 1. a | 2. d | 3. d | 4. b | 5. c | 6. a | 7. d | 8. a | 9. a | 10. d |
| 11. a | 12. a | 13. b | 14. c | 15. d | 16. b | 17. a | 18. d | 19. b | 20. b |
| 21. c | 22. a | 23. d | 24. a | 25. c | 26. a | 27. c | 28. c | 29. a | 30. a |
| 31. b | 32. a | 33. b | 34. b | 35. a | 36. c | 37. c | 38. a | 39. b | 40. d |
| 41. a | 42. a | 43. b | 44. a | 45. d | 46. b | 47. b | 48. a | 49. d | 50. d |
| 51. d | 52. d | 53. d | 54. a | 55. b | 56. d | 57. c | 58. a | 59. b | 60. c |
| 61. c | 62. a | 63. b | 64. d | 65. b | 66. c | 67. d | 68. b | 69. b | 70. a |
| 71. b | 72. b | 73. a | 74. b | 75. c | 76. b | 77. b | 78. a | 79. a | 80. a |
| 81. a | 82. b | 83. b | 84. c | 85. b | 86. c | 87. a | 88. c | 89. c | 90. c |
| 91. b | 92. a | 93. c | 94. c | 95. d | 96. a | 97. c | 98. a | 99. b | 100.a |

1. As per FSSAI regulations-2011, mould count in 25 gm / ml of Candied and Crystalised or Glazed Fruit and Peel is

   a) Absent b) 10

   c) 40 d) 100

2. As per FSSAI regulations-2011, percentage of total sugar (w/w)) in Candied, Crystallised and Glazed Fruit / Vegetable / Rhizome / Fruit Peel is not less than

   a) 65% b) 45%

   c) 70% d) 68%

3. As per FSSAI regulations-2011, percentage of reducing Sugar to total sugar in Candied, Crystallised and Glazed Fruit / Vegetable / Rhizome / Fruit Peel is not less than

   a) 45% b) 25%

   c) 30% d) 65%

4. As per FSSAI regulations-2011, total soluble solids (m/m) of Murabba is not less than

   a) 45% b) 65%

   c) 70% d) 55%

5. As per FSSAI regulations-2011, Fruit contents (m/m) of Murabba is not less than

   a) 45% b) 65%

   c) 70% d) 55%

6. As per FSSAI regulations-2011,Total soluble solids (m/m) of Fruit Chutney is not less than

   a) 45% b) 65%

   c) 50% d) 55%

7. As per FSSAI regulations-2011, total soluble solids (m/m) of Vegetable Chutney is not less than

   a) 45% b) 65%

   c) 25 % d) 55%

8. As per FSSAI regulations-2011, total soluble solids (m/m) of Hot and Sour (Spicy Chutney) is not less than

   a) 45% b) 65%

   c) 25 % d) 55%

9. As per FSSAI regulations-2011, Fruits and Vegetable content (m/m) in final Chutney is not less than

a) 45% b) 40%

c) 25% d) 55%

10. As per FSSAI regulations-2011, pH of final Chutney is not more than

a) 5.4 b) 4.6

c) 6.0 d) 2.0

11. As per FSSAI regulations-2011, drained weight of fruits and vegetable Pickled in citrus juice or brine or edible oil or vinegar is not less than

a) 45% b) 40%

c) 25% d) 60%

12. As per FSSAI regulations-2011, Sodium Chloride content when fruits and vegetable pickled packed in Brine is not less than

a) 15% b) 10%

c) 12% d) 16%

13. Black ring in the neck of bottle is a common problem in

a) Sauces b) Ketchup

c) a & b d) None

14. Spice bag method is related to

a) Sauces b) Ketchup

c) a & b d) None

15. _________ is the product prepared by blending tomato juice/Puree/Paste, salt, sugar, vinegar and other food additives is called as

a) Sauces b) Ketchup

c) a & b d) None

16. As per FSSAI regulations-2011, total soluble solids (m/m) Salt free basis of Tomato Ketchup and Tomato Sauce is not less than

a) 15% b) 25%

c) 30% d) 50%

17. As per FSSAI regulations-2011, acidity (in terms of acetic acid) of Tomato Ketchup and Tomato Sauce is not less than

a) 1% b) 1.5%

c) 2% d) 0.5%

18. As per FSSAI regulations-2011, maximum permitted level of Benzoic acid & its salts in Ketchup and Tomato Sauce is not more than

a) 400 ppm
b) 750 ppm
c) 500 ppm
d) 650 ppm

19. As per FSSAI regulations-2011, maximum permitted level of Sorbic acid & its salts in Ketchup and Tomato Sauce is not more than

a) 400 ppm
b) 750 ppm
c) 1000 ppm
d) 650 ppm

20. As per FSSAI regulations-2011, maximum permitted level of thickening agents in production of Ketchup and Tomato Sauce is not more than

a) 1.0%
b) 0.5%
c) 1.5%
d) 2.0%

21. As per FSSAI regulations-2011, Total plate Count in Tomato ketchup and Tomato Sauce is not more than per ml is

a) 5000
b) 10000
c) 15000
d) 2000

22. As per FSSAI regulations-2011, Yeast and spores in Tomato ketchup and Tomato Sauce is not more than ________per 1/60 c.m.m

a) 100
b) 500
c) 125
d) 200

23. Deaeration is removal of

a) Oxygen
b) Vaccum
c) Carbondioxide
d) All

24. An example of filter aid for juice processing

a) Char coal
b) Diatomaceous earth
c) Cellulose pads
d) All

25. As per FSSAI regulations-2011, total soluble solids (m/m) salt free basis in Chilli Sauce is not less than

a) 15%
b) 8%
c) 3%
d) 30%

26. As per FSSAI regulations-2011, total soluble solids (m/m) salt free basis in Fruits/Vegetable Sauces is not less than

a) 15%
b) 8%
c) 3%
d) 30%

27. As per FSSAI regulations-2011, total soluble solids (m/m) salt free basis in Culinary Paste/ Sauce is not less than

a) 15% b) 8%

c) 3% d) 30%

28. As per FSSAI regulations-2011, total soluble solids (m/m) salt free basis in Ginger Paste is not less than

a) 15% b) 8%

c) 3% d) 30%

29. As per FSSAI regulations-2011, TSS (m/m) of soya sauce is not less than

a) 15% b) 8%

c) 3% d) 25%

30. As per FSSAI regulations-2011, TSS (w/w) of Tomato puree is not less than

a) 15% b) 9%

c) 3% d) 25%

31. As per FSSAI regulations-2011, TSS (w/w) of Tomato paste is not less than

a) 15% b) 9%

c) 3% d) 25%

32. As per FSSAI regulations-2011, Total soluble solids (m/m) (on dilution on ready to serve basis) of soup powder is not less than

a) 15% b) 9%

c) 5% d) 25%

33. Spray residues of arsenic and lead on fruit surface are removed by washing with

a) Dilute HCl b) Dilute alkali

c) a & b d) None

34. Clear sparkling juice is called as

a) Squash b) Crush

c) Cordial d) Nector

35. Important filter aids for juice filtration are

a) Supercel b) Kieselguhr

c) Spanish clay and bentonite d) All

36. Clarification of grape juice by freezing is carried out at
    a) 0-4°C  b) -18°C
    c) -40°C  d) -10°C
37. Chemicals used for juice clarification are
    a) Gelatin  b) Tannin
    c) Albumen & Casein  d) All
38. What is the requirement of Gelatin & Tannin mixture for clarification of 455 lit of juice
    a) 42 g of tannin +85 g of gelatin  b) 42 g of tannin+ 65 g of gelatin
    c) 52 g of tannin+ 65 g of gelatin  d) 52 g of tannin+ 85 g of gelatin
39. What is the requirement of Gelatin & Tannin mixture for clarification of 2500 lit of lime juice
    a) 42 g of tannin +85 g of gelatin  b) 213 g of tannin+ 283 g of gelatin
    c) 52 g of tannin+ 65 g of gelatin  d) 283 g of tannin+ 213 g of gelatin
40. An example of enzyme preparations for juice clarification
    a) Pectinol  b) Filtragol
    c) a & b  d) None
41. Ascorbic acid is a
    a) Antioxidant  b) Vitamin
    c) Acid regulator  d) a & b
42. What is the optimal head space in bottling of juices
    a) 1.5 to 2.5 cm  b) 0.5 to 1.0 cm
    c) 1.2 to 1.5 cm  d) 2.0 to 5 cm
43. Sweetened juice contains
    a) 85% juice  b) 10% TSS
    c) a & b  d) 40% TSS
44. ________ is a clear product free from any cellular matter,
    a) Cordial  b) Squash
    c) Nector  d) Crush
45. As per FSSAI regulations-2011, minimum percentage of fruit juice/ puree in the final product of Squash, Crush, Fruit syrup, Fruit sherbats, Cordial & barley water is
    a) 25%  b) 45%
    c) 10%  d) 30%

46. As per FSSAI regulations-2011, total soluble solids of squash is not less than

    a) 40% b) 55%

    c) 65% d) 30%

47. As per FSSAI regulations-2011, total soluble solids of Crush is not less than

    a) 40% b) 55%

    c) 65% d) 30%

48. As per FSSAI regulations-2011, total soluble solids of Fruit syrup/Fruit sherbats is not less than

    a) 40% b) 55%

    c) 65% d) 30%

49. As per FSSAI regulations-2011, total soluble solids of Cordial & barley water is not less than

    a) 40% b) 55%

    c) 65% d) 30%

50. As per FSSAI regulations-2011, acidity expressed as citric acid in the final product of Squash, Crush, Fruit syrup, Fruit sherbats and Cordial is not more than

    a) 1.5% b) 2.5%

    c) 3.5% d) 5%

51. As per FSSAI regulations-2011, acidity expressed as citric acid in the final product of barley water is not more than

    a) 1.5% b) 2.5%

    c) 3.5% d) 5%

52. Any syrup/ sharbats containing a minimum of ________ percent of dry fruits shall also qualify to be called as fruits syrups.

    a) 10% b) 90%

    c) 65% d) 25%

53. As per FSSAI regulations-2011, total soluble solids of Synthetic syrup or sharbat is not less than

    a) 40% b) 55%

    c) 65% d) 30%

54. As per FSSAI regulations-2011, total soluble solids of Carbonated Fruit Beverages or Fruit Drink is not less than

a) 40% b) 15%

c) 10% d) 30%

55. As per FSSAI regulations-2011, Fruit content (m/m) of Carbonated Fruit Beverages or Fruit Drink is not less than

a) 40% b) 15%

c) 10% d) 30%

56. As per FSSAI regulations-2011, Fruit content (m/m) of carbonated Lime or Lemon juice is not less than

a) 15% b) 5%

c) 10% d) 30%

57. As per FSSAI regulations-2011, maximum permitted level of Benzoic acid in RTS beverages is not more than

a) 400 ppm b) 120 ppm

c) 500 ppm d) 650 ppm

58. As per FSSAI regulations-2011, maximum permitted level of Sulphur dioxide in RTS beverages is not more than

a) 70 ppm b) 120 ppm

c) 500 ppm d) 650 ppm

59. As per FSSAI regulations-2011, total soluble solids of Fruit Nectars is not less than

a) 40% b) 25%

c) 15% d) 30%

60. As per FSSAI regulations-2011, Fruit content (m/m) of Fruit Nectars is not less than

a) 15% b) 20%

c) 40% d) 30%

61. As per FSSAI regulations 2011, Fruit content (m/m) of Orange Nectar/ Pineapple Nectar is not less than

a) 15% b) 20%

c) 40% d) 30%

62. As per FSSAI regulations-2011, total soluble solids of Lemon juice is not less than

a) 40% b) 25%
c) 15% d) 6%

63. As per FSSAI regulations-2011, total soluble solids of freshly expressed Grape juice is not less than

a) 40% b) 25%
c) 15% d) 6%

64. As per FSSAI regulations-2011, total soluble solids of Mango, Guava or any other pulp fruit is not less than

a) 40% b) 25%
c) 15% d) 6%

65. As per FSSAI regulations-2011, maximum permitted level of Benzoic acid in Squashes, Crushes, Fruit Syrups, Sharbats, Cordial and Barley Water is not more than

a) 400 ppm b) 600 ppm
c) 500 ppm d) 650 ppm

66. As per FSSAI regulations-2011, maximum permitted level of Sulphur dioxide in Squashes, Crushes, Fruit Syrups, Sharbats, Cordial and Barley Water is not more than

a) 350 ppm b) 120 ppm
c) 500 ppm d) 650 ppm

67. As per FSSAI regulations-2011, maximum permitted level of Sorbic acid and its salts in Squashes, Crushes, Fruit Syrups, Sharbats, Cordial and Barley Water is not more than

a) 350 ppm b) 1000 ppm
c) 500 ppm d) 650 ppm

68. As per FSSAI regulations-2011, maximum permitted level of Xanthan Gum in Squashes, Crushes, Fruit Syrups, Sharbats, Cordial and Barley Water is not more than

a) 1% b) 2%
c) 0.5% d) 2.5%

69. What is the percentage of barley starch in Barley Water

a) 1% b) 0.25%
c) 0.5% d) 2.5%

70. Mango leather is also known as
    a) Mango slab b) Amawat
    c) Amapapar d) All
71. An example of confection type of fruit halwa
    a) Fruit bar b) Fruit cheese
    c) Fruit leather d) Candied fruit
72. Cheese content in fruit cheese is
    a) 10 % b) 50%
    c) 25% d) None
73. What is non fruit ingredient in fruit cheese
    a) Cheese b) Butter
    c) Cream d) SMP
74. An important ingredient in fruit toffee is
    a) Liquid glucose b) Butter
    c) Cheese d) All
75. Potato is
    a) Perishable vegetable b) Semi perishable vegetable
    c) Non perishable vegetable d) None
76. What is the quality characteristic potato for production of chips, French fries or dehydrated products
    a) Low specific gravity b) High specific gravity
    c) High dry matter d) b & c
77. CIPC (Isopropyl-N-chlorophenyl carbamate) is used as sprout inhibitor for
    a) Grains b) Pulses
    c) Potato d) Carrot
78. What are the quality issues in potato products
    a) High reducing sugars b) Enzymatic browning
    c) After cooking discolouration d) All
79. Commercially peeling of potato is done by
    a) Abrasion peelers b) Lye peeling or steam peeling
    c) Brine Peeling or flame peeling d) All

80. What is the role of potato flour In baking industry

a) Reduces protein content in mix b) Increases protein content in mix
c) Reduces staling of bread d) a & c

81. As per FSSAI regulations-2011, total soluble solids of fruit Bar/ toffee is not less than

a) 40% b) 75%
c) 65% d) 60%

82. As per FSSAI regulations-2011, Fruit content (m/m) of fruit Bar/ toffee is not less than

a) 40% b) 35%
c) 25% d) 60%

83. As per FSSAI regulations-2011, total soluble solids of Tomato juice is not less than

a) 10% b) 5%
c) 15% d) 30%

84. As per FSSAI regulations-2011, total soluble solids of Tomato puree is not less than

a) 10% b) 9%
c) 15% d) 30%

85. As per FSSAI regulations-2011, Fruit juice content (m/m) when reconstituted by dilution according to direction for use is not less than

a) 10% b) 5%
c) 15% d) 20%

86. The organism involved in production of acetic acid is

a) *Streptococcuslactis* b) *Achromabacter aceti*
c) *Saccharomyces sps* d) *Acetobacter aceti*

87. The following fermentation method is used to preserve Fruits and vegetables

a) Acetic b) Lactic
c) Alcoholic d) All

88. Coffee cherries are converted to coffee beans by

a) Acetic acid fermentation b) Lactic acid fermentation
c) Alcoholic fermentation d) All

89. Vinegar acts as a

a) Germicidal agent
b) Antiseptic agent
c) a & b
d) Antioxidant

90. Cider vinegar is manufactured from

a) Lemon
b) Apple
c) Potato
d) Chillies

91. The amount of acetic acid in vinegar is expressed as

a) Percentage
b) Grain strength
c) Aliquot
d) Aqueous

92. One grain strength is equalvent to __________ times of the acetic acid present in it

a) 5
b) 15
c) 10
d) 20

93. What is the grain strength of vinegar having 5 % acetic acid

a) 20
b) 50
c) 40
d) 80

94. What is the grain strength of vinegar having 8 % acetic acid

a) 20
b) 50
c) 40
d) 80

95. Artificial vinegars are also called as

a) Synthetic vinegar
b) Non-brewed vinegar
c) a & b
d) None

96. The most favourable temperature for the growth of alcoholic yeast is

a) 10-15°C
b) 25-27°C
c) 20-25°C
d) 30-45°C

97. Vinegar is prepared by

a) Slow proccss
b) Orleans slow process
c) Quick process (Generator or German process)
d) All

98. What is the acetic acid content in vinegar

a) 6% b) 4%

c) 5% d) 10%

99. What is the optimum content of sugar in fruit juice for production of vinegar

a) 6% b) 4%

c) 5% d) 10%

100. Important steps in vinegar production are

a) Alcoholic fermentation b) Vinegar fermentation.

c) a & b d) None

## Answers

| | | | | | | | | | |
|---|---|---|---|---|---|---|---|---|---|
| 1. a | 2. c | 3. b | 4. b | 5. d | 6. c | 7. c | 8. c | 9. b | 10. b |
| 11. d | 12. c | 13. c | 14. c | 15. c | 16. b | 17. a | 18. b | 19. c | 20. b |
| 21. b | 22. c | 23. a | 24. d | 25. b | 26. a | 27. b | 28. c | 29. d | 30. b |
| 31. d | 32. c | 33. a | 34. c | 35. d | 36. b | 37. d | 38. a | 39. b | 40. c |
| 41. d | 42. a | 43. c | 44. a | 45. a | 46. a | 47. b | 48. c | 49. d | 50. c |
| 51. b | 52. a | 53. c | 54. c | 55. c | 56. b | 57. b | 58. a | 59. c | 60. b |
| 61. c | 62. d | 63. c | 64. c | 65. b | 66. a | 67. b | 68. c | 69. b | 70. d |
| 71. b | 72. d | 73. b | 74. a | 75. b | 76. d | 77. c | 78. d | 79. d | 80. d |
| 81. b | 82. c | 83. b | 84. b | 85. b | 86. d | 87. d | 88. b | 89. c | 90. b |
| 91. b | 92. c | 93. b | 94. d | 95. c | 96. b | 97. d | 98. b | 99. d | 100. c |

1. An ideal vinegar should contain______ percent of sugar

   a) 10% b) 0.3%

   c) 0.5% d) 2%

2. As per FSSAI regulations- 2011, Acidity (m/v) of Brewed Vinegar is not less than ______ percent calculated as acetic Acid

   a) 4% b) 3.75%

   c) 5% d) 10%

3. As per FSSAI regulations- 2011, Total Solids (m/v) in Brewed Vinegar is not less than

   a) 4% b) 3.75%

   c) 1.5% d) 10%

4. Why Grapes have been considered as an important source for production of wine

   a) Juice Is extremely rich in natural sugar

   b) Natural association of fermentative yeasts with berries.

   c) High content of nitrogenous matters

   d) All

5. Wines with 7 to 9 per cent alcohol are known as

   a) Light wines b) Medium wines

   c) Strong wines d) Sweet wines

6. Wines with 9 to 16 per cent alcohol are known as

   a) Light wines b) Medium wines

   c) Strong wines d) Sweet wines

7. Wines with 16 to 21 per cent alcohol are known as

   a) Light wines b) Medium wines

   c) Strong wines d) Sweet wines

8. Carbon dioxide present in

   a) Sparkling wine b) Still wines

   c) Strong wines d) Sweet wines

9. Syphoning off the fermented wine from the solid deposits is Known as

   a) Reckoning b) Racking

   c) Filtration d) All

10. Perry is a wine made from
    - a) Peach
    - b) Pears
    - c) Potato
    - d) Plum fruit
11. Berry wines are produced from
    - a) Strawberry
    - b) Blackberry
    - c) a & b
    - d) None
12. Nira is prepared from
    - a) Palm tree
    - b) Pears
    - c) Cashew nut
    - d) Plum fruit
13. Feni is prepared from
    - a) Palm tree
    - b) Pears
    - c) Cashew apple
    - d) Plum fruit
14. What is the alcohol content in Rum
    - a) 75%
    - b) 95%
    - c) 40%
    - d) 50%
15. In India Rum is bottled at ________ proof
    - a) 85.6
    - b) 90.0
    - c) 75.5
    - d) 88.2
16. Vodka is produced from
    - a) Potato
    - b) Grains
    - c) a & b
    - d) Grapes
17. Scotch whiskey and Irish whiskey is produced from
    - a) Potato
    - b) Malt of barley
    - c) Pears
    - d) Grapes
18. Canadian Whisky is produced from
    - a) Potato
    - b) Malt of barley
    - c) Rye
    - d) Grapes
19. What is the sun dried meat product of Africa
    - a) Beef
    - b) Biltong
    - c) Meat manchuria
    - d) All

20. An example of natural preservative in yeast fermented wines
    a) $CO_2$ b) Ethyl alcohol
    c) Sorbic acid d) Benzoic acid
21. Appert had published the following book
    a) The art of preserving animal and vegetable substance for many years
    b) The art of preserving Fruits and vegetable substance for many years
    c) Canning of fruits and vegetables
    d) None
22. What is the temperature & time required for heat peeling of few fruits & vegetables
    a) 60°C/10 to 30 sec b) 40°C/10 to 60 sec
    c) 70°C/10 to 60 sec d) 80°C/10 sec
23. Low acid fruits are mostly spoiled by
    a) Thermophilic organisms b) Mesophilic organisms
    c) Putrefactive anaerobes d) All
24. What is the minimum pressure requirement for processing of low acid fruits and vegetables
    a) 0.70 to 1.05 kg/$cm^3$ b) 10 to 15 lb per sq inch
    c) a & b d) 0.50 to 2.05 kg/$cm^3$

## Answers

1. b 2. b 3. c 4. d 5. a 6. b 7. c 8. a 9. b 10. b
11. c 12. a 13. c 14. b 15. a 16. c 17. b 18. c 19. b 20. b
21. a 22. b 23. d 24. c

3

# Milk and Meat Products Technology

1. As per FSSAI, 2011 regulations, what is the fat and SNF content in buffalo milk

   a) 5 & 9   b) 6 & 8

   c) 6 & 9   d) None

2. As per FSSAI, 2011 regulations, what is the fat and SNF content in cow milk

   a) 4 & 9   b) 4 & 8.5

   c) 6 & 9   d) None

3. As per FSSAI, 2011 regulations, what is the fat and SNF content in standardized milk

   a) 4 & 9   b) 4.5 & 8.5

   c) 6 & 9   d) None

4. As per FSSAI, 2011 regulations, what is the fat and SNF content in recombined milk

   a) 3 & 8.5   b) 4 & 8.5

   c) 6 & 9   d) None

5. As per FSSAI, 2011 regulations, what is the fat and SNF content in toned milk

   a) 1.5 & 9   b) 3 & 8.5

   c) 1.5 & 8.5   d) None

6. As per FSSAI, 2011 regulations, what is the fat and SNF content in double toned milk

   a) 1.5 & 9   b) 3 & 8.5

   c) 1.5 & 8.5   d) None

7. As per FSSAI, 2011 regulations, what is the fat and SNF content in skim milk

   a) 0.5 & 9 b) 0.5 & 8.7

   c) 1.5 & 8.5 d) None

8. ____________________ heat treatment is used for the production of flavoured milk

   a) Pasteurization b) Sterilization

   c) a & b d) None

9. The major constituents of milk are

   a) Water b) Protein & fat

   c) Lactose and Mineral d) All

10. The protein content in buffalo milk is ______ percent

    a) 3.9 b) 4.9

    c) 4.5 d) 5.9

11. The protein content in cow milk is ______ percent

    a) 1.4 b) 2.4

    c) 3.4 d) 4.4

12. The protein content in Human milk is ______ percent

    a) 1.8 b) 2.8

    c) 3.8 d) None

13. Lactose content in buffalo milk ___________ percent

    a) 5.8 b) 5.2

    c) 4.2 d) 6.2

14. Lactose content in cow milk ___________ percent

    a) 4.9 b) 5.2

    c) 5.9 d) 6.9

15. Lactose content in Human milk ___________ percent

    a) 7.8 b) 5.8

    c) 6.8 d) 4.8

16. Fat content in human milk

    a) 3.6 b) 4.6

    c) 5.6 d) 3.0

17. Longer interval of milking is associated with
    a) More milk with a lower fat  b) More milk with higher fat
    c) Less milk with lower fat  d) Less milk with higher fat
18. Milk proteins are complete proteins because of they contains
    a) All essential Amino Acids  b) All Non essential AA
    c) a & b  d) None
19. Milk is an excellent source of _________ minerals
    a) Ca & P  b) Ca & Fe
    c) Ca & Mg  d) Ca & K
20. Milk is deficient in ___________ vitamin
    a) Vitamin C  b) Vitamin A
    c) Vitamin D  d) All
21. Milk fat imparts ____________ to dairy products
    a) Soft body  b) Smooth texture
    c) Rich Taste  d) All
22. Milk sugar is a
    a) Lactose  b) Maltose
    c) Fructose  d) Sucrose
23. Lactose is a disaccharide of
    a) Glucose + Glucose  b) Glucose + Fructose
    c) Glucose + Galactose  d) None
24. The average energy value of cow milk is
    a) 85 C/100g  b) 75 C/100g
    c) 95 C/100g  d) 65 C/100g
25. The average energy value of buffalo milk is
    a) 80 C/100g  b) 100 C/100g
    c) 75 C/100g  d) 65 C/100g
26. Pasteurization causes ________ percent loss of thiamine and _______ percent loss of vitamin C
    a) 10 & 20  b) 10 & 30
    c) 10 & 50  d) 10 &40

27. Sterilization causes ________ percent loss of thiamine and ________ percent loss of vitamin C

a) 30-50 & 50 b) 30-40 & 30

c) 40-50 & 50 d) 30-50 &40

28. Milk is a ___________ type of emulsion

a) Oil-in –Water b) Water –in- Oil

c) a & b d) None

29. Milk proteins are

a) Casein b) β-lactoglobulin

c) α-lactalbumin d) All

30. Casein is _____________ protein

a) Milk protein b) Meat protein

c) Egg protein d) All

31. _____________ protein exists only in milk

a) Casein b) Gluten

c) Alubumin d) All

32. _____________ is the component in casein micelle that is responsible for the stabilization of the micelle in milk

a) α-casein b) β-casein

c) γ- Casein d) All

33. __________ and _______________are called as whey or serum proteins

a) β-lactoglobulins b) α-lactalbumin

c) a & b d) None

34. Milk is an ____________ type of colloidal system

a) Acidic b) Alkali

c) a &b d) None

35. __________ is the natural emulsifier present in milk

a) Lecithin b) Mono and diaceyl glycerides

c) a & b d) None

36. Phospholipids in milk are

a) Lecithin b) Cephalin

c) Sphingomylin d) All

37. ________is fat soluble, which is responsible for yellow color of milk, cream , butter, ghee and etc.,

a) Riboflavin b) Carotene

c) a &b d) None

38. One molecule of â- Carotene yields ________ molecules of Vitamin A

a) 4 b) 3

c) 2 d) None

39. The white of color of milk is due to

a) Colloidal calcium- Caseinate b) Colloidal phosphorus-albumins

c) a & b d) None

40. The characteristic (greenish) color of milk whey is due to

a) Casein

b) Riboflavin (Lactochrome & lactoflavin)

c) a & b

d) None

41. The characteristic taste of milk is due to

a) Lactose b) Sucrose

c ) Maltose d) fructose

42. ___________ is an enzyme that splits starch in milk

a) Analase b) Diastase

c) a & b d) Lipase

43. ___________ is an enzyme that splits fat in milk and leads to a problem of rancidity

a) Lipase b) Amylase

c) Diastase d) None

44. ___________ is an enzyme that is capable of splitting phosphoric acid esters

a) Lipase b) Amylase

c) Protease d) Phosphatase

45. __________ an enzyme is used for checking of pasteurization efficiency of milk

a) Phosphatase b) Peroxidase

c) Protease d) cataylase

46. ____________ and ______________ enzymes decomposes hydrogen peroxide in milk

a) Peroxidase b) Catalase

c) Both d) None

47. Freshly drawn milk is ___________ to litmus

a) Amphoteric b) Ampholetic

c) a & b d) None

48. Titrable acidity of cow milk is ___________ percent

a) 0.14 to 0.16 b) 0.14 to 0.15

c) 0.13 to 0.15 d) 0.13 to 0.14

49. Titrable acidity of buffalo milk is ___________ percent

a) 0.14 to 0.16 b) 0.14 to 0.15

c) 0.13 to 0.15 d) 0.13 to 0.14

50. pH of fresh cow milk is

a) 6.4 to 6.8 b) 6.4 to 6.7

c) 6.4 to 6.6 d) 6.7 to 6.8

51. pH of fresh buffalo milk is

a) 6.7 to 6.8 b) 6.4 to 6.7

c) 6.7 to 6.9 d) 6.4 to 6.8

52. Specific gravity of milk is usually expressed at ___________degree centigrade

a) 14.6 b) 15.6

c) 16.6 d) 17.6

53. The common types of lactometers

a) Zeal b) Quevenne

c) Both a & b d) None

54. Milk is __________ than water

a) Lighter b) Heavier

c) Both a & b d) None

55. The average range of specific gravity of cow milk is___________

a) 1.028–2.50 b) 1.208–1.030

c) 1.028–1.030 d) 1.201–1.208

56. The average range of specific gravity of buffalo milk is__________

a) 1.030 to 1.032 b) 1.30 to 1.32

c) 1.035 to 1.320 d) 1.031 to 1.039

57. The average range of specific gravity of skim milk is___________

a) 1.035 to 1.037 b) 1.35 to 1.37

c) 1.038 to 1.039 d) 1.37 to 1.39

58. The specific gravity of milk fat is

a) 0.93 b) 0.98

c) 0.83 d) 1.0

59. The specific gravity of Solid not fat (SNF) is

a) 1.717 b) 1.616

c) 1.516 d) 1.721

60. Specific gravity of milk is lowered by __________ and __________

a) Water b) cream

c) a & b d) None

61. Specific gravity of milk is increased by __________ and _________

a) Skim milk b) Removal of fat

c) a & b d) None

62. The percentage of total solids in milk is calculated by

a) TS = 0.25D+1.22F+0.72 b) TS =0.25D+1.5F+0.72

c) TS = 0.35D+1.22F+0.72 d) All

63. The percentage of Solid not fat in milk is calculated by

a) 0.25D+0.28F+0.76 b) 0.25D+0.22F+0.72

c) 0.25D+1.22F+0.72 d) All

64. The average freezing point depression of cow milk is ___________ degree centigrade

a) 0.547 b) 0.534

c) 0.586 d) 0.539

65. The average freezing point depression of buffalo milk is __________ degree centigrade

a) 0.547 b) 0.549

c) 0.541 d) 0.543

66. India's total milk production is _________ million tones in the year of 2011-2012

a) 121.8 b) 150.8

c) 100.18 d) 145.8

67. Freezing-point depression of milk

a) 0.23K b) 0.53 K

c) 0.43K d) 0.63K

68. Casein aggregates at _____ pH

a) 5.5 b) 4.5

c) 4.6 d) 3.6

69. All amino acids occuring in milk are in _______

a) L-form b) D-form

c) Cis-form d) Trans- form

70. _______ fat has butyric acid as one of the constituent fatty acids

a) Oil seeds b) Cereal oils

c) Milk d) Animal fat

71. What is the temperature and time to inactivate phosphatase enzyme in milk

a) 61.1°C/30 min b) 71.1°C/15 sec

c) both a & b d) None

72. What is the temperature and time Pasteurize the milk

a) 61.7°C/30 min b) 71.7°C/15 sec

c) Both a & b d) None

73. What is the batch or holding pasteurization temperature (LTLT) for milk

a) 63°C/30min b) 71.7°C/15 sec

c) both a & b d) None

74. What is the HTST or Continues Pasteurization Temperature for milk

a) 63°C/30 min b) 72°C/15 sec

c) 71.1°C/15 sec d) None

75. What is the Curd Tension (CT) Soft curd milk

a) Less than 25 b) Above 25

c) Zero d) 35

76. Soft curd milk is characterized by

a) Low casein  b) Low calcium

c) both a & b  d) None

77. What is the Curd Tension of human milk

a) Zero  b) 25

c) 10  d) 5

78. What is the Curd Tension of Raw cow milk

a) 55  b) 25

c) 10  d) 5

79. What is the Curd Tension of Pasteurized cow milk

a) 55  b) 44.5

c) 10  d) 5

80. ___________ is a self carbonated beverage containing 1% lactic acid and 1% alcohol

a) Kumiss  b) Kefir

c) Yogurt  d) Butter milk

81. __________ is a lactic acid, alcohol fermented milk contains 2.5% alcohol

a) Kumiss  b) Kefir

c) Yogurt  d) Butter milk

82. Tempering of ice cream is generally carried out at

a) -8 to -7°C  b) -15°C to -12°C

c) -10°C to -12°C  d) -5 to - 6°C

83. Rennet contains __________ and __________enzymes

a) Renin  b) Pepsin

c) a & b  d) None

84. Renin is

a) Enzyme  b) Hormone

c) Protein  d) Vitamin

85. What is the fat content in Table Cream, Light cream and coffee cream (TLC)

a) 20 to 25 %  b) 30 to 40 %

c) 65 to 85 %  d) 60-70%

86. What is the fat content in Whipping cream and Heavy cream contains

a) 20 to 25 % b) 30 to 40 %

c) 65 to 85 % d) 60 to 70%

87. What is the fat content in Plastic cream

a) 20 to 25 % b) 30 to 40 %

c) 65 to 85 % d) 60 to 70%

88. Butter starter culture is

a) *Str. lactis* b) *Str. cremoris*

c) *Str. thermophilus* d) Both a & b

89. What is the moisture content in Indian butter

a) 16.3 % b) 12.3 %

c) 14.3 % d) 10.3 %

90. What is the curd content in Indian butter

a) 2.0 b) 1.0

c) 3.0 d) 4

91. Casein to fat ratio in cheese milk is

a) 0.68 to 0.70 b) 0.60 to 0.78

c) 0.58 to 0.70 d) 0.68 to 0.88

92. Whey proteins are

a) α-lactalbumin b) β-lactoglobulin

c) Both a & b d) None

93. ___________ the organism involved in Yoghurt production

a) *Streptococcus thermophilus* b) *Lactobacillusbulgaricus*

c) Both a & b d) None

94. The ratio of cocci to bacilli in Yoghurt production

a) 1:2 and 2:1 b) 1:1 and 2:1

c) 1:2 and 2:2 d) 1:4 and 2:1

95. What is the protein content in defatted whey protein concentrate

a) 80-85% b) 60-85%

c) 50-85% d) 95-98%

96. Hardening and storage temperatures of ice cream is

    a) -15°C to -12°C    b) -23°C to -29°C

    c) 0°C to 16°C    d) None

97. Tempering of ice cream is generally carried out at

    a) -15°C to -12°C    b) -23°C to -29°C

    c) 0°C to 16°C    d) None

98. What is the temperature of curing/ripening/souring/maturing of cheese

    a) -15°C to -12°C    b) -23°C to -29°C

    c) 0°C to 16°C    d) None

99. Specific sweet products prepared from Danedar type of Khoa

    a) Kalakhan    b) Gourd

    c) Burfi    d) All

100. _______ is a specific sweet products prepared from Dhar type of Khoa

    a) Kalakhan    b) Gourd

    c) Burfi    d) Gulabjamun

## Answers

| | | | | | | | | | |
|---|---|---|---|---|---|---|---|---|---|
| 1. c | 2. b | 3. b | 4. a | 5. b | 6. a | 7. b | 8. b | 9. d | 10. a |
| 11. c | 12. a | 13. b | 14. a | 15. c | 16. a | 17. a | 18. a | 19. a | 20. a |
| 21. d | 22. a | 23. c | 24. b | 25. b | 26. a | 27. a | 28. a | 29. d | 30. a |
| 31. a | 32. a | 33. c | 34. a | 35. a | 36. d | 37. b | 38. c | 39. a | 40. b |
| 41. a | 42. c | 43. a | 44. d | 45. a | 46. c | 47. a | 48. d | 49. b | 50. c |
| 51. a | 52. b | 53. c | 54. b | 55. c | 56. a | 57. a | 58. a | 59. b | 60. c |
| 61. c | 62. a | 63. b | 64. a | 65. b | 66. a | 67. b | 68. c | 69. a | 70. c |
| 71. c | 72. c | 73. a | 74. b | 75. a | 76. c | 77. a | 78. a | 79. b | 80. b |
| 81. a | 82. b | 83. c | 84. a | 85. a | 86. b | 87. c | 88. d | 89. a | 90. b |
| 91. a | 92. c | 93. c | 94. b | 95. a | 96. b | 97. a | 98. c | 99. d | 100. d |

1. Milk of normal composition has a freezing point of
   a) -0.54 to -0.59 °C  b) -0.50 to -0.59 °C
   c) -0.54 to -0.69 °C  d) -0.44 to -0.59 °C
2. As per FSSAI (PFA) regulations, what is the level of BHA is allowed in butter
   a) 0.02%  b) 0.01%
   c) 0.2%  d) 0.1%
3. What is the fat content in Danedar Khoa
   a) 21 -26  b) 20-23
   c) 20-25  d) 31-33
4. Specific sweets of pindi khoa are
   a) Burfi  b) Pedda
   c) Pantooa  d) a & b
5. As per FSSAI (PFA) regulations, what is the fat content in Low fat cream
   a) Not less than 25%  b) Not less than 40%
   c) Not less than 60%  d) Not less than 34.0 percent
6. As per FSSAI (PFA) regulations, what is the fat content in Medium fat cream
   a) Not less than 25%  b) Not less than 40%
   c) Not less than 60%  d) Not less than 34.0 percent
7. As per FSSAI (PFA) regulations, what is the fat content in High fat cream
   a) Not less than 25%  b) Not less than 40%
   c) Not less than 60%  d) Not less than 34.0 percent
8. As per FSSAI (PFA) regulations, what is the fat content in cream powder
   a) Not less than 25%  b) Not less than 40%
   c) Not less than 60%  d) Not less than 42.0 percent
9. As per FSSAI (PFA) regulations, what is the fat content in Chhana or paneer
   a) Not more than 25%  b) Not more than 40%
   c) Not more than 15.0%  d) Not less than 42.0 percent

10. As per FSSAI (PFA) regulations, what is the fat content (%) and moisture content (%) in Hard Pressed Cheese

a) 48.0 & 39.0 b) 40.0 & 45.0

c) 45.0 & 52 d) 20.0 & 80

11. As per FSSAI (PFA) regulations, what is the fat content (%) and moisture content (%) in Semi Hard Cheese

a) 48.0 & 39.0 b) 40.0 & 45.0

c) 45.0 & 52 d) 20.0 & 80

12. As per FSSAI (PFA) regulations, what is the fat content (%) and moisture content (%) in Semi Soft Cheese

a) 48.0 & 39.0 b) 40.0 & 45.0

c) 45.0 & 52 d) 20.0 & 80

13. As per FSSAI (PFA) regulations, what is the fat content (%) and moisture content (%) in Soft Cheese

a) 48.0 & 39.0 b) 40.0 & 45.0

c) 45.0 & 52 d) 20.0 & 80

14. As per FSSAI (PFA) regulations, what is the fat content (%) and moisture content (%) in Cheddar Cheese

a) 48.0 & 39.0 b) 48.0 & 39.0

c) 45.0 & 52 d) 20.0 & 80

15. As per FSSAI (PFA) regulations, what is the fat content (%) and moisture content (%) in Mozzarella Cheese

a) 48.0 & 39.0 b) 48.0 & 39.0

c) 35.0 & 60 d) 20.0 & 80

16. As per FSSAI (PFA) regulations, what is the fat content (%) and SNF(%) in Evaporated skimmed milk

a) 1 & 20 b) 15 & 27

c) 9 & 31 d) 1 & 26

17. As per FSSAI (PFA) regulations, what is the fat content (%) and SNF(%) in Evaporated high fat milk

a) 1 & 20 b) 15 & 27

c) 9 & 31 d) 1 & 26

18. As per FSSAI (PFA) regulations, what is the fat content (%) and SNF(%) in Sweetened condensed milk

a) 1 & 20  b) 15 & 27
c) 9 & 31  d) 1 & 26

19. As per FSSAI (PFA) regulations, what is the fat content (%) and SNF(%) in Sweetened condensed skimmed milk

a) 1 & 20  b) 15 & 27
c) 9 & 31  d) 1 & 26

20. As per FSSAI (PFA) regulations, what is the fat content (%) in Whole milk Powder

a) Not less than 26.0%  b) Not less than1.5%
c) Not morethan 1.5%  d) Not more than 26.0%

21. As per FSSAI (PFA) regulations, what is the fat content (%) in Partly skimmedmilk powder

a) Not less than 26.0%  b) Not less than 1.5%
c) Not morethan 1.5%  d) Not less than 20.0%

22. As per FSSAI (PFA) regulations, what is the fat content (%) in Skimmedmilk powder

a) Not less than 26.0%  b) Not less than 1.5%
c) Not morethan 1.5%  d) Not less than 26.0%

23. As per FSSAI (PFA) regulations, what is the fat content (%) in butter

a) 85  b) 80
c) 81.0  d) 82.0

24. As per FSSAI (PFA) regulations, what is the common salt content in butter

a) 2%  b) 3%
c) 2.5%  d) 1%

25. As per FSSAI (PFA) regulations, what is the SNF content in butter

a) Not more than 12%  b) Not more than 1.5%
c) Not more than 2.5%  d) Not more than 14%

26. As per FSSAI (PFA) regulations, what is the fat content (%) in Whey Powder

a) Not more than 2%  b) Not more than 1.5%
c) Not more than 2.5%  d) Not more than 4%

27. As per FSSAI (PFA) regulations, what is the fat content (%) in Butter oil (ghee)
    a) Not less than 89.6% b) Not less than 97.6%
    c) Not less than 92.6% d) Not less than 99.6%
28. As per FSSAI (PFA) regulations, what is the FFA expressed as oleic acid in Butter oil (ghee) is
    a) Not more than 0.8% b) Not more than 0.1%
    c) Not more than 0.2% d) Not more than 0.4%
29. As per FSSAI (PFA) regulations Boudouins test for Butter oil (ghee) is
    a) Positive b) Negative
    c) both a & b d) None
30. As per FSSAI (PFA) regulations, what is the fat content (%) in Yoghurt
    a) Not less than 5.0% b) Not less than 7.6%
    c) Not less than 3.0% d) Not less than 0.6%
31. As per FSSAI (PFA) regulations, what is the total solids content in High Fat Ice Cream
    a) Not less than 36.0% b) Not less than 30.0%
    c) Not less than 26.0% d) None
32. As per FSSAI (PFA) regulations, what is the total solids content in Medium Fat Ice Cream
    a) Not less than 36.0% b) Not less than 30.0%
    c) Not less than 26.0% d) None
33. As per FSSAI (PFA) regulations, what is the total solids content in Low Fat Ice Cream
    a) Not less than 36.0% b) Not less than 30.0%
    c) Not less than 26.0% d) None
34. As per FSSAI (PFA) regulations, what is Weight/Volume (gms/l) for High Fat Ice Cream is
    a) Not less than 525 b) Not less than 475
    c) Not less than 470 d) None
35. As per FSSAI (PFA) regulations, what is Weight/Volume (gms/l for Medium Fat Ice Cream
    a) Not less than 525 b) Not less than 475
    c) Not less than 470 d) None

36. As per FSSAI (PFA) regulations, what is Weight/Volume (gms/l for Low Fat Ice Cream

a) Not less than 525
b) Not less than 475
c) Not less than 470
d) None

37. As per FSSAI (PFA) regulations, what is the fat content in High Fat Ice Cream

a) Not less than 10.0%
b) More than 2.5% less than 10
c) Not more than 2.5%
d) Noine

38. As per FSSAI (PFA) regulations, what is the fat content in Medium Fat Ice Cream

a) Not less than 10.0%
b) More than 2.5% but less than 10
c) Not more than 2.5%
d) None

39. As per FSSAI (PFA) regulations, what is the fat content in Low Fat Ice Cream

a) Not less than 10.0%
b) More than 2.5% less than 10
c) Not more than 2.5%
d) None

40. How many parts by weight of 40 % cream and 3% milk must be mixed to make testing 5% fat ?

a) 2 parts of cream & 37 parts of milk
b) 4 parts of cream & 37 parts of milk
c) 2 parts of cream & 27 parts of milk
d) 7 parts of cream & 37 parts of milk

41. The organism involved in production of Acidophilus milk is

a) *Lactobacillus acidophilus*
b) *Lactobacillus bulgaricus*
c) *Str.thermophilus*
d) None

42. Over run in butter is caused by

a) Moisture
b) Curd
c) Salt
d) All

43. As per ISI specifications, what is the acidity (expressed in lactic acid) of ice cream

a) 0.15% b) 0.25%

c) 0.16% d) 0.13%

44. As per ISI specifications, what is the standard plate count (per g.) in ice cream

a) Not more than 250000 b) Not more than 200000

c) Not more than 150000 d) Not more than 100000

45. SNF imparts ____________ to ice cream

a) Improves texture b) Gives body

c) Higher over run d) All

46. In ice cream sugar acts as a

a) Improves texture b) Enhances flavour

c) Gives body d) a & b

47. As per FSSAI specifications, what is the standard plate count (per g.) in Pasteurised Milk/Cream or Flavoured Milk

a) 30,000/g b) 20,000/g

c) 40,000/g d) 10,000/g

**Meat and Meat Products**

48. Meat from sheep of less than 12 months of age is referred as

a) Ham b) Lamb

c) Pork d) Chevon

49. ____________ muscles are involuntary in nature

a) Smooth b) Cardiac

c) Skeletal d) a & b

50. Myofibrillar proteins are

a) Soluble in dilute salt solution

b) Soluble in water or very dilute salt solution

c) Almost insoluble

d) None

51. Sarcoplasmic proteins are
    a) Soluble in dilute salt solution
    b) Soluble in water or very dilute salt solution
    c) Almost insoluble
    d) None
52. What is the protein content in typical animal muscle
    a) 15% b) 17%
    c) 19% d) 20%
53. What are the Myofibrillar proteins
    a) Myosin
    b) Actin
    c) Troiponins C, I and T
    d) Connectins and Desmin
    e) All
54. __________ is a conjugated protein consisting of a prosthetic heme moiety and a protein moiety (globin).
    a) Myoglobin b) Actin
    c) Troiponin d) Connectins and Desmin
55. What are the color contributing compounds in meat
    a) Myoglobin b) Cytochrome enzyme
    c) Flavin d) All
56. ______________ is the only protein possessing a fair amount of hydroxyproline (approximately 14%) and low concentratin of hydroxylysine.
    a) Myosin b) Actin
    c) Collagen d) Desmin
57. What is the boiling temperature used to convert collagen to gelatin
    a) 60°C b) 50°C
    c) 90°C d) 100°C
58. _________ is a process employed to create state of immobility or unconsciousness at the time of slaughter to 1 minute.
    a) Slaughtering b) Stunning
    c) Soring d) None

59. __________ meat must be sold and consumed within three days of slaughter

a) Kosher
b) Halal
c) Fresh
d) None

60. What is the temperature and RH of meat cutting room

a) 15-20°C & 40%
b) 15-20°C & 80%
c) 15-20°C & 90%
d) 5 to 10°C & 80%

61. Scalding temperature of pig is

a) 60-62°/6 min
b) 50-52°/6 min
c) 40-52°/6 min
d) 40-55°/6 min

62. pH of Pale, Soft and Exudative (PSE) condition of meat is

a) 4.5-5.7
b) 5.5-5.7
c) 6.5-6.7
d) 5.0-7.0

63. Dark, Firm and Dry (DFD) condition of meat is associated with

a) Low pH
b) High pH
c) Neutral pH
d) None

64. Typical colour of Mutton and chevon is

a) Light to dark red
b) Greyish pink
c) Grey white to dull red
d) Cherry red

65. Typical colour of Pork is

a) Light to dark red
b) Greyish pink
c) Grey white to dull red
d) Cherry red

66. Typical colour of Poultry is

a) Light to dark red
b) Greyish pink
c) Grey white to dull red
d) Cherry red

67. Typical colour of Buffalo meat and beef is

a) Light to dark red
b) Greyish pink
c) Grey white to dull red
d) Cherry red

68. Oxymyoglobin imparts ________ color to meat

a) Light to dark red
b) Greyish pink
c) Bright red colour
d) Cherry red

69. _________ is refers to the intramuscular fat which can be visibly detected when the muscle surface is cut
    a) Marbling b) Mayonnaise
    c) Freeze burn d) All
70. Water activity of fresh meat is
    a) 0.8 b) 0.9
    c) 1 d) 0.99
71. _________ refers to the volume of aqueous extract released by a meat homogenate when it is passed through a filter for a given period of time
    a) Marbling b) Extract release volume
    c) Extension release volume d) All
72. Storage of fresh meat is done at a refrigeration temperature of
    a) 2 to 5°C b) 10 to 15°C
    c) 20 to 25°C d) 5-10°C
73. What is the refrigerated temperature for storage of fresh meat to maintain in good condition for a period of 5-7 days
    a) 5±1°C b) 4±1°C
    c) 6±1°C d) 10±1°C
74. Nitroso haemochromogen _______________ in cured meat
    a) Stable pink pigment b) Greyish pink
    c) Grey white to dull red d) Cherry red
75. _________ is a unstable cured pigment
    a) Nitroso haemochromogen b) Nitirc oxide myoglobin
    c) both a & b d) None
76. What is the moisture content in Intermediate Moisture Meats (IMM).
    a) 10-25% b) 20-50%
    c) 20-40% d) 20-30%
77. An example of humectants
    a) Glycerol b) Propylene glycol
    c) Salt & Sugar d) Sorbitol
    e) All

78. The total weight of egg shell is

a) 1% b) 11%

c) 10% d) 15%

79. The white or albumen portion of egg constitutes _____ percent of the total weight of an egg.

a) 50% b) 58%

c) 40% d) 55%

80. The yolk of egg constitutes _____ percent of the total weight of an egg.

a) 50% b) 58%

c) 31% d) 55%

81. What is the protein content in whole egg

a) 11.8% b) 11.0%

c) 17.5% d) None

82. What is the protein content in egg albumen

a) 11.8% b) 11.0%

c) 17.5% d) None

83. What is the protein content in egg yolk

a) 11.8% b) 11.0%

c) 17.5% d) None

84. What is the fat content in whole egg

a) 11.8% b) 11.0%

c) 17.5% d) None

85. What is the fat content in egg yolk

a) 11.8% b) 11.0%

c) 17.5% d) 32.5%

86. What is the protein, which inhibits several Proteases in egg

a) Ovalbumin b) Ovoinhibitor

c) Avidin d) Ovomucin

87. What is the protein, which Binds biotin in egg

a) Ovalbumin b) Ovoinhibitor

c) Avidin d) Ovomucin

88. What is the protein, which Lyses some bacteria in egg
    a) Ovalbumin b) Ovoinhibitor
    c) Avidin d) Lysozyme
89. What is the protein, which Binds riboflavin in egg
    a) Ovalbumin b) Flavoprotein
    c) Avidin d) Ovomucin
90. What is the protein, which binds metals especially iron in egg
    a) Ovalbumin b) Flavoprotein
    c) Conalbumin d) Ovomucin
91. What is the protein, which Inhibits trypsin in egg
    a) Ovalbumin b) Flavoprotein
    c) Conalbumin d) Ovomucoid
92. _________ is a reference protein
    a) Milk b) Egg
    c) Meat d) Fruits
93. _________is the function of height of thick albumin and weight of the egg.
    a) Honsons unit b) Haugh's unit
    c) Howary Units d) None
94. What is the Haugh's unit of good quality egg
    a) 80 b) 72
    c) 50 d) 60
95. _________ is the ratio of weight the albumin and width of the albumin
    a) Albumin Area Index b) White index
    c) Yolk index d) All
96. __________ is the measurement of height of thickest portion of white divided by diameter of an egg.
    a) Albumin Area Index b) White index
    c) Yolk index d) All
97. ___________ is a Height of yolk to width of yolk
    a) Albumin Area Index b) White index
    c) Yolk index d) All

98. _________ is a refers to immersion of birds in hot water for loosening the feathers

a) Scalding  b) Boiling
c) Soaking  d) None

99. Broiler and young birds are scalded at

a) 55°C /1.5 min  b) 60°C/ 2 min
c) 65°C/ 2 min  d) 56°C/ 1.2 min

100. Culled birds and spent hens are scalded at

a) 55°C /1.5 min  b) 60°C/ 2 min
c) 65°C/ 2 min  d) 56°C/ 1.2 min

## Answers

| | | | | | | | | | |
|---|---|---|---|---|---|---|---|---|---|
| 1. a | 2. b | 3. c | 4. d | 5. a | 6. b | 7. c | 8. d | 9. c | 10. a |
| 11. b | 12. c | 13. d | 14. b | 15. c | 16. a | 17. b | 18. c | 19. d | 20. a |
| 21. b | 22. c | 23. b | 24. b | 25. b | 26. a | 27. d | 28. d | 29. b | 30. c |
| 31. a | 32. b | 33. c | 34. a | 35. b | 36. b | 37. a | 38. b | 39. c | 40. a |
| 41. a | 42. d | 43. b | 44. a | 45. d | 46. d | 47. a | 48. b | 49. d | 50. a |
| 51. b | 52. c | 53. e | 54. a | 55. d | 56. c | 57. a | 58. b | 59. a | 60. b |
| 61. a | 62. b | 63. b | 64. a | 65. b | 66. c | 67. d | 68. c | 69. a | 70. d |
| 71. b | 72. a | 73. b | 74. a | 75. b | 76. b | 77. e | 78. b | 79. b | 80. c |
| 81. a | 82. b | 83. c | 84. b | 85. d | 86. b | 87. c | 88. d | 89. b | 90. c |
| 91. d | 92. b | 93. b | 94. b | 95. a | 96. b | 97. c | 98. a | 99. a | 100. b |

1. Grading of eggs will be done by
   a) Air cell b) Haugh's unit
   c) Strength of Vitelline membrane d) All
2. Giblet consisting of
   a) Heart b) Liver
   c) Gizzard d) All
3. Green rot in eggs is caused by
   a) *Pseudomonas Fluorescens* b) *Proteus* sp.
   c) *Serratia* sp. d) *Sporotrichium*
4. Black rot (Blackening of yolk with "faecal) in eggs caused by
   a) *Pseudomonas Fluorescens* b) *Proteus* sp.
   c) *Serratia* sp. d) *Sporotrichium*
5. Red rot in eggs is caused by
   a) *Pseudomonas Fluorescens* b) *Proteus* sp.
   c) *Serratia* sp. d) *Sporotrichium*
6. Pink spots on egg contents is caused by
   a) *Pseudomonas Fluorescens* b) *Proteus* sp.
   c) *Serratia* sp. d) *Sporotrichium*
7. Black spots on egg contents is caused by
   a) *Pseudomonas Fluorescens* b) *Cladosporium*
   c) *Serratia* sp. d) *Sporotrichium*
8. Yellow or green spots on egg contents is caused by
   a) *Pseudomonas Fluorescens* b) *Cladosporium*
   c) *Penicillium* d) *Sporotrichium*
9. Thermostabilization of eggs is carried out at
   a) 55°C for 15 min b) 58°C for 10 minutes
   c) Both a & b d) None
10. In egg powder production desugaring is done to prevent
    a) Browning b) Maillard reaction
    c) Blackening d) a & b

11. In egg powder production desugaring is done by
    a) Addition of 0.5% yeast
    b) Addition of 5% yeast
    c) Addition of 0.1% yeast
    d) None
12. In egg powder production desugaring is done by
    a) Drying
    b) Fermentation
    c) Concentration
    d) Evaporation
13. Broilers at ____weeks of age are preferred for tandoori chicken because of their tender meat and ability to sustain roasting.
    a) 5 weeks
    b) 6 weeks
    c) 8 weeks
    d) 4 weeks
14. The resolution of rigor mortis is known to occurs due to disintegration of ______
    a) Z-line structure
    b) M –line structure
    c) D –line structure
    d) None
15. Protein content in most fish
    a) 15-24%
    b) 10-20%
    c) 5-15%
    d) None
16. Fish proteins slightly low in
    a) Methionine
    b) Cystine
    c) a & b
    d) None
17. Fish protein is very rich source of __________ amino acid
    a) Methionine
    b) Lysine
    c) Cystine
    d) Leucine
18. Sardine oil is rich source of
    a) Eicospentaenoic acid (5double bonds)
    b) Docosahexaeonic acid (6 double bonds)
    c) a & b
    d) None
19. Quality characteristics of fresh fish
    a) Bright convex eyes with dark pupil
    b) Transparent and slimy
    c) Bright red gills with no slime
    d) All

20. The recommended cold storage temperature for frozen fish is

a) -40°C b) -50°C

c) -30°C d) -18°C

21. __________ is a sprouting inhibitor in onion bulbs

a) Maleic hydrazide b) Cytokinin

c) Alar (B9) d) Gibberlic acid

22. Queen of fruit crops

a) Pine apple b) Apple

c) Mango d) Bananna

23. _______is an antibiotic commonly found in milk, curd, cheese and other fermented milk products

a) Pimaracin b) Nisin

c) Subtilin d) All

24. What are the low cost storage structures for fruits and vegetables

a) In situ b) Sand or coir

c) Pits or trenches d) All

25. ___________system is working on evaporative cooling system for fruits and vegetables storage

a) In situ b) Sand or coir or coir

c) Zero energy cool chamber d) None

26. Blanching of fruits and vegetables is done for

a) To inactivate the enzymes

b) To reduce moisture content

c) To reduce volume to surface area

d) All

27. __________ are the water absorbents used in the package system

a) CaCl2 b) Sorbitol

c) Xylitol d) All

28. Treatment of fruits and vegetables with boiling water or steam for short periods followed by cooling prior to canning is called"

a) Exhausting b) Blanching

c) Sterilization d) Curing

29. What is the inhibitory concentration of $CO_2$ in beverages

a) 14.6 g $CO_2$/ l
b) 24.6 g $CO_2$/ l
c) 10.6 g $CO_2$/ l
d) 6 g $CO_2$/ l

30. ____________ is an antifungal antibiotic can be used for treating fruits and fruits juices

a) Nisin
b) Pimaricin
c) Subtilin
d) All

31. Who was the first person to describe a method of canning of foods in 1807

a) Nicholas Apart
b) Saddington
c) William under Wood
d) None

32. Lye peeling is done for

a) Peaches and carrots
b) Apricots and quinces
c) Sweet oranges and sweet potatoes
d) All

33. Flame peeling is done for

a) Garlic
b) Onions
c) a & b
d) None

34. Brine and syrups can be added to can at ______________temperature

a) 79°C to 82°C
b) 89°C to 92°C
c) 69°C to 72°C
d) 59°C to 62°C

35. The allowed head space in caned products is

a) 0.42 cm to 0.57cm
b) 0.32 cm to 0.47cm
c) 0.62 cm to 0.77cm
d) None

36. The exhausting temperature of canning process is

a) 79°C to 82°C
b) 82-87°C
c) 69°C to 72°C
d) 59°C to 62°C

37. In canning all fruits can be processed satisfactorily at _________ temperature,

a) 100°C
b) 115°C
c) 121°C
d) 105°C

38. What is the pH of Low acid (Non-acid) foods

a) 4.5-5.0 b) >5

c) 3.7-4.5 d) < 3.7

39. What is the pH of Medium acid foods

a) 4.5-5.0 b) >5

c) 3.7-4.5 d) < 3.7

40. What is the pH of acid foods

a) 4.5-5.0 b) >5

c) 3.7-4.5 d) < 3.7

41. What is the pH of high acid foods

a) 4.5-5.0 b) >5

c) 3.7-4.5 d) < 3.7

42. Examples of Low acid (Non-acid) foods

a) Peas, corn & lima beans b) Asparagus & cauliflower

c) Potato & spinach d) All

43. Examples of Medium acid foods

a) Okra & cabbage b) Pumpkin & beet green

c) Soups and sauces d) All

44. Examples of Acid foods

a) Tomato and apple b) Pear, Peach and Banana

c) Mango, jackfruit & pineapple d) All

45. Examples of High Acid foods

a) Citrus juice b) Saukeraut

c) Pickle & chutney d) All

46. What are the chemicals used in Self heating can

a) $Fe_2O_3$ (Ferrous oxide) b) $Fe_3O_4$ (Ferric oxide)

c) a & b d) None

47. A mild swell at one or both ends of a can is called as

a) Springer b) Flipper

c) Leakage d) Breather

48. A can when struck against a table top one or both ends become convex and spring or flip out is a due to ______ swell

a) Springer b) Flipper
c) Leakage d) Breather

49. Flat sour in non-acid foods like vegetables is caused by

a) *B. coagulens* b) *B. sterothermophilus*
c) Both a & b d) None

50. A leaking can is known as

a) Springer b) Flipper
c) Leakage d) Breather

51. A very tiny leak in the can is known as

a) Springer b) Flipper
c) Leakage d) Breather

52. A clear sparkling juice is called as

a) Squash b) Cordial
c) Nectar d) Sherbat

53. ________ is clear sugar syrup which has been artificially flavored

a) Sherbat b) Cordial
c) Nectar d) Squash

54. Preservation of food in common salt or in vinegar is called as

a) Curing b) Salting
c) Pickling d) Nonc

55. Examples of Non climacteric fruits

a) Lemon & Grape
b) Pineapple & Citrus
c) Cherry, Cucumber & straw berry
d) All

56. Examples of climacteric fruits

a) Apple &Banana b) Mango & Sapota
c) Papaya & Tomato d) All

57. Suspending of shreds of citrus fruit peels in jelly or jam
    a) Marmalade  b) Jam
    c) Jelly  d) None
58. Preservation of food in common salt or in edible oil is called as
    a) Curing  b) Salting
    c) Pickling  d) None
59. At what concentration salt acts as preservative
    a) 15-20%  b) 5-10%
    c) 2-10%  d) 1-5%
60. Spontaneous exudation of fluid from a gel is called
    a) Serum loss  b) Syneresis
    c) Weeping  d) b&c
61. Vinegar made from apple juice is known as
    a) Potato vinegar  b) Spirit vinegar
    c) Cider vinegar  d) None
62. Vinegar made from distilled alcoholic is known as
    a) Potato vinegar  b) Spirit vinegar
    c) Cider vinegar  d) None
63. Syphoning off the fermented wine to separate it from the solid deposits is Known as
    a) Racking  b) Separation
    c) Filtration  d) Reverse Osmosis
64. __________ is a measure of how much alcohol (ethanol) is contained in an alcoholicbeverage.
    a) Alcohol content  b) Alcohol proof
    c) both a & b  d) None
65. Pectin is a water soluble polymer of
    a) Glucose molecules  b) Galacturonic acids
    c) D-manose molecules  d) Galactose molecules
66. Sharp/slow freezing temperature
    a) 0 to - 40°C  b) -15 to -29°C
    c) -10 to -25°C  d) -5 to -29°C

67. Cryogenic freezing liquids are
    a) Liquid nitrogen
    b) Liquid carbon- di- oxide
    c) Both a & b
    d) None
68. Vinegar is
    a) 2 % acetic acid
    b) 3 % acetic acid
    c) 4% acetic acid
    d) 6% acetic acid
69. What is prepared with spice bag method
    a) Jam
    b) Jelly
    c) Ketchup/sauce
    d) All
70. ____________is prepared after draining and drying of sugar syrup from preserve.
    a) Jam
    b) Candied fruit
    c) Glazed fruit
    d) Syrup
71. Heat resistance of C. *batulinum* spores are killed at
    a) 100°C/100-300 min
    b) 100°C/1.7 min
    c) 100°C/15-30 min
    d) 100°C over 1000 min
72. Heat resistance of *B. anthracis* spores are killed at
    a) 100°C/100-300 min
    b) 100°C/1.7 min
    c) 100°C/15-30 min
    d) 100°C over 1000 min
73. Heat resistance of Flat sour bacteria spores are killed at
    a) 100°C/100-300 min
    b) 100°C/1.7 min
    c) 100°C/15-30 min
    d) 100°C over 1000 min
74. Typical levels of pectin used as a food additive in products
    a) 0.5 - 1.0%
    b) 0.5 - 3.0%
    c) 0.5 - 4.0%
    d) 0.5 - 5.0%
75. Pectin content is determined by
    a) Alcohol test
    b) Jelmeter test
    c) Both a & b
    d) None
76. Thc optimum pH for a jelly containing 1 per cent pectin is approximately 3.0, 3.2 and 3.4 for ____________, __________ and ______ per cent TSS, respectively
    a) 60, 65 & 71
    b) 60, 65 & 70
    c) 65, 60 & 70
    d) 75, 60 & 65

77. What is the optimum pH value for jelly of 65% TSS and 1% pectin content

a) 3.0 b) 3.2

c) 4.0 d) 4.5

78. Syneresis or weeping of jelly is caused by

a) Excess of acid b) Insufficient pectin

c) Premature gelation d) All

79. Salt content in Sauerkraut (acid cabbage) is

a) 2-3% b) 2-5%

c) 2-10% d) 10-15%

80. ___________ and _____________ g of Tannin and gelatin are generally required for every 455 litres of juice clarification.

a) 45 & 85 b) 42 & 85

c) 52 & 85 d) 32 & 68

81. Red wines from grapes is prepared by

a) With skin & seeds b) Without skin

c) Added with color d) None

82. Alcohol content in light wines

a) 7 to 20% b) 7 to 9%

c) 9 to 16% d) 16-21%

83. Alcohol content in medium wines

a) 7 to 20% b) 7 to 9%

c) 9 to 16% d) 16-21%

84. Alcohol content in strong wines

a) 7 to 20% b) 7 to 9%

c) 9 to 16% d) 16-21%

85. Sparkling wine contains

a) Oxygen b) $CO_2$

c) Nitrogen d) Ar

86. Wine made from pears is known as

a) Champaigne b) Port

c) Perry d) Feni

87. Wine made from cashew apple in Goa is known as

a) Champaigne b) Port

c) Perry d) Feni

88. Juice made from palm tree is known as

a) Champaigne b) Nira

c) Perry d) Feni

89. _______can be made from any types of grains and even potatoes

a) Vodka b) Rum

c) Champaigne d) Port

90. Lycopene is an antioxidant present in

a) Tomatoes b) Oranges and related fruits

c) Red/purple berries d) Citrus

91. ____________ is a plant hormone that plays a key role in the ripening and senescence of fruits and vegetables.

a) Acetylene b) Ethylene

c) Ziberilic acid d) None

92. ______________is a skin disorder of certain apple cultivars which develops during storage and is due to the oxidation of a natural compound in the skin called α-farnesene.

a) Superficial scald b) Bitter pit

c) Both a & b d) Greening

93. What are the chemicals used to control Superficial scald

a) Diphenylamine b) Ethoxyquin

c) Both a & b d) None

94. Bitter pit in apple controlled by spraying

a) Diphenylamine b) Ethoxyquin

c) Calcium d) None

95. As per FSSAI regulations, what is the minimum (%) of TSS of RTS beverages

a) 15 b) 10

c) 12 d) 8

96. As per FSSAI regulations, what is the minimum (%) of TSS of lemon juice

a) 15 b) 10

c) 6 d) 8

97. As per FSSAI regulations, what is the minimum (%) of TSS of Orange Nectar

a) 15 b) 10

c) 6 d) 8

98. As per FSSAI regulations, what is the minimum (%) of TSS of Unsweetened (Natural Mango Pulp) mango pulp

a) 15 b) 10

c) 6 d) 12

99. As per FSSAI regulations, what is the minimum (%) of TSS of Tomato puree

a) 15 b) 10

c) 9 d) 8

100. As per FSSAI regulations, what is the minimum (%) of TSS of Tomato paste

a) 15 b) 25

c) 9 d) 8

101. As per FSSAI regulations, what is the minimum (%) of TSS of Squash

a) 15 b) 25

c) 40 d) 10

102. As per FSSAI regulations, what is the minimum (%) of TSS of Soup Powders

a) 10 b) 5

c) 9 d) 8

103. As per FSSAI regulations, what is the minimum (%) of TSS of Tamarind Pulp/Puree

a) 32 b) 65

c) 9 d) 8

104. As per FSSAI regulations, what is the minimum (%) of TSS of Tamarind Concentrate

a) 32 b) 65
c) 9 d) 8

105. As per FSSAI regulations, what is the minimum (%) of TSS of Fruit Bar/ Toffee

a) 32 b) 65
c) 75 d) 8

106. As per FSSAI regulations, what is the minimum (%) of TSS of Fruit Crush

a) 25 b) 65
c) 55 d) 8

107. As per FSSAI regulations, what is the minimum (%) of TSS of Fruit Syrup/FruitSharbats

a) 25 b) 65
c) 55 d) 8

108. As per FSSAI regulations, what is the minimum (%) of TSS of Cordial

a) 40 b) 65
c) 55 d) 25

109. As per FSSAI regulations, what is the minimum (%) of TSS of Barley Water

a) 40 b) 65
c) 30 d) 25

110. As per FSSAI regulations, what is the minimum (%) of TSS of Murabba

a) 40 b) 65
c) 30 d) 25

111. As per FSSAI regulations, what is the minimum (%) of TSS of Tomato Ketchup and Tomato Sauce

a) 40 b) 65
c) 30 d) 25

112. As per FSSAI regulations, what is the minimum (%) of TSS of Chilli Sauce /Culinary Paste/ Sauce

a) 40 b) 8
c) 30 d) 25

113. As per FSSAI regulations, what is the minimum (%) of TSS of Soya bean Sauce

a) 40 b) 65

c) 30 d) 25

114. As per FSSAI regulations, what is the minimum (%) of TSS of Carbonated Fruit Beverages or Fruit Drink

a) 40 b) 8

c) 10 d) 25

115. As per FSSAI regulations, what is the minimum (%) of TSS of Fruit Chutney

a) 40 b) 65

c) 50 d) 25

## Answers

| | | | | | | | | | |
|---|---|---|---|---|---|---|---|---|---|
| 1. d | 2. d | 3. a | 4. b | 5. c | 6. d | 7. b | 8. c | 9. c | 10. d |
| 11. a | 12. b | 13. b | 14. a | 15. a | 16. c | 17. b | 18. c | 19. d | 20. c |
| 21. a | 22. a | 23. b | 24. d | 25. c | 26. d | 27. d | 28. b | 29. a | 30. b |
| 31. b | 32. d | 33. c | 34. a | 35. b | 36. b | 37. a | 38. b | 39. a | 40. c |
| 41. d | 42. d | 43. d | 44. d | 45. d | 46. c | 47. a | 48. b | 49. c | 50. c |
| 51. d | 52. b | 53. a | 54. c | 55. d | 56. d | 57. a | 58. c | 59. a | 60. d |
| 61. c | 62. b | 63. a | 64. b | 65. b | 66. b | 67. c | 68. c | 69. c | 70. b |
| 71. a | 72. b | 73. d | 74. a | 75. c | 76. b | 77. b | 78. d | 79. a | 80. b |
| 81. a | 82. b | 83. c | 84. d | 85. b | 86. c | 87. d | 88. b | 89. a | 90. a |
| 91. b | 92. a | 93. b | 94. c | 95. b | 96. c | 97. a | 98. d | 99. c | 100. b |
| 101. c | 102. b | 103. a | 104. b | 105. c | 106. c | 107. b | 108. d | 109. c | 110. b |
| 111. d | 112. b | 113. d | 114. c | 115. c | | | | | |

# 4

# Food Preservation

1. Fortification of common salt is done with
   a) Potassium iodide
   b) Potassium iodate
   c) Both a & b
   d) None
2. Vinegar is a
   a) Class-I preservative
   b) Class-II preservative
   c) Natural preservative
   d) a & c
3. As per FSSAI regulations, the permitted level of Sulphur dioxide in RTS beverages is not more than ______ ppm
   a) 100
   b) 70
   c) 150
   d) 80
4. Sodium nitrite is a
   a) Preservative
   b) Buffering agent
   c) Curing agent
   d) Flavour enhancer
5. Surfactants reduce
   a) Pressure
   b) Surface tension
   c) Interfacial tension
   d) b & c
6. One of the following is an oxidizing agent for dough improvement
   a) $KBrO_3$
   b) $KIO_3$
   c) $CaO_2$
   d) All
7. Natural flavours include
   a) Spices
   b) Essential oils
   c) Oleoresins
   d) All
8. One of the following is a pro-oxidant
   a) Fe
   b) Mg
   c) Na
   d) None
9. An example of chelating agent is
   a) BHA
   b) BHT
   c) TBHQ
   d) EDTA

10. Amorphous Silicon Dioxide
    a) Buffering agent b) Anti caking agent
    c) Stabilizer d) None
11. $TD_{50}$ refers to
    a) Lethal dose b) Tumor dose
    c) a & b d) None
12. Diacetyl Tartaric acid Esters of Mono and Diglyceride (DATEM) is a
    a) Emulsifier b) Buffering agent
    c) Curing agent d) Preservative
13. Thiodipropionic acid is a
    a) Acid regulator b) Antioxidant
    c) Preservative d) Flavouring agent
14. The natural emulsifying agent present in soya bean
    a) Lecithin b) Isoflavones
    c) Soya oil d) Soya protein concentrate
15. The yeast extract powder imparts
    a) Umami taste b) Sweet taste
    c) Bitter taste d) Sour taste
16. Moisture retaining agents are
    a) Stabilizers b) Emulsifiers
    c) Humectants d) Anti-caking agents
17. Gelatin is obtained from hydrolysis of
    a) Collagen b) Gluten
    c) a & b d) None
18. One of the following is a protein based stabilizer and thickener
    a) Pectin b) CMC
    c) Gelatin d) All
19. "GRAS" Additives are categorized under
    a) Category I b) Category II
    c) Category III d) None
20. Teriarybutyl hydroquinone is a
    a) Acid regulator b) Antioxidant
    c) Stabilizer d) All

21. Chlorine dioxide is
    a) Bleaching agent b) Flour improver
    c) a & b d) None
22. Propylene glycol is
    a) Water binding agent b) Acid regulator
    c) Antioxidant d) None
23. Sucralose is a
    a) Artificial Sweetener b) Natural color
    c) Flavour d) None
24. Which of the following is an example for curing agent
    a) Sodium nitrite b) NaCl
    c) a & c d) None
25. Hydrolyzed vegetable Protein (HVP) is
    a) Acid regulator b) Stabilizer
    c) Taste enhancer d) All
26. Acesulphame K is a
    a) Buffering agent b) Color
    c) Flavour d) Artificial Sweetener
27. The characteristic flavour of Grapes is due to
    a) Amyl acetate b) Ethyl butyrate
    c) Methyl anthranilate d) None
28. As per FSSAI regulations-2011, maximum permitted level of Sorbic acid and its sodium or potassium salts in Jam/Jelly or Marmalades is not more than _______ ppm
    a) 300 b) 500
    c) 400 d) 1500
29. Dimethyl Polysiloxane is a
    a) Antifoaming agent b) Buffer
    c) a & b d) None
30. Erythrosine imparts__________ color to foods
    a) Blue b) Green
    c) Yellow d) Red

31. Dextrose is a

a) Class-I preservative
b) Class-II preservative
c) Natural preservative
d) a & c

32. As per FSSAI regulations the permitted level of BHA in butter and ghee

a) $\geq 0.01\%$
b) $\geq 0.02\%$
c) $\geq 0.05\%$
d) $\geq 0.03\%$

33. Resin Guaiace is a

a) Acid regulator
b) Antioxidant
c) Preservative
d) Stabilizer

34. An example of Chelating agent

a) Ammonium Carbonate
b) Phosphoric acid
c) EDTA+CA
d) Malic acid

35. Sodium nitrite is a

a) Preservative
b) Buffering agent
c) Curing agent
d) Flavour enhancer

36. Diacetyl Tartaric acid Esters of Mono and Diglyceride (DATEM) is a

a) Emulsifier
b) Buffering agent
c) Curing agent
d) Preservative

37. The characteristic flavour of Grapes is due to

a) Amyl acetate
b) Ethyl butyrate
c) Methyl anthranilate
d) None

38. The characteristic meat flavour in exotic seasonings is produced by

a) MSG
b) I+G
c) Yeast Extract
d) All

39. HVP imparts ________________ taste to foods

a) Sour taste
b) Bitter taste
c) Umami
d) Salt taste

40. What are the chemicals used for flour bleaching and improving

a) Chlorine gas
b) Chlorine dioxide
c) Nitrogen di and tetra oxides
d) All

41. Dough improvement agents are
    a) Potassium bromate    b) Potassium iodate
    c) Calcium peroxide    d) All
42. Biological leavening agent is
    a) Baking powder    b) Ammonium bicarbonate
    c) Yeast    d) All
43. Acesulfame Potassium
    a) Artificial Sweetener    b) Artificial flavour
    c) Artificial colour    d) All
44. As per FSSAI regulations ________ppm of Sulphur dioxide is allowed in preservation of Ready to serve beverages
    a) 60    b) 500
    c) 70    d) 100
45. As per FSSAI regulations ________ppm of Benzoic Acid is allowed in Jam , marmalade, preserve canned cherry and fruit jelly
    a) 450    b) 250
    c) 200    d) 500
46. As per FSSAI regulations ________ppm of Sorbic acid and its sodium or potassium or calcium salts is allowed in preservation of Paneer or channa
    a) 6000    b) 5000
    c) 2000    d) 1000
47. EDTA is used as ________________in latex processing
    a) Preservative    b) Anticoagulant
    c) Both a & b    d) None
48. The following spice acts as an Antioxidant
    a) Coriander    b) Cumin
    c) Turmeric    d) Fenugreek
49. As per FSSAI regulations, the moisture content in chilli powder is not more than _____________percent by weight
    a) 8    b) 6
    c) 10    d) 11

50. King of Spice is
    a) Black pepper b) Green bell pepper
    c) Red bell pepper d) All
51. Queen of Spice is
    a) Chillies b) Cardamom
    c) Celery d) Capsicum
52. The oil content of coconut is
    a) 60-67 % b) 50-57%
    c) 70-75% d) None
53. India is considered as a __________ of spice
    a) Producer b) Home
    c) Consumer d) All
54. ________is an edible part of *Myristica Fragrans* is used as Mace spice
    a) Stigma b) Aril
    c) Nut d) Bud
55. Caraway belongs to the family of
    a) *Apiaceae* b) *Brassicaceae*
    c) *Lauraceae* d) *Solanaceae*
56. Which of the following is Sulpher containing spice
    a) Annise seed b) Star anise
    c) Asafoetida d) All
57. Which of the following statement is true
    a) Volatile oils are lighter than water
    b) Volatile oils are heavier than water
    c) a & b
    d) None
58. What is the moisture content in commercial garlic powder
    a) 10.5% b) 6.5%
    c) 13.5% d) 15%

59. Green tea is a
    a) Fermented tea leaves  b) Non fermented tea leaves
    c) Rich in antioxidants  d) b & c
60. Following one which is not a major spice
    a) Turmeric  b) Zinger
    c) Chillies  d) Garlic
61. Quills, quillings, featherings and scraped chips are graded products of.
    a) Nutmeg  b) Cinnamon
    c) Chillies  d) Vanilla
62. Princess of spice
    a) Cardamom  b) Mint
    c) Clove  d) Vanilla
63. In solvent extraction method, residual solvent should be below
    a) 100 ppm  b) 200 ppm
    c) 300 ppm  d) None
64. Coconut is commonly called as
    a) Tree of Heaven  b) Kalpa vriksha
    c) Wonder palm  d) All
65. Latex of rubber tree contains__________ % of rubber
    a) 20-30%  b) 30-40%
    c) 40-50%  d) None
66. Chilli oleoresin is extracted by using solvent
    a) Ethyl acetate  b) Acctonc
    c) Ethyl alcohol  d) All
67. Cocoa butter (CB) is obtained from the nibs of _____________beans
    a) Cocoa beans  b) Coffee beans
    c) Milk cream  d) None
68. Commercially identified cocoa beans are
    a) Basic/bulk beans  b) Flavor beans
    c) a & b  d) None

69. Most expensive spice is
    a) Big cardamom b) Mace
    c) Saffron d) Asafoetida
70. Leafy Biennial is
    a) Fenugreek b) Garlic
    c) Parsley d) Coriander
71. Queen of spice is
    a) Cardamom b) Coriander
    c) Clove d) Cinnamon
72. As per FSSAI regulations the *Salmonella* should __________ in 25g of the spice sample
    a) Present b) Absent
    c) 10 CFU d) a & c
73. The Greeks called Marjoram as
    a) Joy of Rivers b) Joy of mountain
    c) Joy of leaves d) All
74. Decaffeination of Coffee is done by
    a) Ethyl acetate b) Methylene chloride
    c) Supercritical or liquid $CO_2$ d) All
75. Lead chromate test for turmeric
    a) Negative b) Positive
    c) a & b d) None
76. Withering is a process involved in
    a) Coffee b) Tea
    c) Cocoa d) Cinnamon
77. King of Spice is
    a) Black pepper b) Green bell pepper
    c) Red bell pepper d) All
78. The following spice acts as an Antioxidant
    a) Coriander b) Cumin
    c) Turmeric d) None

79. Commercially Garlic is dried to a moisture content of ______ percent

a) 8 b) 6

c) 10 d) 11

80. One gram of Onion oil is equivalent to how many Kgs of fresh onions

a) 5.4 b) 4.4

c) 6.4 d) 3.4

81. Queen of Spice is

a) Chillies b) Cardamom

c) Celery d) Capsicum

82. Spice sterilizing agent is

a) Water b) Steam

c) ETO d) b & c

83India is considered as a __________ of spice

a) Producer b) Home

c) Consumer d) All

84. __________ is an edible part of *Myristica Fragrans* is used as baking spice

a) Stigma b) Aril

c) Nut d) Bud

85. Caraway belongs to the family of

a) *Apiaceae* b) *Brassicaceae*

c) *Lauraceae* d) *Solanaceae*

86. Which of the following is Sulpher containing spice

a) Annise seed b) Star anise

c) Asafoetida d) All

87. Which of the following statement is true

a) Volatile oils are lighter than water

b) Volatile oils are heavier than water

c) a & b

d) None

88. As per FSSAI regulations, the moisture content in Soluble Coffee Powder is not more than ______________ percent by weight

a) 5 b) 6

c) 4 d) 8

89. Decaffeination of Coffee is done by____________method

a) Ethyl acetate b) Methylene chloride

c) Supercritical or liquid $CO_2$ d) All

90. Following one which is not a major spice

a) Turmeric b) Zinger

c) Chillies d) Garlic

91. Finest quality of essential oil is obtained by__________ distillation method

a) Steam b) Water

c) Water-Steam d) All

92. Princess of spice

a) Cardamom b) Mint

c) Clove d) Vanilla

93. What is the suitable packaging material for storing of Herbs and Spice oils and Oleoresins

a) Laminated films b) PET bottles

c) HDPE drums d) LDPE pouches

94. As per FSSAI regulations, the volatile oil content (db) in Black pepper not less than ______________ percent by weight

a) 2 b) 6

c) 4 d) 3

95. Cinnamon is also known as

a) Baking spice b) Pungent spice

c) Sweet wood d) Cassia

96. One of the following is a natural antioxidant

a) Ascorbic acid b) Malic acid

c) a & b d) None

97. Chelating agents are otherwise referred as

a) Saponins b) Sesquitarpens

c) Sequestrants d) Sesquins

98. As per FSSAI regulations, the permitted level of Sulphur dioxide in RTS beverages is not more than ____________ ppm

a) 100 b) 70

c) 150 d) 80

99. As per FSSAI regulations, the permitted level of BHA in edible oils or fats is not more than ____________ percent by weight

a) 0.01 b) 0.03

c) 0.04 d) 0.02

100. Surfactants reduce

a) Pressure b) Surface tension

c) Interfacial tension d) b & c

## Answers

| | | | | | | | | | |
|---|---|---|---|---|---|---|---|---|---|
| 1. c | 2. d | 3. b | 4. c | 5. d | 6. d | 7. d | 8. a | 9. d | 10. b |
| 11. b | 12. a | 13. b | 14. a | 15. a | 16. c | 17. a | 18. c | 19. a | 20. b |
| 21. c | 22. a | 23. a | 24. c | 25. c | 26. d | 27. c | 28. b | 29. a | 30. d |
| 31. d | 32. b | 33. b | 34. c | 35. c | 36. a | 37. c | 38. d | 39. c | 40. d |
| 41. d | 42. c | 43. a | 44. c | 45. c | 46. c | 47. c | 48. c | 49. d | 50. a |
| 51. b | 52. a | 53. d | 54. b | 55. a | 56. c | 57. c | 58. b | 59. d | 60. d |
| 61. b | 62. d | 63. c | 64. d | 65. b | 66. d | 67. a | 68. c | 69. c | 70. c |
| 71. a | 72. b | 73. c | 74. d | 75. a | 76. b | 77. a | 78. c | 79. b | 80. b |
| 81. b | 82. d | 83. d | 84. b | 85. a | 86. c | 87. c | 88. c | 89. d | 90. d |
| 91. b | 92. d | 93. c | 94. a | 95. c | 96. a | 97. c | 98. b | 99. d | 100. d |

1. One of the following is an oxidizing agent for dough improvement
   a) $KBrO_3$ b) $KIO_3$
   c) $CaO_2$ d) All
2. Natural flavours include
   a) Spices b) Essential oils
   c) Oleoresins d) All
3. One of the following is a pro-oxidant
   a) Cu b) Mg
   c) Na d) None
4. An example of chelating agent is
   a) BHA b) BHT
   c) TBHQ d) EDTA
5. Amorphous Silicon Dioxide
   a) Buffering agent b) Anti caking agent
   c) Stabilizer d) None
6. $TD_{50}$ refers to
   a) Lethal dose b) Tumor dose
   c) a & b d) None
7. As per FSSAI regulations, the permitted level of Benzoic acid in Tomato and other sauces is not more than ______________ ppm
   a) 550 b) 650
   c) 750 d) 850
8. As per FSSAI regulations, the permitted level of Benzoic acid in Pickles and chutneys is not more than ______________ ppm
   a) 250 b) 350
   c) 520 d) 490
9. The natural emulsifying agent present in soya bean
   a) Lecithin b) Isoflavones
   c) Soya oil d) Soya protein concentrate
10. The MSG imparts
    a) Umami taste b) Sweet taste
    c) Bitter taste d) Sour taste

11. Moisture retaining agents are
    a) Stabilizers  b) Emulsifiers
    c) Humectants  d) Anti-caking agents
12. Gelatin is obtained from hydrolysis of
    a) Collagen  b) Gluten
    c) a & b  d) None
13. One of the following is a protein based stabilizer and thickener
    a) Pectin  b) CMC
    c) Gelatin  d) All
14. "GRAS" Additives are categorized under
    a) Category I  b) Category II
    c) Category III  d) None
15. Teriarybutyl hydroquinone is a
    a) Acid regulator  b) Antioxidant
    c) Stabilizer  d) All
16. Chlorine dioxide is
    a) Bleaching agent  b) Flour improver
    c) a & b  d) None
17. Propylene glycol is
    a) Water binding agent  b) Acid regulator
    c) Antioxidant  d) None
18. Sucralose is a
    a) Artificial Sweetener  b) Natural color
    c) Flavour  d) None
19. Which of the following is an example for curing agent
    a) Sodium nitrite  b) NaCl
    c) a & c  d) None
20. Hydrolyzed vegetable Protein (HVP) is
    a) Acid regulator  b) Stabilizer
    c) Taste enhancer  d) All

21. Acesulphame K is a
    a) Buffering agent b) Color
    c) Flavour d) Artificial Sweetener
22. Sugar syrup is a
    a) Class –I Preservative b) Class –II Preservative
    c) a & b d) None
23. As per FSSAI regulations, the permitted level of Sorbic acid and its sodium or potassium salts in Jam/Jelly or Marmalades is not more than ___ ppm
    a) 300 b) 500
    c) 400 d) 1500
24. Dimethyl Polysiloxane is a
    a) Antifoaming agent b) Buffer
    c) a & b d) None
25. Erythrosine imparts__________ color to foods
    a) Blue b) Green
    c) Yellow d) Red
26. The dough properties of wheat flour effects on the quality of
    a) Geometry of backed products b) Taste of baked goods
    c) a & b d) None
27. Dough-forming properties of flours are commonly evaluated by
    a) Alveograph b) Mixograph
    c) Farinograph d) All
28. _____________measures air pressure inside of a dough bubble as it is inflated until it bursts.
    a) Alveograph b) Mixograph
    c) Farinograph d) All
29. Mixograph and Farinograph measures
    a) Dough water absorption
    b) Strength
    c) Mixing time & tolerance to over mixing
    d) All

30. In which stage dough is at maximum stiffness and comes together as one mass.

a) Pickup b) Clean Up

c) Final development d) Letdown

31. In which stage dough will begin to liquefy

a) Breakdown b) Clean Up

c) Final development d) Letdown

32. The following spice imparts colour to food

a) Coriander b) Cumin

c) Turmeric d) None

33. As per FSSAI regulations, the moisture content in chilli powder is not more than _________ percent by weight

a) 8 b) 6

c) 10 d) 11

34. As per FSSAI regulations, the volatile oil content (db) in whole Clove is not less than __________ percent by weight

a) 15 b) 16

c) 18 d) 17

35. Queen of Spice is

a) Chillies b) Cardamom

c) Celery d) Capsicum

36. The oil content of coconut is

a) 60-67 % b) 50-57%

c) 70-75% d) None

37. India is considered as a __________ of spice

a) Producer b) Home

c) Consumer d) All

38. __________ is an edible part of *Myristica Fragrans* is used as Mace spice

a) Stigma b) Aril

c) Nut d) Bud

39. Caraway belongs to the family of
    a) *Apiaceae* b) *Brassicaceae*
    c) *Lauraceae* d) *Solanaceae*

40. Which of the following is Sulpher containing spice
    a) Annise seed b) Star anise
    c) Asafoetida d) All

41. Which of the following statement is true
    a) Volatile oils are lighter than water
    b) Volatile oils are heavier than water
    c) a & b
    d) None

42. As per FSSAI regulations, the moisture content in Soluble Coffee Powder is not more than ______ percent by weight
    a) 5 b) 6
    c) 4 d) 8

43. As per FSSAI regulations, the Coffee content in Coffee - Chicory Mixture is shall not be less than ______ percent by mass
    a) 50 b) 51
    c) 52 d) 49

44. Following one which is not a major spice
    a) Turmeric b) Zinger
    c) Chillies d) Onion

45. Anti-coagulants used in the processing of latex
    a) Formalin b) Zinc Oxide
    c) Lauric acid d) Boric acid

46. Princess of spice
    a) Cardamom b) Mint
    c) Clove d) Vanilla

47. In solvent extraction method, residual solvent should be below
    a) 100 ppm b) 200 ppm
    c) 300 ppm d) None

48. Coconut is commonly called as

a) Tree of Heaven b) Kalpa vriksha

c) Wonder palm d) All

49. Latex of rubber tree contains_____ % of rubber

a) 20-30% b) 30-40%

c) 40-50% d) None

50. Chilli oleoresin is extracted by using solvent

a) Ethyl acetate b) Acetone

c) Ethyl alcohol d) All

51. Cocoa butter (CB) is obtained from the nibs of ___________ beans

a) Cocoa beans b) Coffee beans

c) Milk cream d) None

52. Commercially identified cocoa beans are

a) Basic/bulk beans b) Flavor beans

c) a & b d) None

53. Most expensive spice is

a) Big cardamom b) Mace

c) Saffron d) Asafoetida

54. Leafy Biennial is

a) Fenugreek b) Garlic

c) Parsley d) Coriander

55. Queen of spice is

a) Cardamom b) Coriander

c) Clove d) Cinnamon

56. As per FSSAI regulations the *Salmonella* should __________ in 25g of the spice sample

a) Present b) Absent

c) 10 CFU d) a & c

57. As per FSSAI regulations headless cloves should not be more than __________ percent by weight in given Clove sample

a) 2 b) 3

c) 4 d) 5

58. As per FSSAI regulations light berries should not be more than __________ percent by weight in given Pepper sample

a) 2 b) 5

c) 4 d) 8

59. Lead chromate test for checking of adulteration of turmeric should be

a) Negative b) Positive

c) a & b d) None

60. As per FSSAI regulations, Acid Insoluble Ash (AIA) content of compounded asafoetida should not be more than __________ percent by weight

a) 1.5 b) 2.0

c) 1.0 d) 2.5

61. Life blood of food companies are

a) Existing products b) New Products

c) New packaging material d) Marketing

62. Date fruits packed in vacuum packaging is an example of

a) Repositioning of existing products

b) New form or size of existing products

c) New packing of existing products

d) All

63. Bring into existence of new product into market is called as

a) Innovative product

b) Creative product

c) New form or size of existing products

d) a & b

64. Risk of failure of new product may be high in

a) Creative products

b) Innovative products

c) New form or size of existing products

d) All

65. Who buys in a market place is called as

a) Seller b) Consumer

c) Customer d) All

66. The decline in product launches may decrease due to
   a) Consolidating of many food companies
   b) Potential product failures
   c) Private label products.
   d) All
67. Operating a food plant year round is
   a) More profitable  b) Less profitable
   c) No change in profits  d) More losses
68. In any market place, profile of customers and consumers are altering due to
   a) Population movement  b) Population age
   c) Ethnic makeup  d) All
69. What are the governmental influences in pushing for new product development
   a) Limits on food additives usage  b) Population age
   c) Technological advances  d) All
70. New product is developed to capture vacuum (needs) of
   a) Vendors  b) Customers
   c) Whole sellers  d) All
71. Role of Marketing people in new product development
   a) Preparation of labels
   b) Consumer analysis & Marketing strategy
   c) Promotional activities
   d) All
72. The manufacturing department determines
   a) Production capabilitics  b) Manpower requirements
   c) a & b  d) Develop new processing
73. Food technologists designs a new product based on
   a) Concept statement  b) Consumer research
   c) Quality of PM  d) a & b
74. In Brazil, diesel fuel was developed from
   a) Pea nut oil  b) Soya oil
   c) Caster oil  d) Mustard oil

75. The following one is technical source of databases for new product development ideas

a) FSTA b) FROST
c) AFSTI d) a & b

76. Fluidity as an organization tool in new product development related to

a) Shifting personnel between departments
b) R&D team
c) Marketing team
d) Finance team

77. Product concepts and prototypes are developed by R&D team by applying

a) Knowledge b) Experience
c) Technical Training d) All

78. New product ideas screening must consider _________ of the company

a) Financial planning b) Strategic planning
c) Tactical planning d) All

79. The following facilities are required for new product development team

a) PD lab & Test kitchen b) Commercial plants
c) Pilot plant facilities d) a & c

80. Transfer of technology from global R&D centers to state of art manufacturing locations of the company can result in failure due to

a) None interested here (NIH) syndrome
b) Not interested here (NIH) syndrome
c) a & b
d) No one interested here (NIH) syndrome

81. When a product fails, least affected personnel in the company are

a) R&D b) Marketing
c) Production d) All

82. The following category of consumers in market place are called as pace setters

a) Inner directed b) Outer directed
c) Sustenance driven d) All

83. According to British brew Bass Tavern's research, consumers of Night clubbers are called as
   a) Young affluent and adventures
   b) Unskilled laborers
   c) With money and want entertainment
   d) Pint and pension group
84. Quantitative market information is based on
   a) Interpretation of focus groups
   b) Interviews requiring subjective interpretation
   c) Statistical analysis of data
   d) a & b
85. Neuromarketing is carried out by
   a) Functional marketing resonance imaging
   b) Functional marketing research imaging
   c) Functional magnetic resonance imaging
   d) Functional magnetic revolution imaging
86. The generation of product ideas is based on the needs and desires of
   a) Consumers b) Customers
   c) a & b d) Marketing team
87. The following technology helps in to develop tailored made food products to meet the specific nutritional taste and safety specifications
   a) Extrusion Technology b) Biotechnology
   c) Hurdle technology d) All
88. What are the major concerns of Manufacturing plant
   a) Space & facilities b) Labour and disruptiveness
   c) a & b d) Location
89. What is considered as a buffer between the end of production line and distribution system?
   a) Purchasing department b) Warehouse
   c) R&D team d) Marketing
90. The following variables will be measured in Full sale method of test marketing
   a) Competition b) Trade
   c) Advertisement d) All

91. No evidence of swelling for the blown cans name is

a) Flat
b) Springer
c) Hard swell
d) Flipper

92. A non-nutritive substance added intentionally to food, to improve its appearance, flavour, texture or storage properties is called as

a) Food
b) Food additive
c) Preservative
d) All

93. Placing the food below or above or in between a red hot surface

a) Grilling
b) Roasting
c) Sauteing
d) All

94. What is the temperature of deep fat frying

a) 100-150°C
b) 180-200°C
c) 180-220°C
d) 100-180°C

95. Vermicelli payasam is prepared by combined cooking of

a) Roasting
b) Simmering
c) a & b
d) Roasting and boiling

96. Which is less penetrating frequency for Domestic microwave ovens

a) 2450MHz
b) 915 MHz
c) A and B
d) None

97. The microwave region of the e.m. spectrum occupies frequencies between

a) 29 Hz up to 4012 Hz
b) 1019 Hz up to 4000 Hz
c) 10 Hz up to 1012 Hz
d) $10^9$ to $10^{12}$ HZ

98. Function of Sugar as a ingredient

a) Provides flavor, volume and texture
b) Acts as food for yeast
c) Acts as a preservative in jams, jellies and pickles
d) All

99. When a food is in moisture equilibrium with its environment, then the water activity (aw) of the food will be quantitatively equal to

a) RH/100
b) RH-100
c) RH*100
d) RH+100

100. Most enzymatic reactions are slowed down at water activities.

a) Below 0.8
b) Below 0.5
c) Below 0.9
d) Below 0.6

## Answers

| | | | | | | | | | |
|---|---|---|---|---|---|---|---|---|---|
| 1. d | 2. d | 3. a | 4. d | 5. b | 6. b | 7. c | 8. a | 9. a | 10. a |
| 11. c | 12. a | 13. c | 14. a | 15. b | 16. c | 17. a | 18. a | 19. c | 20. c |
| 21. d | 22. a | 23. b | 24. a | 25. d | 26. a | 27. d | 28. a | 29. d | 30. b |
| 31. a | 32. c | 33. d | 34. d | 35. b | 36. a | 37. d | 38. b | 39. a | 40. c |
| 41. c | 42. c | 43. b | 44. d | 45. a | 46. d | 47. c | 48. d | 49. b | 50. d |
| 51. a | 52. c | 53. c | 54. c | 55. a | 56. b | 57. a | 58. b | 59. a | 60. a |
| 61. b | 62. c | 63. b | 64. a | 65. c | 66. d | 67. a | 68. d | 69. a | 70. b |
| 71. d | 72. c | 73. d | 74. b | 75. d | 76. a | 77. d | 78. d | 79. d | 80. b |
| 81. c | 82. a | 83. c | 84. c | 85. c | 86. c | 87. b | 88. c | 89. b | 90. d |
| 91. a | 92. b | 93. a | 94. c | 95. c | 96. a | 97. d | 98. d | 99. a | 100. a |

1. Antimicrobial agent in egg
   a) Trypsin b) Lypase
   c) amylase d) Avidin
2. Most resistant organism for irradiation
   a) Bacteria b) Yeast
   c) Mould d) Virus
3. Membrane pore sizes for RO
   a) 0.1 μ to 0.5 μ b) 1,000 to 1,00,000 $M_W$
   c) 250 to 400 $M_W$ d) 125 $M_W$
4. Membrane pore sizes for UF
   a) 0.1 μ to 0.5 μ b) 1,000 to 1,00,000 $M_W$
   c) 250 to 400 $M_W$ d) 125 $M_W$
5. Ohmic heating is also called as
   a) Resistance heating b) Electroheating
   c) a & b d) None
6. Advantage of microwave and D.E heating
   a) Uniform heating
   b) Selective heating may occur
   c) Efficient of energy conservation
   d) All
7. The following material heat well by Microwave and D.E Heating
   a) Water acid anhydrides, b) Alcohols, aldehydes,ketones,
   c) Amides, amines d) All
8. The energy conversion in microwave and Dielectric heating is mainly by
   a) Ionic conduction b) Dipole rotation.
   c) a & b d) None
9. The principle involved in freeze drying is
   a) Sublimation b) Submission
   c) Sun drying d) Sun filling
10. Slow freezing formss
    a) Large crystals b) Small crystals
    c) Both d) None

11. One of the following is a fermented product
    a) Cheese b) Curd
    c) Wine d) All
12. Freeze drying is also termed as
    a) Lipolysis b) Lyophilization
    c) Concentration d) Dehydration
13. Cold sterilization of food is so called as
    a) Freezing b) Irradiation
    c) Waxing d) Chilling
14. One of the following is a masonry storage structure
    a) Gade b) Kotlu
    c) Pathra d) Puri
15. Mediums used for dry heat methods
    a) Air or fat b) Water of fat
    c) Water or air d) Water or steam
16. The peeling method is only used for garlic and onion is____________
    a) Flame peeling b) Lye peeling
    c) Hand peeling d) Steam peeling
17. Acid resistant lacquer is called as
    a) R-enamel b) C-enamel
    c) a & b d) None
18. The process of bringing back the temperature of the frozen food to room temperature is called as
    a) Shrinkage b) Thawing
    c) Icing d) None
19. Spray drying is based on
    a) Atomization b) Aromization
    c) hydrogenation d) a & b
20. 1 Gray rad =
    a) $10^2$ rad b) 10 rad
    c) $10^3$ rad d) $10^4$ rad

21. Low acid fruits are mostly spoiled by
    a) Thermophilic organisms b) Mesophilic organisms
    c) Putrefactive anaerobes d) All
22. What is the minimum pressure requirement for processing of low acid fruits and vegetables
    a) 0.70 to 1.05 $kg/cm^3$ b) 10 to 15 lb per sq inch
    c) a & b d) 0.50 to 2.05 $kg/cm^3$
23. Strength of brine is measured by
    a) Brix hydrometer b) Salometer
    c) Salt meter d) All
24. What is the temperature & time required for heat peeling of few fruits & vegetables
    a) 60°C/10 to 30 sec b) 40°C/10 to 60 sec
    c) 70°C/10 to 60 sec d) 80°C/10 sec
25. An example of natural preservative in yeast fermented wines
    a) $CO_2$ b) Ethyl alcohol
    c) Sorbic acid d) Benzoic acid
26. Vodka is produced from
    a) Potato b) Grains
    c) a & b d) Grapes
27. Nira is prepared from
    a) Palm tree b) Pears
    c) Cashew nut d) Plum fruit
28. Feni is prepared from
    a) Palm tree b) Pears
    c) Cashew apple d) Plum fruit
29. An ideal vinegar should contain______ percent of sugar
    a) 4% b) 0.3%
    c) 0.5% d) 2%
30. One grain strength is equalvent to __________ times of the acetic acid present in it
    a) 5 b) 15
    c) 10 d) 20

31. The following fermentation method is used to preserve Fruits and vegetables

a) Acetic b) Lactic

c) Alcoholic d) All

32. What is the role of potato flour in baking industry

a) Reduces protein content in mix b) Increases protein content in mix

c) Reduces staling of bread d) a & c

33. What is non fruit ingredient in fruit cheese

a) Cheese b) Butter

c) Cream d) SMP

34. What is the percentage of barley starch in Barley Water

a) 1% b) 0.25%

c) 0.5% d) 2.5%

35. As per FSSAI regulations-2011, total soluble solids of Fruit Nectars is not less than

a) 40% b) 25%

c) 15% d) 30%

36. An example of enzyme preparations for juice clarification

a) Pectinol b) Filtragol

c) a & b d) None

37. What is the requirement of Gelatin & Tannin mixture for clarification of 455 lit of juice

a) 42 g of tannin +85 g of gelatin b) 42 g of tannin+ 65 g of gelatin

c) 52 g of tannin+ 65 g of gelatin d) 52 g of tannin+ 85 g of gelatin

38. Important filter aids for juice filtration are

a) Supercel b) Kieselguhr

c) Spanish clay and bentonite d) All

39. As per FSSAI regulations-2011, TSS (m/m) of soya sauce is not less than

a) 15% b) 8%

c) 3% d) 25%

40. Spice bag method is related to
    a) Sauces b) Ketchup
    c) a & b d) None
41. Softness and slipperiness of pickle is due to
    a) Weak brine b) Strong brine
    c) Strong acid d) All
42. The organism involved in end of Sauerkraut production is
    a) *Leuconostac mesentroids* b) *Lactobacillus brevis*
    c) *Lactobacillus plantarum* d) *Lactobacillus delbrukii*
43. What is the optimum concentration of salt to inhibit most of the bacteria
    a) 5% b) 10%
    c) 15% d) 25%
44. Suspension of shredded peels of orange, lemon in product is called as
    a) Murrabba b) Marmalade
    c) Preserve d) Candy
45. The optimum pH for a jelly containing 1 per cent pectin is approximately 3.0, 3.2 and 3.4 for _________ per cent TSS, respectively.
    a) 60, 65 & 70 b) 65, 60 & 70
    c) 70, 65 & 60 d) 70,60 & 65
46. What are the essential ingredients in jelly production
    a) Pectin b) Sugar
    c) Acid & water d) All
47. Surface graining and shrinkage of jam is caused by
    a) Loss of moisture b) Microbial action
    c) Low acid d) All
48. What are the source of irradiation
    a) Radioisotopes of $^{60}Co$ or $^{137}Cs$ b) Machine-generated X-rays
    c) Accelerated electrons d) All
49. The water activity of semi moist textured products is in between
    a) 0.70 &0.85 b) 0.91 &0.99
    c) 0.85& 0.90 d) None

50. A very tiny leak in can is called as
   a) Springer  b) Leakage
   c) Breathing  d) Flipper
51. The following one is major producing vegetable in India
   a) Tomato  b) Chillies
   c) Potato  d) Beans
52. What is the concentration of Sodium hydroxide used in caustic peeling of fruits and vegetables
   a) 10-20%  b) 0.1-2%
   c) 5-10%  d) 1- 2%
53. A canned product having pH more than 4.5 can be processed at
   a) 121°C  b) 100°C
   c) 105°C  d) 150°C
54. Sulphur dioxide acts effectively against
   a) Bacteria  b) Moulds
   c) a & b  d) Yeast
55. The following one is not a physical method for determination of maturity of fruits and vegetable
   a) Weight  b) Specific gravity
   c) Flesh firmness  d) Bio electrical conductance
56. What is the Minimum juice content in mature Mandarins
   a) 30%  b) 28%
   c) 33%  d) 35%
57. The following one is a right statement
   a) Calcium increases firmness of fruits
   b) Calcium decreases firmness of fruits
   c) Calcium increases size, weight of fruits
   d) All
58. Blanching of fruits and vegetables helps in
   a) Inactivates enzymes
   b) Reduces area to surface volume
   c) Enhances the green colour of vegetables
   d) All

59. Which is major producing plantation crop in India
    a) Cashew nut b) Arc nut
    c) Coconut d) Rubber
60. The following spice acts as an Antioxidant
    a) Coriander b) Cumin
    c) Turmeric d) Fenugreek
61. The oil content of coconut is
    a) 60-67% b) 50-57%
    c) 70-75% d) None
62. Which of the following statement is true
    a) Volatile oils are lighter than water
    b) Volatile oils are heavier than water
    c) a & b
    d) None
63. What is the moisture content in commercial garlic powder
    a) 10.5% b) 6.5%
    c) 13.5% d) 15%
64. Green tea is a
    a) Fermented tea leaves b) Non fermented tea leaves
    c) Rich in antioxidants d) b & c
65. Quills, quillings, featherings and scraped chips are graded products of
    a) Nutmeg b) Cinnamon
    c) Chillies d) Vanilla
66. The following test is used to detect meat spoilage
    a) Dye reduction test b) Phenolphthalein test
    c) Hydrogen test d) None
67. During pre-freezing the meat is frozen at °C
    a) -40 b) -50
    c) -60 d) -70
68. When the meat depicts sign of decomposition and putrefaction is referred as
    a) Shrinking b) Rigormortis
    c) Spoilage d) All

69. Veal is a type of meat obtained from
    a) Sheep b) Goat
    c) Pig d) Calf
70. The sausages are classified basically on the basis of wheather meat is
    a) Fermented b) Cooked
    c) Smoked d) All
71. The following is an example of sarcoplasmic protein
    a) Collagen b) Elastin
    c) Myoglobin d) None
72. The following chemical acts as humectants
    a) Glycerol b) Nacl
    c) a & b d) None
73. The involuntary muscles are
    a) Smooth muscles b) Cardiac muscles
    c) a & b d) None
74. The superior method of slaughter of animal as far efficiency of bleeding is consider is
    a) Halal b) Jhatka
    c) Kosher d) None
75. For the artificial tenderization of meat muscles, which of the following could be the source of the enzymes?
    a) Figs b) Papaya
    c) Pinc applc d) All
76. The phase where the bacterial growth is maximum
    a) Lag b) Log
    c) Stationary d) None
77. Most spoilage bacteria can grow up to a water activity of
    a) 0.99 b) 0.10
    c) 0.15 d) 0.91
78. A sharp decline in muscle PH, the muscle depicts following condition
    a) PSE (pale, soft & exudative) b) DFD (dark,firm&dry)
    c) Both a & b d) None

79. A sharp increase in muscle PH, the muscle depicts following condition
    a) PSE (pale, soft & exudative) b) DFD (dark,firm&dry)
    c) Both a & b d) None
80. The colour of pork meat is
    a) Grayish Pink b) Cherry Red
    c) Black d) All
81. The salami is a
    a) Sausage b) Frozen meat
    c) Meat burger d) Grilled meat
82. Dressing percentage of desi pigs
    a) 65-70 b) 20-30
    c) 10-20 d) None
83. Smooth muscles are found in
    a) Gastro intestinal tract b) Blood vessels
    c) Lymphatics d) All
84. Shank and brisket is the primal cut of
    a) Goat b) Sheep
    c) Buffalo d) All
85. Packaging of meat can be done in
    a) Overwraps b) Shrink film overwraps
    c) Both a & b d) None
86. Single screw operation depends on
    a) Pressure requirement of die b) Slit at the barrel wall
    c) Degree of screw fill d) All of the above
87. TVP is prepared from which type of extruder?
    a) Collet extruder b) Pasta press
    c) High shear cooking extruder d) Low shear cooking extruder
88. Velocity of drag flow in an extruder is directly proportional to
    a) Motor power b) In put energy
    c) Screw speed d) Length of screw

89. Pre-conditioner is used in conjugation with the extruders to
   a) Increase the residence time
   b) Increase capacity
   c) Reduce mechanical power consumption
   d) All of the above
90. Puffed snack is prepared from
   a) Collet Making extruder   b) High shear cooking extruder
   c) Low shear cooking extruder   d) All of the above
91 The processing zone of a single screw barrel consists of
   a) Feeding zone   b) Kneading zone
   c) Final cooking zone   d) All of the above
92. In the production of potato sticks the screw speed should be maintained at
   a) 400 rpm   b) 600 rpm
   c) 390 rpm   d) 650 rpm
93. The fallowing factors influence on the selection of an extruder
   a) Product type   b) Packaging
   c) Finishing   d) None
94. Co-rotating twin screw extruder are most popularly used in food industry due to
   a) High capacity and enhanced mixing capacity
   b) Less energy consumption
   c) Economical price
   d) Easy handling
95. Modified starches can be prepared from
   a) Low shear cooking extruder   b) High shear cooking extruder
   c) Collet extruder   d) High pressure forming extruder
96. Pasta products are prepared from
   a) Hard wheat   b) Soft wheat
   c) Durum wheat   d) Buck wheat flour
97. Noodles were originated in
   a) China   b) Japan
   c) Korea   d) Taiwan

98. The pellet frying quality depends on
    a) Pellet thickness b) Pellet shape
    c) Pellet composition d) All

99. What is the moisture content range in single twin screw extruder?
    a) 20-28% b) 21-28%
    c) 22-28% d) 23-28%

100. Expanded ready to eat cereals require cooking moisture
    a) >5% b) >10%
    c) >15% d) >20%

## Answers

1. d 2. c 3. d 4. b 5. c 6. d 7. d 8. c 9. a 10. a
11. d 12. b 13. b 14. b 15. a 16. a 17. a 18. b 19. a 20. a
21. d 22. c 23. b 24. b 25. b 26. c 27. a 28. c 29. b 30. c
31. d 32. d 33. b 34. b 35. c 36. c 37. a 38. d 39. d 40. c
41. a 42. c 43. c 44. b 45. a 46. d 47. a 48. d 49. a 50. c
51. b 52. d 53. a 54. c 55. d 56. c 57. a 58. d 59. c 60. c
61. a 62. c 63. b 64. d 65. b 66. a 67. a 68. d 69. d 70. d
71. c 72. c 73. c 74. a 75. d 76. b 77. d 78. a 79. b 80. a
81. a 82. a 83. d 84. c 85. c 86. d 87. c 88. c 89. d 90. a
91. d 92. a 93. a 94. a 95. b 96. c 97. a 98. d 99. a 100. d

1. Structure forming component in extruded product is
   a) Flour b) Starches
   c) Protein d) All
2. An extruded products with no expansion in volume is called as
   a) Collet b) Dellet
   c) Pellet d) Sellet
3. It is the combination of two or more preservation methods to increase the shelf life of food products
   a) Microwave heating b) Hurdle technology
   c) HPP d) PEF
4. In general packaging materials used should have
   a) Protective strength b) Attractive
   c) Economy d) All
5. It is considered as a poor man's food
   a) Bajra b) Barley
   c) Kodo d) Finger millet
6. Blue revolution aims to increase in ____________ production
   a) Fish b) Milk
   c) Rice d) Coffee
7. Factors effecting food demand
   a) Population b) Life style
   c) Consumption d) All
8. Indian institute of horticultural research is located at
   a) Kassargud b) Anand
   c) Karnal d) None
9. Central social Welfare Board: set up by
   a) WHO b) FAO
   c) Govt. of India d) a & b
10. Cow pea is commonly known as
    a) Lobia b) Soybean
    c) Til d) Pend

11. Sesame is commonly known as
    a) Lobia b) Soybean
    c) Til d) Pend
12. By product of sugar cane is used for manufacturing of methyl alcohol, citric acid, rum.
    a) Bagasses b) Molasses
    c) Sugar beet d) a & b
13. By product of sugar cane is used for the preparation of paper and plastics.
    a) Bagasses b) Molasses
    c) Sugar beet d) a & b
14. The oil of ________ is used for the manufacturing of baby protein
    a) Sunflower oil b) Ground nut oil
    c) Palm oil d) All
15. Queen of cereals
    a) Rice b) Maize
    c) Wheat d) Sorghum
16. Niacin deficiency causes
    a) Beri –beri b) Pellagra
    c) Anemia d) None
17. Vitamin $B_{12}$
    a) Cynacobalamine b) Pyridoxine
    c) Folic acid d) Niacin
18. It is an important trace element as it forms part of vitamin B12
    a) Calcium b) Copper
    c) Zinc d) Cobalt
19. In India first Food Processing Industry was established in
    a) 1942 b) 1952
    c) 1962 d) 1972
20. Vitamin acts as an antioxidant
    a) Vitamin E b) Vitamin K
    c) Vitamin A d) None

21. Vitamin D deficiency causes
    a) Scurvy b) Blindness
    c) Pellagra d) Rickets
22. pH is 3.7 and below food are
    a) Highly acidic b) Acid foods
    c) Low acid d) None
23. Example for acid foods
    a) Meat b) Soups
    c) Figs d) Fish
24. It is considered that pH of _______ is dividing line between acid and non acid foods
    a) 4.5 b) 5
    c) 6 d) 3.7
25. Example for ready to cook (RTC)
    a) Noodles b) Papads
    c) Wafers d) All
26. Example for ready to use foods
    a) Garam masala b) Ginger/garlic paste
    c) Butter chicken mix d) All
27. Fermented foods are produced by action of
    a) Bacteria b) Moulds
    c) Yeast d) All
28. An example of non nutritive substance
    a) Carbohydrates b) Protein
    c) Fiber d) Fat
29. Foods provides health benefits beyond basic nutrition is called as
    a) Fortified foods b) Supplemented foods
    c) Functional foods d) All
30. The following one is an example of empty calories food
    a) Pure carbohydrates b) Fats
    c) Protein d) All

31. Aflatoxin in ground nut is produced by

a) *A. flavus* b) *A. niger*

c) *A. ochraceus* d) All

32. Insect infestations in grains, dry fruits and spices are controlled by fumigating with

a) Methyl bromide b) Ethylene oxide

c) Propylene oxide d) All

33. The dividing line of pH in between medium acid and acid products is

a) 3.5 b) 4.5

c) 5.3 d) 3.0

34. Functions of sugar in foods

a) Enhances flavor, volume & texture

b) Food for yeast

c) Preservative in jams jellies & pickles

d) All

35. When a food is in moisture equilibrium with its environment, then the water activity (aw) of the food will be quantitatively equal to

a) RH/100 b) RH-100

c) RH*100 d) RH+100

36. What are the refrigerants used in cryogenic freezing

a) Liquid nitrogen b) Liquid carbon dioxide

c) a & b d) CFC

37. A very tiny leak in can is called as

a) Springer b) Leakage

c) Breathing d) Flipper

38. Most resistant organism for irradiation

a) Bacteria b) Yeast

c) Mould d) Virus

39. Membrane pore sizes for RO

a) 0.1μ to 0.5μ b) 1,000 to 1,00,000 MW

c) 250 to 400 MW d) 125 MW

40. Membrane pore sizes for UF
    a) 0.1μ to 0.5μ
    b) 1,000 to 1,00,000 MW
    c) 250 to 400 MW
    d) 125 MW
41. Ohmic heating is also called as
    a) Resistance heating
    b) Electro heating
    c) a & b
    d) None
42. Advantage of microwave and D.E heating
    a) Uniform heating
    b) Selective heating may occur
    c) Efficient of energy conservation
    d) All
43. Most commonly used microwave frequencies for food applications are
    a) 2450MHZ & 915 MHZ
    b) 2540MHZ & 915 MHZ
    c) 2405MHZ & 915 MHZ
    d) 2540MHZ & 951 MHZ
44. The energy conversion in microwave and Dielectric heating is mainly by
    a) Ionic conduction
    b) Dipole rotation.
    c) a & b
    d) None
45. The principle involved in freeze drying is
    a) Sublimation
    b) Submission
    c) Sun drying
    d) Sun filling
46. Slow freezing forms
    a) Large crystals
    b) Small crystals
    c) Both
    d) None
47. An example of fermented product
    a) Cheese
    b) Curd
    c) Winc
    d) All
48. Freeze drying is also termed as
    a) Lipolysis
    b) Lyophilization
    c) Concentration
    d) Dehydration
49. Passing of a food through a fine dry or powdery substance in order to coat it is called as
    a) Breading
    b) Dredging
    c) Battering
    d) Coating

50. One of the following is a masonry storage structure
    a) Gade b) Kotlu
    c) Pathra d) Puri
51. Mediums used for dry heat methods
    a) Air or fat b) Water of fat
    c) Water or air d) Water or steam
52. Exhausting of cans helps in
    a) Minimizes discoloration b) Avoids corrosion of the tinplate
    c) Reduces chemical reaction d) All
53. Acid resistant lacquer is called as
    a) R-enamel b) C-enamel
    c) a & b d) None
54. The process of bringing back the temperature of the frozen food to room temperature is called as
    a) Shrinkage b) Thawing
    c) Icing d) None
55. The percentage of volume of inter-grain space to the total volume of grain bulk is
    a) Bulk density b) Porosity
    c) Angle of repose d) Spherecity
56. According to mode of heat transfer, drying method is of type
    a) Conduction b) Convection
    c) Radiation d) All of these
57. Sun drying is an example for which of the following drying method
    a) Conduction b) Convection
    c) Radiation d) All of these
58. The Dobhapa method of parboiling involves ___________ steaming
    a) Single b) Double
    c) Triple d) None
59. Instant rice needs __________ minutes to cook
    a) 20-30 b) 10-20
    c) 5-10 d) 1-5

60. Rice starch is used in ______________

a) Pudding b) Ice cream

c) Custard powder d) All of these

61. Kilning process involves _________________

a) Drying (Roasting) b) Soaking

c) Cleaning d) Steaming

62. The process of water absorption is known as

a) Soaking b) Steeping

c) Imbibition d) All of these

63. Which millet is a rich source of calcium and iron?

a) Ragi b) Bajra

c) Wheat d) All of these

64. A method of rice fortification is

a) Hot extrusion b) Cold extrusion

c) Coating d) All of these

65. Which is a component of wheat mill?

a) Break roll b) Reduction roll

c) Purifier d) All of these

66. Porosity depends upon

a) Shape b) Dimensions

c) Roughness d) All of these

67. Parboiling method developed at I.I.T., Kharagpur is

a) Brine solution method b) Pressure parboiling method

c) R P E C method d) Sodium chromate method

68. Which is a major step in parboiling?

a) Soaking b) Steaming

c) Drying d) All of these

69. A single or double-decked sieve which is given a swinging motion produced by an eccentric drive is

a) Trieur b) Destoner

c) Plansifter d) Tray separator

70. Cereals are deficient in
    a) Lysine b) Methionine
    c) Tryptophan d) Tannin
71. The principal food grain of the rural population in India, especially in the southern region is
    a) Ragi b) Bajra
    c) Sorghum d) Barley
72. The 'minute rice' is also known as
    a) Instant rice b) Breakfast rice
    c) Aged rice d) Fortified rice
73. Engelberg type huller is
    a) Sheller b) Polisher
    c) Sheller-cum-polisher d) Modern mill
74. Which among the following is ready-to-serve breakfast cereal?
    a) Flaked type b) Puffed type
    c) Shredded type d) All of these
75. An emery roller also called as
    a) Gota machine b) Miller
    c) Husker d) Ghani
76. Monkey nuts are
    a) Ground nuts b) Peas
    c) Walnuts d) Arica nuts
77. Saponins found in
    a) Coconut b) Soya bean
    c) Bengal gram d) All of these
78. Parboiling of legumes means
    a) Steam boiling b) Chemical boiling
    c) Chemical heating d) None
79. The germination process leads to the increase of ________ content in legumes.
    a) Vitamin C b) Mineral
    c) Phosphate d) Zink

80. Which oil is safe for human health?

a) Cotton seed oil b) Sunflower oil

c) Lard oil d) Palm oil

81. During storage the damage of grain may be caused by

a) Insects b) Moulds

c) Rodents d) All of these

82. The anti-nutrients are removed in legumes by

a) Cooking b) Germination

c) Both a & b d) Polishing

83. The cooking quality of dhal depends on

a) Soaking in water b) Hardness of water

c) Both a & b d) Seed coat

84. What are the best oil extraction techniques?

a) Solvent extraction technique b) Expelling

c) Refining d) Both a & b

85. Peanut butter is made from

a) Ground nuts b) Coconuts

c) Kidney beans d) Sunflower

86. Husk is a_____ product of legumes

a) Edible product b) By-product

c) End product d) None of these

87. Which types of legumes require shorter cooking time?

a) Quick cook b) Parboiled

c) Canned d) Polished

88. Separating meal from the oil solvent solution is called as

a) Miscella b) Clarification

c) Solvent d) Retentiate

89. The oldest existing unit of milling machine having multipurpose use is

a) Electric mill b) Hand operated flour mill

c) Diesel mil d) All

90. The process by which oils are hardened is known as
    a) Winterization b) Randomization
    c) Hydrogenation d) None of these
91. During conditioning pulses are finally dried to have moisture about ________ %.
    a) 6 to 8 b) 8 to 10
    c) 10 to 12 d) 100
92. Generally pulse crops are grown mixed with
    a) Coarse cereals b) Sugar cane
    c) Rice d) Banana
93. Solvent extraction process is also called
    a) Sublimation b) Evaporation
    c) Leaching d) Distillation
94. Cotton seed contains _______ % of oil.
    a) 5 to 8 b) 15 to 20
    c) 40 to 60 d) 80
95. The following will not a cause of off flavor in baked products
    a) Inferior ingredients b) Balanced formula
    c) Wrong pH d) Faulty baking
96. Pectin as a setting agent used in the production of
    a) Jams b) Chewing gums
    c) Crackers d) Toffee
97. The following dough yield less number of breads and are hard to bake
    a) Under fermented b) Over fermented
    c) Both a & b d) None
98. Good quality Garibaldi will have _________ of the product as fruit.
    a) Two-third b) Half
    c) One-third d) One-fourth
99. An unfermented dough is known as
    a) Old dough b) Poor dough
    c) Strong dough d) Young dough

100. Nuts with bland flavor

a) Cashew nuts
b) Walnuts
c) Almonds
d) Coco Nuts

## Answers

| | | | | | | | | | |
|---|---|---|---|---|---|---|---|---|---|
| 1. d | 2. c | 3. b | 4. d | 5. a | 6. a | 7. d | 8. d | 9. c | 10. a |
| 11. c | 12. b | 13. a | 14. a | 15. b | 16. b | 17. a | 18. d | 19. a | 20. a |
| 21. d | 22. a | 23. c | 24. a | 25. a | 26. d | 27. d | 28. c | 29. c | 30. a |
| 31. a | 32. d | 33. b | 34. d | 35. a | 36. c | 37. c | 38. c | 39. d | 40. b |
| 41. c | 42. d | 43. a | 44. c | 45. a | 46. a | 47. d | 48. b | 49. b | 50. b |
| 51. a | 52. d | 53. a | 54. b | 55. b | 56. d | 57. c | 58. b | 59. c | 60. d |
| 61. a | 62. d | 63. a | 64. d | 65. d | 66. d | 67. c | 68. d | 69. c | 70. a |
| 71. a | 72. a | 73. c | 74. d | 75. a | 76. a | 77. b | 78. a | 79. a | 80. b |
| 81. d | 82. c | 83. c | 84. d | 85. a | 86. b | 87. a | 88. a | 89. b | 90. c |
| 91. c | 92. a | 93. c | 94. b | 95. b | 96. a | 97. b | 98. c | 99. d | 100. a |

1. Annatto is used for colouring
   a) Butter b) Margarine
   c) Cheese d) All the above
2. If the cake is reddish brown in colour, it indicates too much addition of
   a) Cocoa b) Soda
   c) Sugar d) Fat
3. In the manufacture of toffee, fudge and chocolate type of granulate used
   a) Cubes b) Industrial granulated
   c) Nibs d) Granulated
4. The main ingredient of lozenges is
   a) Icing sugar b) Gelatin
   c) Gum Arabic d) None
5. Second step in the preparation of fondant is
   a) Mixing b) Dissolving
   c) Cooling d) Boiling
6. This method of cake preparation is suitable for making high ratio cakes
   a) Boiled b) Blending
   c) Flour-batter d) Sugar-batter
7. Sugar alcohol used in manufacture of diabetic and sugar free confections
   a) Maltitol b) Xylitol
   c) Sorbitol d) Both b & c
8. Popular variety of baked products on west coast of India are
   a) Macaroons b) Surti Butters
   c) Dough nuts d) Biscuits
9. These are made out of seedless yellow grapes
   a) Sultanas b) Raisins
   c) Currants d) Angelica
10. During this stage of mixing the dough is too warm and becomes sticky
    a) Let down b) Break down
    c) Final development d) Clean up

11. In the production of flour for baking ________ are removed during milling process
    a) Bran & endosperm    b) Endospem & germ
    c) Bran & germ    d) Only bran
12. If flavours are dissolved in ethyl alcohol they are called as
    a) Non-alcoholic    b) Alcoholic
    c) Emulsifiers    d) Powdered
13. In yeast raised doughs, the rate of fermentation is controlled by
    a) Sugar    b) Water
    c) Baking powder    d) Salt
14. The process of building up the sugar layer is
    a) Hard panning    b) Engrossing
    c) Crystallization    d) Liquorice
15. The plants whose fruit is enclosed in a pod is called
    a) Pulses    b) Legumes
    c) Fruit    d) None
16. A subgroup of the legume family; refers only to the dried seed is called
    a) Pulses    b) Legumes
    c) Fruit    d) None
17. Average water efficiency of pulses is
    a) 1 kg/1250 lits of water    b) 1kg/4325 lit of water
    c) 1kg/5 520 litres of water    d) 1kg/ 13 000 litres of water
18. Average water efficiency of Chicken is
    a) 1 kg/1250 lits of water    b) 1kg/4325 lit of water
    c) 1kg/5 520 litres of water    d) 1kg/ 13 000 litres of water
19. Average water efficiency of Mutton is
    a) 1 kg/1250 lits of water    b) 1kg/4325 lit of water
    c) 1kg/5 520 litres of water    d) 1kg/ 13 000 litres of water
20. Average water efficiency of Beef is
    a) 1 kg/1250 liters of water    b) 1kg/4325 litres of water
    c) 1kg/5 520 litres of water    d) 1kg/ 13 000 litres of water

21. The following ‘anti nutrients present in raw pulses
    a) Phytate b) Tannin
    c) Phenol d) All
22. Glycemic Index of Chick pea is
    a) 42 b) 39
    c) 80 d) 100
23. Glycemic Index of white rice is
    a) 42 b) 39
    c) 80 d) 100
24. Glycemic Index of white bread is
    a) 42 b) 39
    c) 80 d) 100
25. The following percentage of oil is used in dry milling method of pigeon pea
    a) 0.5-2.0 per cent linseed oil or any edible oil
    b) 5-20 per cent linseed oil or any edible oil
    c) 3-5 per cent linseed oil or any edible oil
    d) None
26. The extent of post harvest losses in case of pulses is
    a) 30% b) 20%
    c) 45% d) 35%
27. The extent of post harvest losses in case of cereals is
    a) 30% b) 20%
    c) 45% d) 35%
28. The following citrus fruit varieties is mainly used for processing purpose
    a) Kagzi Lime b) Rangpur Lime
    c) Sweet Lime d) All
29. The following one is right statement
    a) The period of maturity is shortest in lime (5-6 months)
    b) The period of maturity is longest in smandarins (9-10 months)
    c) The period of maturity is longest in sweet oranges (9-10 months)
    d) All

30. Most apples maintain their quality when stored at

    a) 5 to 10°C with RH 90-95% b) 1.8 to 0°C with RH 90-95%

    c) 10.8 to 20°C with RH 90-95% d) None

31. Recommended atmosphere for storage of Apples

    a) 1-2% $O_2$ and 2-4% $CO_2$ b) 16 % $O_2$ and 0.03% $CO_2$

    c) 10% $O_2$ and 15% $CO_2$ d) 5% $O_2$ and 2-4% $CO_2$

32. In the cold storage, bananas are stored at

    a) 5-10°C & RH 90-95% b) 13-14°C & RH 90-95%

    c) 27°C & RH 90-95% d) All

33. The following variety of Grapes can be used for table purpose

    a) Anab-e-Shahi b) Bangalore Blue

    c) Bhokri & Gulabi d) a & c

34. The following variety of Grapes can be used for Juice and wine

    a) Anab-e-Shahi b) Bangalore Blue

    c) Bhokri & Gulabi d) a & c

35. The following variety of Grapes can be used for raisin making

    a) Anab-e-Shahi b) Sharad Seedless

    c) Thompson Seedless d) a & c

36. The following variety of Grapes has good export potential

    a) Arka Majestic b) Arka Krishna

    c) Arka Soma d) Arka Shweta

37. The following variety of Grapes is good for preparing white desert wine

    a) Arka Majestic b) Arka Krishna

    c) Arka Soma d) Arka Shweta

38. The ideal storage temperatures for Nagpur mandarin is

    a) 6-7°C at 85-90% RH b) 10-12°C at 85-90% RH

    c) 16-27°C at 85-90% RH d) None

39. The ideal storage temperatures for lime & lemon is

    a) 6-7°C at 85-90% RH b) 10-12°C at 90-95% RH

    c) 16-27°C at 85-90% RH d) None

40. Shelf life of Nagpur mandarin can be extended by three weeks on wrapping the fruits with
    a) Heat shrinkable polythene and Cryovate film (D 955 and BDF 2001)
    b) Waxing with carnuba oils
    c) Keeping PE pouches
    d) All
41. The following is right statement
    a) The disk separator separates materials on the basis of difference in length of various constituents
    b) The indented cylinder separator separates the materials on the basis of relative lengths
    c) The spiral separator separates the grain as per their roundness
    d) All these above
42. Instant coffee is packed in flexible laminates of
    a) LDPE/Al/PET
    b) PET/Al/LDPE
    c) PET/LDPE
    d) Both b & c
43. In recent years, biscuits, cookies and crackers are packed in
    a) RCF coated with LDPE/PVE
    b) PET/Al foil/LDPE
    c) OPP either plane or coextruded
    d) RCF coated with acrylic
44. Cheddar cheese is originated in
    a) India
    b) England
    c) Egypt
    d) Japan
45. When a ball mill is centrifuging
    a) Maximum grinding takes place
    b) Normal grinding takes place
    c) Little or no grinding takes place
    d) None
46. Lard is obtained from
    a) Sheep
    b) Buffalo
    c) Goat
    d) Hogs

47. White deposition on the surface of the chocolate is due to faulty
    a) Winterization b) Plasticization
    c) Tempering d) None
48. Which of the following oil do you expect to have more free fatty acids?
    a) Sunflower oil b) Plasticization
    c) Tempering d) None
49. Polystyrene is used for
    a) Acid foods b) Neutral foods
    c) Basic foods d) All
50. Antimicrobial agent in egg
    a) Trypsin b) Lypase
    c) Amylase d) Avidin
51. Salami is prepared from
    a) Soybean b) Pork
    c) Meat d) Beef
52. Surimi is prepared from
    a) Fish b) Pork
    c) Meat d) Beef
53. Dhokla is prepared from
    a) Rice and bengalgram b) Wheat and rye
    c) Wheat and rice d) Rice only
54. The protein of egg white known for its antibiotic activity
    a) Conalbumin b) Lysozyme
    c) Ovalbumin d) Avidin
55. Pork never aged because
    a) It is not consumed largely
    b) Ageing of pork is not liked by the consumers
    c) It has high fat content
    d) None of the above
56. Essential components of contraction relaxation cycle is
    a) Iron&phospherous b) Ca and Mg
    c) Na and K d) Zn and Fe

57. Many quick–freezing processes freeze fish under pressure to
    a) Prevent denaturation of protein
    b) Prevent desiccation
    c) Enhance freezing
    d) None

58 Fat of fish prone to oxidation because of
    a) More saturation
    b) Less saturation
    c) Higher amount of lipoxidase enzyme
    d) Oxidation has nothing to do with saturation

59. An agent used for glazing is
    a) Edible wax b) Layers of ice
    c) Sodium salt d) Bee wax

60. Smoking is done
    a) After slaughtering b) Before curing
    c) After curing d) At any time

61. Cured meats are called
    a) Beef b) Bacon
    c) Ham d) Marinades

62. Which of the following statement is not correct
    a) Bromelin from pine apple can be used for tenderization of meat
    b) Electrical stimulation of carcasses after slaughtering of animals can cause tenderization
    c) Ficin is a protolytic enzyme, obtained from faba beans can be used for tenderization of meat
    d) None of the above

63. The first kvk in India is at
    a) Pondicherry b) Pattambi
    c) Jalna d) Varanasi

64. Golden rice is genetically modified rice with yellow colored seeds rich in
    a) Vitamin A b) Vitamin C
    c) Vitamin E d) Vitamin K

65. Pungency of onion is due to

a) Allyl propyl disulphide
b) Allicin
c) Allyl teriarybutyl sulphide
d) Propanal

66. Milk is deficient in

a) Na
b) K
c) Fe
d) None of the above

67. AGMARK was promulgated in

a) 1954
b) 1934
c) 1937
d) 1935

68. The first aseptic packaging of food was carried out by

a) Dunkley
b) Neilson
c) Nicholas
d) Martin

69. From the packaging point of view very important property of coffee is

a) Particle size
b) Bulk density
c) Density
d) Moisture

70. The possible migrant from paperboard carton is

a) Biphenyls
b) Dioxin
c) Nitrosamines
d) Trichloroanisole

71. Costliest spice is

a) Vanilla
b) Pepper
c) Turmeric
d) Saffron

72. Which of the following is sulphur containing spice

a) Dill
b) Asafoetida
c) Clove
d) Vanilla

73. Which of the following spices have highest amount of fiber in them

a) Ginger
b) Rosemary
c) Coriander
d) Clove

74. Vanilla flavor is naturally obtained by curing of which plant of the palnt

a) Seed
b) Pod
c) Leaves
d) Buds

75. Uniform black color of black pepper is obtained by
    a) Oven drying b) Using cooker
    c) Enzymic treatment d) None
76. Rolling is done in the process of
    a) Tea b) Coffee
    c) Cashew nut d) Coccoa
77. Minimal heat treatment of market milk at 137.8°C is for atleast
    a) 1 sec b) 2 sec
    c) 3 sec d) 4 sec
78. Tin cans was patented by
    a) Nicholas Appert b) Peter Durand
    c) Louis Pasteur d) Flemming
79. "C" enamels contains
    a) FeS b) ZnS
    c) MgS d) CaS
80. Mustard oil is effective against
    a) Mold b) Yeast
    c) Bacteria d) All of the above
81. Heating of the frozen food is termed as
    a) Dehydrofreezing b) Thawing
    c) Freeze drying d) None
82. Boiling of water will leads to
    a) Sterilization b) Pasteurization
    c) Disinfection d) None
83. Both extra cellular and intracellular crystallization takes place in
    a) Slow freezing only b) Fast freezing only
    c) Both a & b d) a & b
84. What is the strength of the brine solution for the canning of vegetables?
    a) 2% b) 8%
    c) 15% d) 67%
85. In standard salinometer 100° is equal to ________% of brine
    a) 22.5 b) 26.5
    c) 27.5 d) 29.5

86. Freeze crack is seen when foods are frozen by
    a) Air freezing  b) Compressed freezing
    c) Contact freezing  d) Cryogenic freezing
87. Corn gluten is rich in the corn protein known as
    a) Zein  b) Glutenin
    c) Glutelin  d) Prolamin
88. Marine products export development authority rules was formulated in
    a) 1941  b) 1951
    c) 1961  d) 1971
89. How many central food laboratories are there in India?
    a) 4  b) 8
    c) 16  d) 20
90. Reynolds number for laminar flow
    a) Less than 2100  b) Above 4000
    c) 2100-4000  d) None
91. Reynolds number for turbulent flow
    a) Less than 2100  b) Above 4000
    c) 2100-4000  d) None
92. Reynolds number for either laminar or turbulent flow
    a) Less than 2100  b) Above 4000
    c) 2100-4000  d) None
93. Lard is obtained from
    a) Sheep  b) Buffalo
    c) Goat  d) Hogs
94. Tooth paste is
    a) Bingham plastic  b) Pseudoplastic
    c) Newtonian fluid  d) Dilatant
95. Cooking fats are
    a) Bingham plastic  b) Pseudoplastic
    c) Newtonian fluid  d) Dilatant
96. Tomato sauce is the example for
    a) Casson plastic  b) Pseudoplastic
    c) Newtonian fluid  d) Dilatant

97. Veal is obtained from
    a) Sheep b) Buffalo
    c) Goat d) Calf
98. Detergent slurries
    a) Bingham plastic b) Pseudoplastic
    c) Newtonian fluid d) Dilatant
99. Blood is example for
    a) Casson plastic b) Pseudoplastic
    c) Newtonian fluid d) Dilatant
100. ISO standards are
    a) Mandatory orders b) Mandatory regulations
    c) Non mandatory regulations d) Non mandatory orders

## Answers

| | | | | | | | | | |
|---|---|---|---|---|---|---|---|---|---|
| 1. d | 2. b | 3. b | 4. a | 5. d | 6. b | 7. d | 8. b | 9. a | 10. a |
| 11. c | 12. b | 13. d | 14. b | 15. b | 16. a | 17. a | 18. b | 19. c | 20. d |
| 21. d | 22. b | 23. c | 24. d | 25. a | 26. b | 27. a | 28. a | 29. d | 30. b |
| 31. a | 32. b | 33. d | 34. b | 35. c | 36. a | 37. c | 38. a | 39. b | 40. a |
| 41. d | 42. b | 43. c | 44. b | 45. c | 46. d | 47. c | 48. b | 49. a | 50. d |
| 51. c | 52. a | 53. a | 54. b | 55. c | 56. c | 57. a | 58. a | 59. a | 60. d |
| 61. d | 62. c | 63. a | 64. a | 65. a | 66. c | 67. c | 68. b | 69. d | 70. b |
| 71. d | 72. b | 73. c | 74. b | 75. c | 76. a | 77. a | 78. b | 79. b | 80. d |
| 81. b | 82. a | 83. a | 84. a | 85. b | 86. d | 87. a | 88. d | 89. a | 90. a |
| 91. b | 92. c | 93. d | 94. a | 95. a | 96. a | 97. d | 98. a | 99. a | 100. d |

1. The following describes what enterprises do to meet environmental impacts, work conditions etc.

   a) SA 80001: 2008 b) ISO 22000: 2005

   c) ISO 14001: 2007 d) Both a & c

2. The following is odd one

   a) SEIRI = Sorting out

   b) SEITION = Systematic arrangement

   c) SEIARI = Serene Atmosphere

   d) SEISO = Spic and spam

3. The following deals with continuous improvement or tomorrow is better than today

   a) 5S-conecpt b) ISO 22000

   c) Kaizen d) Quality Control Circles

4. The following quality system consisting of officers from different departments should be formed to work on solving operational problems

   a) 5S-conecpt b) CFT

   c) Kaizen d) Quality Control Circles

5. The following Quality management model meant for a systematic approach to improvement and problem-solving

   a) PDCA b) CFT

   c) Kaizen d) Quality Control Circles

6. In 7 QC tool "fishbone diagram meant for

   a) To determine the causes of a problem

   b) To understand the correlation of two sets of variable parameters

   c) To take action on those items which are identified as the vital few

   d) To understand the distribution and variation of the data

7. In 7 QC tool "Histogram meant for

   a) To determine the causes of a problem

   b) To understand the correlation of two sets of variable parameters

   c) To take action on those items which are identified as the vital few

   d) To understand the distribution and variation of the data

8. In 7 QC tool “Pareto chart meant for
   a) To determine the causes of a problem
   b) To understand the correlation of two sets of variable parameters
   c) To take action on those items which are identified as the vital few
   d) To understand the distribution and variation of the data
9. In 7 QC tool Scatter diagram meant for
   a) To determine the causes of a problem
   b) To understand the correlation of two sets of variable parameters
   c) To take action on those items which are identified as the vital few
   d) To understand the distribution and variation of the data
10. The following are the principles of Auditing
    a) Ethical Conduct
    b) Fair Presentation and Independence
    c) Due Professional Care and Evidence-based approach
    d) All
11. The process when an independent and competent third party certifies that a product, service, system, process, or material conforms to specific requirements.
    a) Certification b) Verification
    c) Accreditation d) None
12. Confirmation, through the provision of objective evidence, that specified requirements have been fulfilled
    a) Certification b) Verification
    c) Accreditation d) None
13. Third-party attestation related to a conformity assessment body conveying formal demonstration of its competence to carry out specific conformity assessment tasks.
    a) Certification b) Verification
    c) Accreditation d) None
14. The following disinfectant used for all food contact surfaces, spray, combines sanitising and acid rinse into one operation
    a) Acid anionics b) Iodophors
    c) QAC d) Per acetic acid

15. The following disinfectant used for all food contact surfaces, usually restricted to CIP, especially cold temperature and carbon dioxide environments

a) Acid anionics
b) Iodophors
c) QAC
d) Per acetic acid

16. The following disinfectant used for all food contact surfaces, approach as a handdip

a) Acid anionics
b) Iodophors
c) QAC
d) Per acetic acid

17. The following disinfectant used for all food contact surfaces, mostly used for Environmental control; walls, drains, tiles etc.

a) Acid anionics
b) Iodophors
c) QAC
d) Per acetic acid

18. The following disinfectant used for all food contact surfaces, spray, CIP, fogging

a) Acid anionics
b) Iodophors
c) Chlorine
d) Per acetic acid

19. The following construction material is used when food contact surfaces requires good to resistance against salt and alkalis at moderate temperatures

a) AISI 304
b) AISI 316L
c) AISI 316l
d) AISI 316 Ti

20. The following construction material is used, when food contact surfaces requires good resistance against organic acids at moderate temperatures

a) AISI 304
b) AISI 316L
c) AISI 316l
d) AISI 316 Ti

21. The following construction material is used, when food contact surfaces requires increased resistance against non-oxidising acids such as acetic acid, tartaric acid, phosphoric acid

a) AISI 304
b) AISI 316L
c) AISI 316l
d) AISI 316 Ti

22. The following construction material is used, when food contact surfaces requires increased resistance against pitting and intercrystalline corrosion

a) AISI 304
b) AISI 316L
c) AISI 316l
d) AISI 316 Ti

23. The following construction material for food contact surfaces having better corrosion resistance than Type 1.4404
    a) AISI 304 b) AISI 316L
    c) AISI 316l d) AISI 316 Ti
24. The following construction material is used, when food contact surfaces requires increased corrosion resistance against particular acids and salt water
    a) AISI 304 b) AISI 316L
    c) AISI 316l d) AISI 316 Ti
25. The following construction material is used, when food contact surfaces requires Resistance against pitting corrosion
    a) AISI 304 b) AISI 316L
    c) AISI 316l d) AISI 316 Ti
26. Typical operating cycle for a CIP re-use system
    a) Pre-rinse → Detergent wash → Intermediate water rinse → Acid wash → Final water rinse → Final flush
    b) Pre-rinse → Acid wash → Intermediate water rinse → Detergent wash → Final water rinse → Final flush
    c) Pre-rinse → Acid wash → Detergent wash → Intermediate water rinse → Final water rinse → Final flush
    d) Detergent wash → Pre-rinse → Intermediate water rinse → Acid wash → Final water rinse → Final flush
27. Recommended disinfectant for Hand-dip (production)
    a) Quat b) Iodophor
    c) QAC d) Chlorhexadine
28. Recommended disinfectant for Odour control
    a) Quat b) Iodophor
    c) QAC d) Chlorhexadine
29. Recommended disinfectant for Wood crates
    a) Quat b) Iodophor
    c) QAC d) Chlorine
30. Recommended disinfectant for Porous surfaces
    a) Quat b) Iodophor
    c) QAC d) Chlorine

31. Recommended disinfectant for Plastic crates
    a) Amphoteric b) Iodophor
    c) QAC d) All
32. Recommended disinfectant for Hand sanitiser (washroom)
    a) Chlorhexadine b) Iodophor
    c) QAC d) All
33. Recommended disinfectant for Fogging, atmosphere
    a) Amphoteric b) Chlorine
    c) QAC d) All
34. Recommended disinfectant for Concrete floors
    a) Amphoteric b) Chlorine
    c) QAC d) b & c
35. Recommended disinfectant for Processing equipment (aluminium)
    a) Iodophor b) Amphoteric
    c) QAC d) All
36. Recommended disinfectant for Rubber belts
    a) Iodophor b) Amphoteric
    c) QAC d) All
37. Recommended disinfectant for Tile walls
    a) Iodophor b) Amphoteric
    c) QAC d) All
38. Recommended disinfectant for Processing equipment (stainless steel)
    a) Acid sanitiser b) Acid-quat
    c) chlorine, iodophor, amphoteric d) All

## Answers

1. d 2. c 3. c 4. b 5. a 6. a 7. d 8. c 9. b 10. d
11. a 12. b 13. c 14. a 15. d 16. b 17. c 18. c 19. a 20. a
21. b 22. b 23. c 24. d 25. d 26. a 27. b 28. a 29. d 30. d
31. d 32. d 33. d 34. d 35. d 36. d 37. d 38. d

# 5

# Food Product Development

1. Life blood of food companies are
   - a) Existing products
   - b) New Products
   - c) New packaging material
   - d) Food technologists
2. New Products are developed to meet
   - a) Health & Nutritional requirements
   - b) Environmental & Social requirements
   - c) Convenience & handling
   - d) All
3. What are the characteristics of a line extension in a variant of an established line of food products
   - a) Little research
   - b) Minimum manufacturing changes
   - c) No change in logistics and supply chain
   - d) All
4. The following one is not a characteristic of new product development through line extension
   - a) New purchasing skills
   - b) No change in logistics and supply chain
   - c) No change in storage of RM &PM
   - d) Minimum R&D
5. An example of new product developed through a line extension of RTE potato chips
   - a) Different seasonings coated chips
   - b) New size & shapes of chips
   - c) Different exotic flavored chips
   - d) All

6. New flavoured coated bread crumbs are an example of new product developed through
   a) Repositioning of existing products
   b) New form or size of existing products
   c) Line extension
   d) Reformulating of the existing products
7. A coarse or more natural peanut butter is new product developed through
   a) Repositioning of existing products
   b) New form or size of existing products
   c) Line extension
   d) Reformulating of the existing products
8. New varieties canned RTS soups, are new products developed through
   a) Repositioning of existing products
   b) Line extension
   c) New packing of existing products
   d) Reformulating of the existing products
9. Launching of new seasoning variant of ACT-II Popcorn by Conagrafoods is an example of
   a) Repositioning of existing products
   b) Line extension
   c) New packing of existing products
   d) Reformulating of the existing products
10. Launching of new seasoning variant of Kurkure (Extruded product) by PEPSICO is an example of
    a) New form or size of existing products
    b) Line extension
    c) New packing of existing products
    d) Reformulating of the existing products
11. Finding of new application for an existing product by consumers may leads to development of new product by
    a) Repositioning of existing products
    b) Line extension
    c) New packing of existing products
    d) Reformulating of the existing products

12. Arm & Hammer baking soda an example of
    a) New form or size of existing products
    b) Repositioning of existing products
    c) Innovative products
    d) Reformulating of the existing products
13. Repositioning of existing products requires
    a) Highly level of R&D
    b) Extensive marketing
    c) Brand extension
    d) Maximum modifications in process line
14. An example of new products developed under repositioning of existing products
    a) Oat meal rich foods have high dietary fiber
    b) Soya containing foods are functional health foods
    c) Millets containing foods are prebiotic foods
    d) All
15. Conversion of existing product into a new form like converting a paste product into a sauce or powder or tablet form is called as development of
    a) Repositioning of existing products
    b) New form or size of existing products
    c) New packing of existing products
    d) Reformulating of the existing products
16. The following general characters are observed in development of New form or size of existing products
    a) Extensive R&D is required
    b) New Manufacturing lines are required
    c) Vigorous marketing is required
    d) All
17. An example of New form or size of existing products
    a) Instant Foods & Quick meals b) Fruit juice powders
    c) Fat powders or sauce powders d) All

18. Reformulation in existing product is required due to

    a) Changes organoleptic attributes & availability of New technologies

    b) Nutritional & health benefits

    c) Unavailability of RM or availability of alternative ingredients

    d) All

19. The following authority policies may lead to Reformulation in existing products

    a) FAO
    b) FSSAI
    c) MOFPI
    d) APEDA

20. Development of low calories foods is an example of

    a) Repositioning of existing products

    b) New form or size of existing products

    c) New packing of existing products

    d) Reformulating of the existing products

21. Functional food products can be developed through

    a) Repositioning of existing products

    b) New form or size of existing products

    c) New packing of existing products

    d) Reformulating of the existing products

22. General characters of the reformulation of an existing product are

    a) Moderate R&D

    b) Slight changes in process line & machinery

    c) Minimum impact on marketing

    d) All

23. The following one does not come under reformulating of the existing products

    a) New flavor variant in extruded balls

    b) Lactose free milk products

    c) Gluten free

    d) Organic foods

24. Packing of the bulk produce into units of package suitable for easy handling comes under
    a) Repositioning of existing products
    b) New form or size of existing products
    c) New packing of existing products
    d) Reformulating of the existing products
25. New packing of existing products may utilizes the following technologies for better shelf life of foods
    a) Controlled atmosphere
    b) Modified atmosphere
    c) Vacuum packaging & Shrink wrapping
    d) All
26. The following one is an example of new packing of existing products
    a) Replacing of metal can with pouch and semi rigid tray
    b) Use of plastic squeeze bottles withsnap cap lids for dispersing
    c) Use of laminated films
    d) All
27. Availability of Kissan Ketchup in squeeze bottle rather than in glass bottle is an example of
    a) Repositioning of existing products
    b) New form or size of existing products
    c) New packing of existing products
    d) Reformulating of the existing products
28. Date fruits packed in vacuum packaging is an example of
    a) Repositioning of existing products
    b) New form or size of existing products
    c) New packing of existing products
    d) Reformulating of the existing products
29. What is the technology utilized in Surimi development and it become a basis for manysea food dishes.
    a) Wet salting
    b) Freeze drying
    c) Hurdle concept
    d) Pickling

30. General characters of innovative products
    a) R&D depends on nature of product
    b) Heavy impact on marketing
    c) High impact on manufacturing facilities
    d) All
31. Bring into existence of new product into market is called as
    a) Innovative product b) Creative product
    c) a & b d) None
32. The following one is not a creative products
    a) Tofu b) Bean curd
    c) Bottled beer d) Limed corn meal
33 The following one is creative technology
    a) Extrusion b) Drying
    c) Freezing d) All
34. The general characters of creative products are
    a) Expensive R&D b) New machinery
    c) New branding d) All
35. Risk of failure may be high in
    a) Creative products
    b) Innovative products
    c) New packing of existing products
    d) Reformulating of the existing products
36. An example of creative products
    a) Reformed meat cuts
    b) Surimi & Kamaboka based products
    c) Short chain FA containing products
    d) All
37. The character many new products are purported to have from single raw material is called as
    a) Creative products b) Added value
    c) Value added d) b & c

38. Added value provide to potatoes by

a) Pre-peeling
b) Dicing
c) Slicing
d) All

39. A conceptual representation of need discovered in customers and consumers, making marketing personal hope to develop into a want, i.e. a potential to sell.

a) Market place
b) Consumers
c) Customers
d) Market

40. Who buys in a market place is called as

a) Seller
b) Consumer
c) Customer
d) All

41. Customer is attracted by

a) Sales material
b) Promotions
c) Tastings in market place
d) All

42. Who described the customer as a 'gate keeper"

a) Dr. Kurt
b) Fuller
c) Bogaty
d) Woods

43. Gatekeeper is a

a) Family member
b) Purchasing dept.
c) Chefs & Owners
d) All

44. In three dimensional six numbered cubes in the figure of marketing characteristics of product, y-axis is a measure of

a) Increasing market place complexity
b) Increasing technical complexity
c) Measure of the volatility and elusiveness
d) All

45. In three dimensional six numbered cubes in the figure of marketing characteristics of product, x-axis is a measure of

a) Increasing market place complexity
b) Increasing technical complexity
c) Measure of the volatility and elusiveness
d) All

46. In three dimensional six numbered cubes in the figure of marketing characteristics of product , z-axis is a measure of
    a) Increasing market place complexity
    b) Increasing technical complexity
    c) Measure of the volatility and elusiveness
    d) All
47. The decline in product launches may decrease due to
    a) Consolidating of many food companies
    b) Potential productfailures
    c) Private label products.
    d) All
48. Driving forces beyond new product development are
    a) Business goals
    b) New markets
    c) Advanced technologies & health sciences
    d) All
49. Different reasons for new product development are
    a) Corporate & Governmental reasons
    b) Market place reasons
    c) Technological pressures
    d) All
50. Growth of food companies can be achieved by
    a) Business expansion in to new geographic markets
    b) Greater market penetration through promotional activities
    c) Developing new products & acquiring the rivals or smaller companies
    d) All
51. Management takes the following measures to meet the financial expenses associated with business development & growth
    a) Reducing staff
    b) Cutting cost
    c) Improving processing efficiency & QC
    d) All

52. Operating a food plant year round is
   a) More profitable
   b) Less profitable
   c) No change in profits
   d) More losses
53. Thrift measures for successful food business operation are
   a) Year round operation of food plant
   b) Production of new products in slack season
   c) Keeps the trained workers employed throughout the year
   d) All
54. Market place reasons for new product development are
   a) Ware house stores like retail & Restaurant giants
   b) Mail catalogue shopping andteleshopping.
   c) Supermarkets & Hyper markets with 24 ×7 hr
   d) All
55. In any market place, profile of customers and consumers are altering due to
   a) Population movement
   b) Population age
   c) Ethnic makeup
   d) All
56. Food products are now recognized as functional health foods by
   a) Low sodium content
   b) Low cholesterol & Saturated fats
   c) Enriched vitamins & Minerals
   d) All
57. What are the technological pressure on new product development
   a) Improved packaging materials properties
   b) Advanced preservation technologies
   c) Advances in nutritional studies
   d) All
58. What are the governmental influcences in pushing for new product development
   a) Limits on food additives usage
   b) Population age
   c) Advanced preservation technologies
   d) All

59. What are the governmental influcences in pushing for new product development
    a) Standards of composition of foods
    b) Cost affordability of customer.
    c) Labeling standards
    d) All

60. What are the quasi governmental regulatory bodies influcences in pushing for new product development
    a) ISO
    b) Codex
    c) a & b
    d) All

61. What are the factors influences on food business activities by government and quasi governmental regulatory bodies
    a) Fiscal policy & Environment protection
    b) Consumer protection & Health
    c) Trade barriers & IPR
    d) All

62. Four major developmental stages in product development are
    a) Development of business strategy that describes the project.
    b) The research and development phase of the project including the manufacturing design.
    c) Development of programs for marketing, production and quality assurance.
    d) Organization of production and distribution for launch and analysis of the initial sales data.
    e) All the above

63. New product is developed to capture vacuum ( needs) of
    a) Vendors
    b) Customers andConsumers
    c) Whole sellers
    d) All

64. Objectives of technical skills of research and development department are
    a) To create bench top prototypes
    b) To develop standards for RM & PM
    c) To develop process design and equipment requirements
    d) All

65. Business plan for new product development is refined by

a) R&D department b) Financial department

c) Operational department d) Marketing department

66. Role of Marketing people in new product development

a) Preparation of labels

b) Consumer analysis & Marketing strategy

c) Promotional activities

d) All

67. The manufacturing department determines

a) Production capabilities b) Manpower requirements

c) a & b d) Develop new processing

68. The final phase of new product development is an evaluation

a) Process design b) Test marketing

c) Quality control of RM d) All

69. Food technologists designs a new product based on

a) Concept statement

b) Sensory and consumer evaluation

c) a & b

d) Quality of raw materials

70. An example of unusual products developed are

a) Whey powder from whey b) Jam from fruits

c) Sugar from corn cobs d) All

71. In Brazil, diesel fuel was developed from

a) Pea nut oil b) Soya oil

c) Caster oil d) Mustard oil

72. New product development idea must fit into the following criteria

a) Needs and desires of consumers

b) Within the skill level, technical capabilities

c) Core competenciesof the company

d) All

73. Sources for ideas for new product development are
    a) Market place b) Within the company
    c) Outside the market place d) All
74. The following technologies helps in generation of new food products
    a) Hurdle concept b) Water activity
    c) a & b d) Drying
75. The following aids helps in getting to know customers and consumers requirements for new product development
    a) Census data b) Statistical data
    c) a & b d) None
76. An example of religiously responsible foods
    a) No sugar foods b) No onion & garlic foods
    c) Low salt foods d) Low non vegan items
77. The following one is technical source of databases for new product development ideas
    a) FSTA b) FROST
    c) a & b d) FSSAI
78. Mission oriented and is directed towards specific goal is a
    a) Applied research b) Fundamental research
    c) Basic research d) All
79. Fluidity as an organization tool in research involves
    a) Shifting personnel between departments
    b) R&D team
    c) Marketing team
    d) None
80. The new product development team should be an embody of skills necessary to evaluate team resources should include
    a) Management & Financial skills
    b) Engineering skills
    c) R&D and Marketing and sales data and skills
    d) All

81. Product concepts and prototypes are developed by R&D team by applying
    a) Knowledge b) Experience
    c) Training d) All

82. The following one is not right statement regarding new product ideas screening
    a) Satisfy the goals for new product success set by senior management
    b) Need not be in within Technical & Manufacturing capability set by senior management
    c) Lead to profitable products based on criteria established by senior management
    d) Satisfy the needs and wants of customers and consumers

83. New product ideas screening must consider ______________ of the company
    a) Financial planning b) Strategic planning
    c) Tactical planning d) All

84. The following facilities are required for new product development team
    a) PD lab & Test kitchen b) Pilot plant facilities
    c) a & b d) Commercial plants

85. Transfer of technology from global R&D centers to state of art manufacturing locations of the company can result in failure due to
    a) Not invented here (NIH) syndrome
    b) Not interested here (NIH) syndrome
    c) a & b
    d) None interested here (NIH) syndrome

86. Following one is not a constraint in innovation of new product
    a) Marketing research b) Communication problem
    c) Job security d) Management's attitude

87. When a product fails, least affected personnel in the company are
    a) R&D b) Marketing
    c) Production d) All

88. Who is to be blamed? If marketing, production and technical personnel disparities are not defused.
    a) Management b) Marketing
    c) R&D d) Production

89. Three primary functions of marketing people are
    a) Development of marketing strategies
    b) Development of sales and promotional materials
    c) Monitoring of customers and consumers needs & satisfaction
    d) All
90. Market research is an organized and unbiased investigation to
    a) Measure all the factors qualitatively and quantitatively
    b) To an analyze data and interpretation
    c) Sensory evalution
    d) a & b
91. Mac Nutty categorized the consumers in market place into
    a) Inner directed b) Outer directed
    c) Sustenance driven d) All
92. What category of consumers in market place are called as pace setters
    a) Inner directed b) Outer directed
    c) Sustenance driven d) All
93. According to British brew Bass Tavern's research, consumers of first time tasters are called as
    a) Young affluent and adventures
    b) Unskilled laborers
    c) Singles on prowl and out for enjoyment
    d) Pint and pension group
94. According to British brew Bass Tavern's research, consumers of Blue collar hunters are called as
    a) Young affluent and adventures
    b) Unskilled laborers
    c) Singles on prowl and out for enjoyment
    d) Pint and pension group
95. According to British brew Bass Tavern's research, consumers of Premium wanders are called as
    a) Young affluent and adventures
    b) Unskilled laborers
    c) Singles on prowl and out for enjoyment
    d) Pint and pension group

96. According to British brew Bass Tavern's research, consumers of Night clubbers are called as
    a) Young affluent and adventures
    b) Unskilled laborers
    c) With money and want entertainment
    d) Pint and pension group
97. Qualitative market information is based on
    a) Interpretation of focus groups
    b) Interviews requiring subjective interpretation
    c) Statistical analysis of data
    d) a & b
98. Quantitative market information is based on
    a) Interpretation of focus groups
    b) Interviews requiring subjective interpretation
    c) Statistical analysis of data
    d) a & b
99. ZMET stands for
    a) Zaltman Metaphor Evaluation Technique
    b) Zaltman Metaphor Elicitation Technique
    c) Zaltman Methodology Elicitation Target
    d) Zaltman Metaphor Elicitation Target
100. Neuromarketing is carried out by
    a) Functional marketing resonance imaging
    b) Functional marketing research imaging
    c) Functional magnetic resonance imaging
    d) Functional magnetic revolution imaging

## Answers

| | | | | | | | | | |
|---|---|---|---|---|---|---|---|---|---|
| 1. b | 2. d | 3. d | 4. a | 5. d | 6. c | 7. c | 8. b | 9. b | 10. b |
| 11. a | 12. b | 13. b | 14. d | 15. b | 16. d | 17. d | 18. d | 19. b | 20. d |
| 21. d | 22. d | 23. a | 24. c | 25. d | 26. d | 27. c | 28. c | 29. c | 30. d |
| 31. b | 32. c | 33. a | 34. d | 35. a | 36. d | 37. d | 38. d | 39. d | 40. c |
| 41. d | 42. a | 43. d | 44. a | 45. b | 46. c | 47. d | 48. d | 49. d | 50. d |
| 51. d | 52. a | 53. d | 54. d | 55. d | 56. d | 57. d | 58. a | 59. d | 60. c |
| 61. d | 62. e | 63. b | 64. d | 65. b | 66. d | 67. c | 68. b | 69. c | 70. c |
| 71. b | 72. d | 73. d | 74. c | 75. c | 76. b | 77. c | 78. a | 79. a | 80. d |
| 81. d | 82. b | 83. d | 84. c | 85. c | 86. a | 87. c | 88. a | 89. d | 90. d |
| 91. d | 92. a | 93. a | 94. b | 95. c | 96. c | 97. d | 98. c | 99. b | 100. c |

1. The generation of product ideas is based on the needs and desires of
   a) Consumers  b) Customers
   c) a & b  d) Marketing team
2. The following technology helps in to develop tailored made food products to meet the specific nutritional taste and safety specifications.
   a) Extrusion Technology  b) Biotechnology
   c) Hurdle technology  d) All
3. The following packaging technology has impact on shelf life and quality of various products
   a) Laminated film packaging  b) Aseptic packaging
   c) PET bottles  d) All
4. Demands of consumers for new food products
   a) Quality & convenience
   b) Nutrition & health
   c) Safety & environmentally sound
   d) All
5. Functions of engineering personnel in new product development are
   a) Modifying process line
   b) Process development & utilities requirements
   c) Assessing the cost of new equipment
   d) All
6. New Process line is designed to produce new products based on
   a) Mathematical modeling
   b) Experimenting and analyzing data statistically
   c) a & b
   d) Pilot plant design
7. What are the major concerns of Manufacturing plant
   a) Space & facilities  b) Labour and disruptiveness
   c) a & b  d) Location
8. The following one is not associated with benefits of Co-pack partnerships
   a) No capital expenses  b) Extra vigilance
   c) No utilities cost  d) All

9. Production costs can be lowered if purchasing department can obtain materials and ingredients at
   a) High cost
   b) Low cost
   c) Debit basis
   d) All
10. What are the points to be consider for vendor development
    a) Better price & quality
    b) Availability & delivery
    c) Reliability of supply and supplier
    d) All
11. What is considered as a buffer between the end of production line and distribution system?
    a) Purchasing department
    b) Warehouse
    c) Marketing department
    d) All
12. Test marketing is carried out by
    a) Pseudo sale method
    b) Controlled sales method
    c) Full sale methods
    d) All
13. The following one is not associated with Pseudo sale method of test marketing
    a) Product is real
    b) Advertising
    c) Price is real
    d) All
14. The central idea of Pseudo sale method of test marketing is
    a) Estimation of trialrates
    b) Estimation of repeat rates
    c) a & b
    d) None
15. Long term market share is calculated by
    a) $s = p \times r$
    b) $s = p / r$
    c) $s = p - r$
    d) $s = r/p$
16. Sale of new products through one or several outlets by firm is known as
    a) Pseudo sale method
    b) Minimarket
    c) Full sale methods
    d) All
17. Sale of new products in two or three cities is known as ___________ of test marketing
    a) Pseudo sale method
    b) Minimarket
    c) Full sale method
    d) All

18. What are variables will be measured in Full sale method of test marketing
    a) Competition b) Trade
    c) Advertisement d) All
19. The following department acts as the internal policing unit in food factory
    a) R&D b) NPD
    c) Quality control d) All
20. Shelf life of packaged foods, under normal marketing conditions are determined by
    a) Static tests b) Accelerated tests
    c) Abuse temperature d) All
21. The following ISO-9000, series describes the necessary quality system requirements for the end product inspection and detection
    a) ISO 9003 b) ISO 9002
    c) ISO 9001 d) ISO 9004
22. The following ISO-9000, series describes guideline for establishing Total Quality Management (TQM) in company
    a) ISO 9003 b) ISO 9002
    c) ISO 9001 d) ISO 9004
23. Pricing decision of new food products are dictated by
    a) Product costs b) Competitive price levels
    c) Objectives of company d) All
24. The following one is not a right statement
    a) Fixed costs don't vary with production or sales level
    b) Variable costs vary directly with level of production
    c) Fixed costs vary with production or sales level
    d) Total costs sum of fixed and variable costs
25. a) Prefernce tests supply information about people's likes and dislikes of a product. Theyare not intended to evaluate specific characteristics, such as crunchiness or smoothness
    b) Discrimination Tests aim to evaluate specific attributes, i.e. characteristics of products (crunchiness).
    c) Prefernce tests are subjective tests and include hedonic, paired comparison and scoring.
    d) Discrimination Tests are objective tests and include triangle, duo trio, ranking and paired comparison

a) A only correct
b) a & c correct
c) c & d correct
d) a,b,c & d correct

26. The following is a right statement in case of sensory evaluation of foods
    a) Maintain a product with same sensorial characteristics so consumers of that product continue to buy it
    b) Release production batch for sale because the products inspected meet the standards set by the products inspected meet the standards set by the manufacturer
    c) Reject production batch because the products inspected are below the acceptable level of quality set by inspected are below the acceptable level of quality set by the manufacturer the manufacturer
    a) a only correct
    b) b & c correct
    c) b only correct
    d) All are correct

27. In following test sensory panelists evaluates three samples, two are Panelists evaluates three samples, two are identical and one if different
    a) Duo –Trio test
    b) Triangle test
    c) Hedonic test
    d) All

28. In following test sensory Panelists evaluates three samples, and one of them Panelists evaluates three samples, and one of them is marked as reference
    a) Duo –Trio test
    b) Triangle test
    c) Hedonic test
    d) All

29. A savoury and subtle taste that is associated with a soupy or brothy note is called
    a) Sweet
    b) Salt
    c) Umami
    d) Bitter

30. The product Aroma and Flavour is perceived by
    a) Auditory
    b) Olfactory
    c) Visual
    d) All

## Answers

1. c 2. b 3. b 4. d 5. d 6. c 7. c 8. b 9. b 10. d
11. b 12. d 13. b 14. c 15. a 16. b 17. c 18. d 19. c 20. b
21. a 22. d 23. d 24. c 25. d 26. d 27. b 28. a 29. c 30. b

1. Food control systems are essential to protect for
   a) Health b) Safety
   c) a & b d) None
2. It is a mandatory regulatory activity of enforcement
   a) Food control b) Food safety
   c) Food Quality d) Food audit
3. SPS stands for
   a) Special Program service
   b) Sanitary protect Schemes
   c) Sanitary and protect standards
   d) Sanitary and phyto sanitary measures
4. Food Legislation includes
   a) Health Protection b) Legal security
   c) Increase consistency d) All
5. GAP stands for
   a) Green agro products b) Good agricultural practices
   c) Good Agro practices d) Good agmark scheme
6. Expand GHP
   a) Good Health Practices b) Good hydrocarbon products
   c) Good hygienic Practices d) None
7. RIA stands for
   a) Regular Investment Assessment
   b) Regular Increase Assessment
   c) Regular Impact Assessment
   d) Regulated Incentive Assessment
8. The CEO is the legal representative of
   a) Food Authority b) Food safety
   c) Food standards d) None
9. The chairperson of the food authority is appointed by
   a) State Government b) Central Government
   c) Joint secretary d) Joint Commissioner

10. Export Act
    a) 1982 b) 1984
    c) 1963 d) 1962
11. Insecticide Act
    a) 1962 b) 1968
    c) 1964 d) 1970
12. FPO stands for
    a) Fruit Pulp Order b) Fresh Product Order
    c) Food product Order d) Fruit product Order
13. The MMPO is exercised under
    a) Export Act b) PFS
    c) Essential commodities Act d) Environment Act
14. The edible oils packaging order came into force on
    a) 17$^{th}$ September b) 12$^{th}$ October
    c) 28$^{th}$ November d) 19$^{th}$ July
15. The PFS order came to effect on
    a) 11$^{th}$ march b) 24$^{th}$ may
    c) 27$^{th}$ October d) 30$^{th}$ April
16. Standards of weight and measures act is based on the
    a) Regular system b) Metric system
    c) Unit system d) Numeric system
17. Standards of weight and measures act was introduced in
    a) 1954 b) 1976
    c) 1972 d) 1986
18. IILMs
    a) Indian Institute of Legal Metrology
    b) Indian Institute of Law Metrology
    c) Indian Industry of Legal Metrology
    d) Indian Impact of Legal Metrology
19. The main objective of ECA act is to
    a) High Price b) Deflation
    c) Good Food d) Black Marketing

20. The most familiar ISO standard 14000 series is

a) ISO 14001 b) ISO 14002

c) ISO 14003 d) ISO 14004

21. It is defined as a mandatory regulatory activity of enforcement by national or local authorities to provide consumer protection and ensure that all foods during production, handling, storage, processing, and distribution are safe, wholesome and fit for human consumption.

a) Food Safety b) Food control

c) Food chain d) All

22. A process consisting of three components: risk assessment, risk management and risk communication

a) Risk Characterization b) Risk Analysis

c) Regulatory impact assessment d) Regulatory impact assessment

23. It is a Voluntary food law

a) AGMARK b) BIS

c) FPO d) A & B

24. According to FSSAI any material which is or could be employed for the purposes of adulteration is called as

a) Adulterant b) Contaminant

c) a & b d) None

25. The Food Authority shall be composed of ________ members excluding the Chairperson

a) 15 b) 12

c) 11 d) 3

26. Who is the Cabinet Secretary in selection committee of FSSAI

a) Chair person b) Joint-secretary

c) Food technologist d) Secretary

27. Prevention of food adulteration act came into force in

a) 1955 b) 1954

c) 1956 d) 2006

28. He may enter and inspect any place where article of food is manufactured, or stored for sale, or stored for the manufacture of any other article of food, or exposed or exhibited for sale and where any adulterant is

manufactured or kept, and take samples of such articles of food or adulterant for analysis.

a) Food Inspector
b) Food commissioner
c) Food authority
d) None

29. The person who is responsible to issue guidelines from time to time with regard to exercise of powers of the Food Inspectors.

a) State Commissioner of Food Safety
b) Central govt. authority
c) Food authority
d) All

30. Consumer Protection Act, 1986 extends to the whole of India except the state of

a) Jammu and Kashmir
b) New Delhi
c) Kerala
d) Tamilnadu

31. FSSAI act, 2006 is regulated by

a) Ministry of Agriculture
b) Ministry of environment and forests
c) Ministry of health and family welfare
d) Ministry of commerce

32. The codex secretariat is located at

a) Delhi
b) Rome
c) Japan
d) USA

33. The kind of legislation for BIS is

a) Voluntary
b) Mandatory
c) Both
d) None

34. The Government of India promulgated a fruit products order in

a) 1946
b) 1956
c) 1955
d) 1954

35. Fruit product order is operated by

a) Ministry of Agriculture
b) Food and nutrition board of the ministry of Food Processing industries
c) Ministry of commerce
d) None

36. The PFA and FPO are enforced by the Department of

a) Agriculture b) Nutrition

c) Commerce d) Health

37. The Milk and milk products order (MMPO) is regulated by

a) Ministry of Agriculture

b) Food and nutrition board of the ministry of Food Processing industries

c) Ministry of commerce

d) None

38. The Plant Food and Seed (Regulation of import into India) order was amended in

a) 1992 b) 1990

c) 1996 d) 1998

39. The Edible Oils Packing (Regulation) Order, came into force on

a) 17th September 1998 b) 18th September 1998

c) 14th September 1999 d) 17th September 1997

40. Meat Food Products Order is regulated by

a) Directorate of Marketing and Inspection at the ministry of Agriculture

b) Ministry of Agriculture

c) Ministry of commerce

d) None

41. The Export (Quality Control and Inspection) Act, 1963 was amended in

a) 1974 b) 1964

c) 1984 d) 1994

42. The main objective of this act is to supply of essential commodities to the public by proper regulation, prevention of black marketing and making it available to the public at reasonable price.

a) Essential Commodities Act b) PFA

c) FPO d) MMPO

43. Environmental Management Systems, Specification with Guidance for Use is series of

a) ISO-9005 b) ISO-14001

c) ISO-9000 d) ISO-9001-2008

44. The Agmark standard was set up by
    a) DMI of Government of India b) Ministry of Agriculture
    c) Ministry of commerce d) None
45. The function of BIS is
    a) Standard formulation
    b) Consumer affairs
    c) Awareness and training program
    d) All
46. Codex Alimentarius Commission was established in
    a) 1982 b) 1992
    c) 1972 d) 1962
47. The meaning of Codex Alimentarius in latin is
    a) Food law b) Food code
    c) A and B d) None
48. The codex commission members meets once in every
    a) 2years b) 3years
    c) 4years d) 5years
49. According to Joint FAO/WHO Expert Committee on Food Additives (JECFA) Sulphite used in dried fruits should not exceed
    a) 1000mg/Kg weight b) 2mg/Kg weight
    c) 10mg/Kg weight d) 1mg/Kg weight
50. How many stages are there before affixing the AGMARK label?
    a) 5 b) 2
    c) 3 d) 4

## Answers

| | | | | | | | | | |
|---|---|---|---|---|---|---|---|---|---|
| 1. c | 2. a | 3. d | 4. d | 5. b | 6. c | 7. c | 8. a | 9. b | 10. c |
| 11. b | 12. d | 13. C | 14. a | 15. c | 16. b | 17. b | 18. a | 19. d | 20. a |
| 21. b | 22. b | 23. d | 24. a | 25. c | 26. a | 27. a | 28. a | 29. a | 30. a |
| 31. c | 32. b | 33. a | 34. a | 35. b | 36. d | 37. a | 38. a | 39. a | 40. a |
| 41. c | 42. a | 43. b | 44. a | 45. d | 46. d | 47. c | 48. a | 49. a | 50. d |

# 6

# Food and Industrial Microbiology

1. Laminar flow hood works on the principle of
   a) High-efficiency particulate air filter
   b) High-energy particulate air filter
   c) High-efficiency pressure air filter
   d) None
2. Most of the organisms are killed
   a) 121°C for 5 min.  b) 15lb/in$^2$
   c) a & b  d) None
3. Autoclaving is not recommended for
   a) Oils  b) Powders
   c) Sensitive fluids  d) All
4. _______ grade AISI (Austeniti Stainless Steels) steels are recommended to use in food industry
   a) AISI 304  b) AISI 316
   c) a & b  d) None
5. _______ are Unicellular, procaryotic, cell multiplication is usually by binary fission
   a) Fungi  b) Bacteria
   c) Viruses  d) Moulds
6. Who was the discoverer of the microbial world
   a) Leeuwenhoek  b) Selman A Waksman
   c) Pasteur  d) *A.bassi*
7. Pure cultures of bacteria were first obtained by
   a) Leeuwenhoek  b) Joseph Lister
   c) Brefeld  d) *A.bassi*

8. Pure cultures of fungi were first obtained by
   a) Leeuwenhoek b) Joseph Lister
   c) Brefeld d) *A.bassi*
9. An optimum incubation temperature for bacteria is
   a) 30-37°C b) 20-27°C
   c) 10-27°C d) 10-37°C
10. Who developed nutrient broth & its solid counterpart nutrient agar.
    a) Leeuwenhoek b) Joseph Lister
    c) Koch d) *A.bassi*
11. Who developed Appertization and is the principle of food canning
    a) Leeuwenhoek b) Francois Appert
    c) Koch d) Pasteur
12. Who discovered the bacteriophages
    a) Twort b) Herelle
    c) Stanley d) a & b
13. _______is the portion of a bacterial cell consisting of the cytoplasmic membrane and the cell material bound by it.
    a) Protoplast b) Sphaeroplast
    c) Bacteriophages d) None
14. Is a protoplast surrounded by the outer membrane of cell wall.
    a) Protoplast b) Sphaeroplast
    c) Bacteriophages d) None
15. Spores produced within the cell are called
    a) Endospores b) Exospores
    c) Both a & b d) None
16. The formation of two cells from one parent cell is called as
    a) Generation time b) Doubling time
    c) a & b d) All
17. ________is the change in cell number or mass per unit time
    a) Generation time b) Doubling time
    c) Growth rate d) All

18. Number of generations can be calculated by
    a) log N- log N (o)/ log 2  b) log N/Nlog 2
    c) log N/1- log N (o)/ log  d) All
19. Growth cycle of bacteria is
    a) Lag phase  b) Log phase
    c) Stationary phase  d) Declined phase
    e) All
20. Secondary metabolites (toxins, antibiotics, sterols) are produced in
    a) Lag phase  b) Log phase
    c) Stationary phase  d) Declined phase
21. Most microbes grow best at pH values of
    a) 7  b) 5
    c) 3  d) 8
22. Moulds and yeasts grow best at pH values of
    a) 7.5  b) 5.5
    c) 3.5  d) 8.5
23. Most of the meats and sea foods have pH of
    a) > 5.6  b) > 3.6
    c) >2.6  d) 8.6
24. The Eh values will be positive for the
    a) Oxidized substances  b) Reduced substances
    c) Both a & b  d) None
25. Antimicrobial compounds in milk are
    a) Lactenin  b) Anticoliform factor
    c) Lysozyme  d) a & b
26. Antimicrobial compound in Egg is
    a) Lactenin  b) Anticoliform factor
    c) Lysozyme  d) a & b
27. Examples of psychrotrophs
    a) *Pseudomonas*  b) *Streptococcus*
    c) *Alcaligenes*  d) All

28. Most thermophilic bacteria that grow on foods belong to the genera
   a) *Bacillus* b) *Clostridium*
   c) *Alcaligenes* d) Both a & b
29. The organisms associated with spoilage of vegetables
   a) *Erwinia* b) *Pseudomonas*
   c) *Botrytis* d) All
30. The organisms associated with spoilage of fruits
   a) *Penicillium* b) *Botrytis*
   c) *Rhizopus and Aspergillus* d) All
31. The temperatures used for canning of high acid foods
   a) 120°C b) 100°C
   c) 121°C d) 111°C
32. The temperatures used for canning of low acid foods
   a) 120°C b) 100°C
   c) 121°C d) 111°C
33. One microorganism produces a substance that is inhibitory to other microbial population results in a negative outcome for later one.
   a) Parasitism b) Antagonism
   c) Symbiosis d) None
34. ______is refers to the killing of one microorganism by another for consumption and sustenance
   a) Parasitism b) Antagonism
   c) Symbiosis d) Predation
35. Bacteria grow fastest in the pH range
   a) 6-8 b) 3-6
   c) 4-9 d) 1-6
36. Yeast grow fastest in the pH range
   a) 6-8 b) 4.5-6
   c) 4-9 d) 1-6
37. Moulds grow fastest in the pH range
   a) 6-8 b) 4.5-6
   c) 2-8.5 d) 1-6

38. The pH of post – rigor mammalian muscle is
    a) 6.4 b) 7.0
    c) 5.6 d) 6.6
39. Yeasts are
    a) Aerobic b) Anaerobic
    c) Facultative d) Both a & c
40. Bacteria is a
    a) Aerobic b) Anaerobic
    c) Facultative d) All
41. High O - R potential favours______organisms
    a) Aerobic b) Anaerobic
    c) Facultative d) Both a & c
42. Low O-R potential favours
    a) Aerobic b) Anaerobic
    c) Facultative d) Both b&c
43. Aerobic microorganisms require __________ Eh
    a) Positive b) Negative
    c) Botha a & b d) None
44. Anaerobic micro organisms require__________Eh
    a) Positive b) Negative
    c) Botha a & b d) None
45. Natural Inhibitory compounds in Cran berries are
    a) Lactenin b) Anticoliform factor
    c) Lysozyme d) Benzoic acid
46. Natural Inhibitory compounds in Garlic, onion, leeks are
    a) Lactenin b) Allicin
    c) Lysozyme d) Benzoic acid
47. Natural Inhibitory compounds in Allspice (pimento), cloves, cinnamon are
    a) Lactenin b) Allicin
    c) Eugenol d) Benzoic acid

48. Natural Inhibitory compounds in Cinnamon and Cassia are
    a) Cinnamic aldehyde  b) Allicin
    c) Eugenol  d) Benzoic acid
49. The bitter principle of green olives, have antimicrobial properties
    a) Cinnamic aldehyde  b) Allicin
    c) Eugenol  d) Oleuropein
50. Lysozyme present in milk, egg is most active against________bacteria
    a) Gram Negative  b) Gram positive
    c) Both a & b  d) None
51. Natural Inhibitory compounds in Egg are
    a) Ovotransferrin  b) Avidin
    c) Ovolflaroprotein  d) All
52. Ovoflavo protein and avidin in egg white which sequester _______ and _____ the growth of those bacteria
    a) Biotin  b) Riboflavin
    c) a & b  d) None
53. _______produced by the propionibacteria in a swiss cheese is inhibitory to molds.
    a) Sorbic acid  b) Propionic acid
    c) Citric acid  d) Acetic acid
54. Salmonellosis associated with
    a) Eggs  b) Poultry
    c) Meat  d) All
55. The optimal temperature for most molds is
    a) 25 to 30°C  b) 15 to 40°C
    c) 10 to 30°C  d) 25 to 50°C
56. _________organisms are involved in Amylo" process for the saccharification of starch
    a) *Mucor racemosus*  b) *M. rouxii*
    c) Both a & b  d) *Aspergillus spp*
57. __________ is a bread mold
    a) *Rhizopus stolonifer*  b) *Mucor racemosus*
    c) *M. rouxii*  d) None

58. Whiskers" on chilled meat is caused by

a) *Rhizopus stolonifer* b) *Thamnidium elegans*

c) *M. rouxii* d) None

59. __________is a blue-green-spored mold, causes soft rots of fruits

a) *Pencillium expansum* b) *P. digitatum*

c) *P. italicum* d) All

60. ________is producing olive, or yellowish-green conidia, and causes a soft rot of citrus fruits

a) *Pencillium expansum* b) *P. digitatum*

c) *P. italicum* d) All

61. ________________ is a blue contact mold in citrus fruit spoilage

a) *Pencillium expansum* b) *P. digitatum*

c) *P. italicum* d) All

62. __________ is useful in the ripening of Camembert cheese

a) *Pencillium camemberti* b) *P. roqueforti*,

c) *P. italicum* d) All

63. __________ is useful in the ripening of blue cheese

a) *Pencillium camemberti* b) *P. roqueforti*

c) *P. italicum* d) All

64. __________ is known as "dairy mold," gives white to cream colored growth.

*a) Geotrichum candidum* *b) Neurospora*

c) *P. italicum* d) All

65. Red bread mold is

*a) Geotrichum candidum* *b) Neurospora*

c) *P. italicum* d) All

66. The organism involved in fermentation of Chinese red rice(ang -khak)

*a) Geotrichum candidum* *b) Neurospora*

c) *Monascus purpureus* d) All

67. *Saccharomyces cerevisiae* is used as a

a) Bread strains b) Beer strains

c) Wine strains d) All

68. _________ is a top yeast

a) *Saccharomyces cerevisiae* b) *Saccharomyces uvarum*

c) a & b d) None

69. _________ is a bottom yeast

a) *Saccharomyces cerevisiae* b) *Saccharomyces uvarum*

c) a & b d) None

70. ____________ is causes gaseous spoilage of canned vegetables

*a) Geotrichum candidum*

*b) Clostridium thermosaccharoolyticum*

c) *P. italicum*

d) All

71. _______ is a "bacterial soft rot

a) *Erwinia carotovora* b) *Geotrichum candidum*

c) *P. italicum* d) All

72. ___________ is involved in thermophilic flat sour spoilage of low acid canned foods

a) *Geotrichum candidum* b) *Bacillus stearothermophilus*

c) *Erwinia carotovora* d) None

73. Extreme halophiles (15-30% salt) are

a) *Halobacterium* b) *Halococcus*

c) *Erwinia carotovora* d) a & b

74. _________ is Osmophilic bacteria

a) *Halobacterium* b) *Halococcus*

c) *Leuconostoc* d) None

75. __is the temperature necessary to kill all the organisms in 10 minutes

a) TDP b) LTLT

c) HTST d) TDT

76. Bovine phosphatase enzyme is always_________in milk pasteurization

a) Negative b) Positive

c) both a & b d) None

77. HTST (High temperature short time) temperature for pasteurization is

a) 71.7°C/15 sec b) 62.8°C/30 min

c) 137.8°C/ 2 sec d) None

78. LTLT (Low temperature long time) temperature for pasteurization is
  a) 71.7°C/15 sec b) 62.8°C/30 min
  c) 137.8°C/ 2 sec d) None
79. Ultra pasteurization temperature for pasteurization is
  a) 71.7°C/15 sec b) 62.8°C/30 min
  c) 137.8°C/ 2 sec d) None
80. *Mycobacterium tuberculosis* is killed at
  a) 71.7°C b) 62.8°C
  c) 61.7°C d) All
81. Q fever is caused by
  a) *Mycobacterium tuberculosis* b) *Coxiella burnetti*
  c) *Erwinia carotovora* d) All
82. *Coxiella burnetti* is killed at
  a) 71.7°C b) 62.8°C
  c) 61.7°C d) 68.2°C
83. Who is the Father of canning
  a) Nicolas Appert b) Louis Pasteur
  c) Frazier d) Robert Koch
84. __________ Enamel contains zinc oxide
  a) R-enamel b) AR-enamel
  c) C- enamel d) All
85. C – Enamel is used for canning of
  a) Acid foods b) S- containing foods
  c) Zn-containing foods d) All
86. _____ is a time required to kill known number of micro organisms at a given temperature.
  a) TDP b) DRT
  c) HTST d) TDT
87. _____ is a time required to reduce the microbial population at specified temperature
  a) TDP b) DRT
  c) HTST d) TDT

88. ________ is a storage, usually in bins or boxes for equalization of moisture or readdition of moisture to a desired level

a) Steeping
b) Sweating
c) Rehydration
d) Drying

89. The moisture content of Intermediate Moisture foods

a) 20-50 %
b) 10-20%
c) 30-40%
d) 25-55%

90. Examples of Intermediate Moisture foods

a) Honey
b) Jelly/Jam
c) Raisins
d) Soft candies
e) All

91. Intermediate moisture products are also referred as

a) Reduced – water activity products
b) Low– water activity products
c) High– water activity products
d) Increased– water activity products

92. Propionic acid is naturally found in

a) Swiss cheese
b) Blue cheese
c) Red Cheese
d) None

93. Sorbic acid and its salts effective against

a) Yeast
b) Mould
c) Bacteria
d) a & b

94. Sodium diacetate used as a flavoring agent in

a) Butter
b) Fat spread
c) Margarine
d) All

95. ______________ is the quantity of gamma or x-radiation which produces one electrostatic unit of electric charge of either sign in a cubic centimetre of air under standard conditions.

a) Roentgen
b) Roentgen-equivalent
c) Megarep
d) All

96. ________ is the quantity of ionizing energy which produces, per gram of tissue, an amount of ionization equivalent to a roentgen.

a) Roentgen
b) Roentgen-equivalent
c) Megarep
d) None

97. _______ is a 1 million rep. One r, or 1 rep, is equivalent to the absorption of 83 to 90 erg per gram of tissue.

a) Roentgen b) Roentgen-equivalent

c) Megarep d) None

98. _____________ is 1 million rad

a) Megarad b) kilorad

c) Both a & b d) None

99. ____________ is a 1,000 rad.

a) Megarad b) kilorad

c) Both a & b d) None

100. Cold sterilization is also called as

a) Aseptic packaging b) Aseptic canning

c) Microwave cooking d) Irradiation

## Answers - Food Microbiology

| | | | | | | | | | |
|---|---|---|---|---|---|---|---|---|---|
| 1. a | 2. c | 3. d | 4. c | 5. b | 6. a | 7. b | 8. c | 9. a | 10. c |
| 11. b | 12. d | 13. a | 14. b | 15. a | 16. c | 17. c | 18. a | 19. e | 20. c |
| 21. a | 22. c | 23. a | 24. a | 25. d | 26. c | 27. d | 28. d | 29. d | 30. d |
| 31. b | 32. c | 33. b | 34. d | 35. a | 36. b | 37. c | 38. c | 39. d | 40. d |
| 41. d | 42. d | 43. a | 44. b | 45. d | 46. b | 47. c | 48. a | 49. d | 50. b |
| 51. d | 52. c | 53. b | 54. d | 55. a | 56. c | 57. a | 58. b | 59. a | 60. b |
| 61. c | 62. a | 63. b | 64. a | 65. b | 66. c | 67. d | 68. a | 69. b | 70. b |
| 71. a | 72. b | 73. d | 74. c | 75. a | 76. a | 77. a | 78. b | 79. c | 80. c |
| 81. b | 82. b | 83. a | 84. c | 85. b | 86. d | 87. b | 88. b | 89. a | 90. e |
| 91. a | 92. a | 93. d | 94. d | 95. a | 96. b | 97. c | 98. a | 99. b | 100. d |

1. A Gray (Gy) equals __________ rads
   a) 200 b) 100
   c) 300 4) 1000
2. ________ is a term used to define "radiation sterilization" which would imply high dose treatments, with the resulting product being shelf-stable.
   a) Radurization b) Radappertization
   c) Radicidation d) Picowaved
3. ____________ is a term used to define "radiation pasteurization" low-dose treatments, where the intent is to extend a product's shelf life.
   a) Radurization b) Radappertization
   c) Radicidation d) Picowaved
4. _____________ is a term used to define "radiation pasteurization" treatment, but with the specific intent being the elimination of a particular pathogen
   a) Radurization b) Radappertization
   c) Radicidation d) Picowaved
5. _______ is a term used to label foods treated with low-level ionizing radiation
   a) Radurization b) Radappertization
   c) Radicidation d) Picowaved
6. Frequency (megacycles) for microwave heating and dielectric heating of processed foods
   a) 2450 & 950 b) 2540 & 950
   c) 2250 & 950 d) 3245 & 950
7. High Hydrostatic Pressure (HHP) process is also called as
   a) Picowaved b) Thermal processing
   c) Pascalization d) None
8. Centrifugal separation of bacteria from milk is called as
   a) Cremation b) Centrifugation
   c) Bactofugation d) None
9. A combination of steam injection and evacuation for processing of cream is called as
   a) Vacreation b) Sterilization
   c) Pascalization d) Vaccum processing

10. *Coxiella burnetii (*Q fever) is effectively killed at

    a) 68.2°C/30 min　　b) 62.8°C/30 min

    c) 61.1°C/30 min　　d) 67.1°C/30 min

11. ________________spoilage organism is responsible for blue Milk

    a) *Pseudomonas syncyanea*　　b) *P. synxantha*

    c) *Serratia*　　d) *P. putrefaciens*

12. ________________spoilage organism is responsible for yellow milk

    a) *Pseudomonas syncyanea*　　b) *P. synxantha*

    c) *Serratia*　　d) *P. putrefaciens*

13. ________________spoilage organism is responsible for Red milk

    a) *Pseudomonas syncyanea*　　b) *P. synxantha*

    c) *Serratia*　　d) *P. putrefaciens*

14. __________ spoilage organism producing a red layer at the top of the milk

    a) *Brevibacterium erythrogenes*　　b) *P. synxantha*

    c) *Serratia*　　d) *P. putrefaciens*

15. ________________spoilage organism is responsible for brown milk

    a) *Pseudomonas syncyanea*　　b) *P. synxantha*

    c) *Serratia*　　d) *P. putrefaciens*

16. What is the concentration of salt solution at 100° salometers

    a) 18.6 %　　b) 25.6%

    c) 26.5%　　d) 100%

17. Bacterial soft rot in fruits and vegetables is caused by

    a) *Pseudomonas*　　b) *Erwinia caratovora*

    c) *Serratia*　　d) *Botrytis cinerea*

18. ___________is a fermenter of pectin substances

    a) *Pseudomonas*　　b) *Erwinia caratovora*

    c) *Serratia*　　d) *Botrytis cinerea*

19. Gray mold rot in fruits and vegetables is caused by

    a) *Pseudomonas*　　b) *Erwinia caratovora*

    c) *Serratia*　　d) *Botrytis cinerea*

20. Rhizopus soft rot in fruits and vegetables is caused by
    a) *Pseudomonas*
    b) *Erwinia caratovora*
    c) *Rhizopus stolonifer*
    d) *Botrytis cinerea*
21. Anthrachose disease in fruits and vegetables is caused by
    a) *Collectotrichum lindemuthianum*
    b) *Collectotrichum coccodes*
    c) Both a & b
    d) *Botrytis cinerea*
22. Alternaria rot in fruits and vegetables is caused by
    a) *Pseudomonas*
    b) *Alternaria tenuis*
    c) *Rhizopus stolonifer*
    d) *Botrytis cinerea*
23. Blue mold rot in fruits and vegetables is caused by
    a) *Pseudomonas*
    b) *Serratia*
    c) *Penicillium digitatum*
    d) *Botrytis cinerea*
24. Downey mildew in fruits and vegetables is caused by
    a) *Pseudomonas*
    b) *Serratia*
    c) *Penicillium digitatum*
    d) *Phytopthora bremia*
25. Watery soft rot in vegetables is caused by
    a) *Collectotrichum lindemuthianum*
    b) *Sclerotinia sclerotiorum*
    c) *Fusarium*
    d) *Botrytis cinerea*
26. Black mold rot in fruits and vegetables is caused by
    a) *Aspergillus niger*
    b) *Cladosporium*
    c) *Fusarium*
    d) *Botrytis cinerea*
27. Green mold rot fruits and vegetables is caused by
    a) *Aspergillus niger*
    b) *Cladosporium*
    c) *Richoderma*
    d) *b&c*
28. What is the safe level of moisture content in cereal flours
    a) 18%
    b) 15%
    c) 13%
    d) 14%

29. ______ Percent of moisture permits the growth of mold in cereal flours.
    a) 18%  b) 15%
    c) 13%  d) 14%
30. ______ Percent of moisture permits the growth of mold and bacteria in cereal flours
    a) 17%  b) 15%
    c) 13%  d) 14%
31. Bread mold is
    a) *Rhizopus stolonifer*
    b) *Rhizopus nigricans*
    c) *Collectotrichum lindemuthianum*
    d) Both a & b
32. Red bread or bloody bread is caused by
    a) *Rhizopus stolonifer*
    b) *Rhizopus nigricans*
    c) *Collectotrichum lindemuthianum*
    d) *Serratia marcescens*
33. Optimum meat chilling temperature is
    a) -1.0 to 2.2°C  b) -1.4 to 2.2°C
    c) -1.6 to 2.2°C  d) -1.8 to 2.2°C
34. Giblet is
    a) Gizzard, heart and liver  b) Gizzard, heart and lungs
    c) Gizzard, Lungs and liver  d) None
35. Flat sour spoilage of acid canned foods is caused by
    a) *Bacillus subtilis*  b) *Bacillus coagulans*
    c) *C. botulinum*  d) All
36. Thermophilic anaerobe (TA) spoilage in canned foods is caused by
    a) *Bacillus subtilis*  b) *B.coagulans*
    c) *C. botulinum*  d) *C.thermosaccharolyticum*
37. What are the diuretic (water-excreting) drinks
    a) Tea  b) Coffee
    c) Cola  d) All

38. What is the most abundant disaccharide,
    a) Sucrose b) Maltose
    c) Starch d) Lactose
39. Which is the correct statement
    a) GI of complex CHO is higher than simple CHO
    b) GI of complex CHO is Lower than simple CHO
    c) Both are true
    d) None
40. _______ is a illness caused both by the ingestion of toxins as well as organism resulting from the infection through the intestinal tract
    a) Food Poisoning b) Intoxication
    c) Infection d) None
41. ________is a illness caused by the consumption of a bacterial toxin formed in the food
    a) Food Poisoning b) Intoxication
    c) Infection d) None
42. What are the organisms causes intoxication of foods
    a) *Clostridium botulinum* b) *Staphylococcus aureus*
    c) Both a & b d) None
43. An enterotoxin is produced by
    a) *Vibrio cholerae* b) *Escherichia coli*
    c) *Shigella dysentriae* d) All
44. Neurotoxin is produced by
    a) *Clostridium botulinum* b) *Escherichia coli*
    c) *Clostridium tetani* d) a & c
45. Cytotoxin is produced by
    a) *Corynebacterium diptheriae* b) *Streptococcus pyogenes*
    c) *Clostridium botulinum* d) a & b
46. *Clostridium botulinum* produces ___________toxin
    a) Botulism Neuro Toxin b) Diptheria toxin
    c) Botulism Cerbo Toxin d) All

47. Gastroenteritis is caused by
    a) *Clostridium botulinum*  b) *Staphylococcus aureus*
    c) *Streptococcus pyogenes*  c) *Shigella dysentriae*
48. Enterotoxins produced by *Staphylococcus aureus* are
    a) A  b) B
    c) C1 & C2  d) d & e
    e) All
49. Vibriosis caused by
    a) *Corynebacterium diptheriae*  b) *Streptococcus pyogenes*
    c) *Clostridium botulinum*  d) *Vibrioparahaemolyticus*
50. What is an indicator for fecal contamination of water
    a) *Clostridium botulinum*  b) *Escherichia coli*
    c) *Salmonella*  d) *Shigella dysentriae*
51. Haemorrhagic colitis and haemolytic uraemic syndrome is caused by
    a) *Escherichia coli*  b) *E. coli* sero type 0157:47
    c) *Salmonella*  d) *Shigella dysentriae*
52. ________is a catalase positive, oxidage negative, fermentative, short Gram negative, non sporeforming rod.
    a) *Clostridium botulinum*  b) *Escherichia coli*
    c) *Salmonella*  d) *Shigella dysentriae*
53. Listeriosis caused by
    a) *Listeria monocytogenes*  b) *Escherichia coli*
    c) *Salmonella*  d) *Shigella dysentriae*
54. Salmonellosis caused by
    a) *Clostridium botulinum*  b) *Escherichia coli*
    c) *Salmonella*  d) *Shigella dysentriae*
55. Widal test is used to detect
    a) *Clostridium botulinum*  b) *Escherichia coli*
    c) *Salmonella*  d) *Shigella dysentriae*
56. Aflatoxin B is the most potent of all toxins and produced by
    a) *Aspergillus flavus*  b) *Aspergillus niger*
    c) *P. expansum*  d) *Aspergillus clavatus*

57. Patulin (Clavicin or expansin) is a mycotoxin produced by
    a) *Aspergillus clavatus* b) *Penicillium patulum*
    c) *P. expansum* d) All
58. Ochratoxin is a potent nephrotoxin and produced by
    a) *Aspergillus clavatus* b) *Aspergillus ochraceus*
    c) *P. expansum* d) All
59. Citrinin is a mycotoxin produced by
    a) *Penicillium citrinum* b) *Penicillium viridicatum*
    c) *P. expansum* d) a & b
60. Fumonisins are produced by
    a) *Fusarium sacchari* b) *Fusarium subglutinans,*
    c) *Fusarium thapsinum.* d) All
61. Sambutoxin is produced by
    a) *Fusarium sacchari* b) *Fusarium subglutinans,*
    c) *Fusarium sambucinum.* d) *Fusarium thapsinum*
62. Zeralenone is a toxin is produced by
    a) *Fusarium graminearum* b) *Fusarium subglutinans,*
    c) *Fusarium tricinctum* d) a & c
63. Ergotism is caused by
    a) *Fusarium graminearum* b) *Caviceps purpurea*
    c) *C. paspali* d) b & c
64. Cyclopiazonic acid is a toxic indole tetramic acid formed in spoiled corn & pea nuts and produced by
    a) *Aspergillus* b) *Penicillium*
    c) a & b d) None
65. Rubratoxin is formed in infected cereals and legumes and produced by
    a) *Penicillum rubrum* b) *Penicillium viridicatum*
    c) *P. expansum* d) *Penicillium citrinum*
66. ____________ is a poisonous species of mushrooms
    a) *Agaricus sps* b) *Amanita muscaria*
    c) Both a & b d) None

67. Paralytic Shellfish Poisoning is caused by
    a) *Penicillum rubrum* b) *Gonyaulax catenella*
    c) *P. expansum* d) *Penicillium citrinum*
68. Saxitoxin is produced by
    a) *Saxidomas giganteus* b) *Gonyaulax catenella*
    c) *P. expansum* d) *Penicillium citrinum*
69. Gonyatoxin is produced by
    a) *Gonyaulax catenella* b) *G.tamarensis*
    c) Both a & b d) None
70. Ciguatera poisoning is caused by ingestion of fish that feed on herbivorous or reef fishes which in turn feed on_____________ dinoflagellates
    a) *Gambierdiscus toxicus* b) *Gonyaulax catenella*
    c) *G.tamarensis* d) *Saxidomas giganteus*
71. Ciguatoxin is produced by
    a) *Gonyaulax catenella* b) *Gambierdiscus toxicus*
    c) *G.tamarensis* d) *Saxidomas giganteus*
72. What is an algal toxin accumulates in the gonads and viscera of the fish that feed on the Calcareous alga of *Jania sp.*
    a) Ciguatoxin b) Gonyatoxin
    c) Tetrodo toxin d) Saxitoxin
73. Amebiasis (amoebic dysentery) caused by
    a) *Agaricus* spp b) *Entamoeba histolytica*
    c) *Amanita muscaria* d) None
74. Sarcocystosis caused by
    a) *Sarcocystis hominis* b) *Entamoeba histolytica*
    c) *S. suihominis* d) a & c
75. Cryptosporidiosis caused by
    a) *Sarcocystis hominis* b) *Entamoeba histolytica*
    c) *S. suihominis* d) *Cryptosporidium partum*
76. As per WHO, maximum acceptable level of Hg in foods other than fish is
    a) 0.01 ppm b) 0.05 ppm
    c) 0.2 ppm d) 0.5 ppm

77. Chlorinated pesticides are
   a) DDT  b) Chlordane
   c) Endosulfan  d) Heptachlor
   e) All
78. Organophosphates (Ops) are
   a) Aliphatic Ops  b) Phenyl Ops
   c) Hetero cyclic  d) All
79. _____is defined as a noncancerous and non inflammatatory enlargement of the thyroid gland
   a) Goiter  b) Favism
   c) Neurolathyrism  d) None
80. Hemolytic anemia, jaundice and hyperbilirubinemia are Syptoms of
   a) Goiter  b) Favism
   c) Neurolathyrism  d) None
81. Neurolathyrism is caused by consumption of peas of the genus
   a) *Lathyruspisum*  b) *Lathyrussativus*
   c) *Lathyrusporna*  d) All
82. Allyl isothiocyanates are present in
   a) Mustard  b) Horse raddish
   c) Both a & b  d) None
83. Glycyrrhyzin is derived from dried roots ___________ and is popularly known as licorice
   a) Ginger roots  b) Horse raddish
   c) *Glycyrrhiza glabra*  d) All
84. Optimum pH for sodium benzoate is
   a) 2.5-4  b) 3.0-4.5
   c) 2.5-8  d) 2.5-9
85. ____________is used in the prevention of mold growth and rope development in backed goods
   a) Sodium benzoate  b) Propionic acid
   c) KMS  d) None

86. The organism involved in the fermentation of Natto (resembling cheese, which is produced from soya bean)

a) *Bacillus spp.*
b) *Aspergillus Oryzae*
c) *Aspergillus niger*
d) All

87. The organism involved in the fermentation of Tempehe

a) *Bacillus spp.*
b) *Aspergillus Oryzae*
c) *Aspergillus niger*
d) All

88. Flat sour is a kind of spoilage and is occurred in

a) Acid foods
b) Non acid foods
c) Cereal foods
d) Milk foods

89. Flat sour is a disease in non –acid foods like vegetables is caused by

a) *Bacillus coagulans*
b) *Bacillus Sterothermophilus*
c) a & b
d) None

90. Thermophilic acid (TA) spoilage is occurred in

a) Low acid foods
b) Medium acid foods
c) Acid foods
d) Both a & b

91. Thermophilic acid (TA) spoilage is caused by

a) *Bacillus coagulans*
b) *B. sterothermophilus*
c) *Clostridium thermosaccharolyticum*
d) All

92. Sulphide spoilage (Sulphur stinker) is caused by

a) *Clostridium nitrificans*
b) *B. sterothermophilus*
c) *Clostridium thermosaccharolyticum*
d) All

93. What species of bacteria is used as test organism for Inoculated pack studies

a) *Clostridium sporogenes* (P.A, 3679)
b) *B. pumpilus* (ATCC 27142)
c) *B. subtilis* (ATCC 6653)
d) *B. Stearothermophillus* (ATCC 7953)

94. What species of bacteria is used as test organism for Hot air and ethylene oxide studies

a) *Clostridium sporogenes* (P.A, 3679)

b) *B. pumpilus* (ATCC 27142)

c) *B. subtilis* (ATCC 6653)

d) *Bacillus subtilis* Var. *niger* (Syn. *B.globigii*, ATCC9372)

95. What species of bacteria is used as test organism for Cobalt or gamma radiation sterilization

a) *Clostridium sporogenes* (P.A, 3679)

b) *B. pumpilus* (ATCC 27142)

c) *B. subtilis* (ATCC 6653)

d) *Bacillus subtilis* Var. *niger* (Syn. *B.globigii*, ATCC9372)

96. What species of bacteria is used as test organism for Steam sterilization

a) *B. subtilis* (ATCC 6653)

b) *B. Stearothermophillus* (ATCC 7953)

c) Both a & b

d) None

97. ________ is the time in minutes required to destroy the organisms in a specified medium at 121.1°C.

a) F-value b) Z-value

c) D-value d) All

98. Sauerkraut is

a) Acid cabbage b) Acid Okra

c) Acid brinjal d) Acid capsicum

99. Sauerkraut contains ________________ of lactic acid

a) Not less than 2% b) Not less than 1.2%

c) Not less than 1.5% d) None

100. The organism initially (first) involved in Sauerkraut production

a) *Leuconostac mesentroids* b) *Lactobacillus brevis*

c) *Lactobacillus plantarum* d) *pediococcuss*

## Answers

|  |  |  |  |  |  |  |  |  |  |
|---|---|---|---|---|---|---|---|---|---|
| 1. b | 2. b | 3. a | 4. c | 5. d | 6. a | 7. c | 8. c | 9. a | 10. b |
| 11. a | 12. b | 13. c | 14. a | 15. d | 16. c | 17. b | 18. b | 19. d | 20. c |
| 21. c | 22. b | 23. c | 24. d | 25. b | 26. a | 27. d | 28. c | 29. b | 30. a |
| 31. d | 32. d | 33. b | 34. a | 35. b | 36. d | 37. d | 38. a | 39. b | 40. a |
| 41. b | 42. c | 43. d | 44. d | 45. d | 46. a | 47. b | 48. e | 49. d | 50. b |
| 51. b | 52. b | 53. a | 54. c | 55. c | 56. a | 57. d | 58. b | 59. d | 60. d |
| 61. c | 62. d | 63. d | 64. c | 65. a | 66. b | 67. b | 68. a | 69. c | 70. a |
| 71. b | 72. c | 73. b | 74. d | 75. d | 76. b | 77. e | 78. d | 79. a | 80. b |
| 81. b | 82. c | 83. c | 84. a | 85. b | 86. a | 87. b | 88. b | 89. c | 90. d |
| 91. c | 92. a | 93. a | 94. d | 95. b | 96. c | 97. a | 98. a | 99. c | 100. a |

1. What are the organisms are involved in Sauerkraut production

   a) *Leuconostac mesentroids*  b) *Lactobacillus brevis*

   c) *Lactobacillus plantarum*  d) *Pediococcuss* and *Cerevisiae*

   e) All

2. The final or last organism involved in Sauerkraut production

   a) *Leuconostac mesentroids*  b) *Lactobacillus brevis*

   c) *Lactobacillus plantarum*  d) *Pediococcuss* and *Cerevisiae*

3. Ergot alkaloid bodies are found in infected grains of

   a) Rye  b) Jowar (sorghum)

   c) Bajra  d) All

4. What are the alkaloids which causes ergotism ?

   a) Erotamine  b) Ergotoxin

   c) Ergometine  d) All

5. The organism which causes ergotism ?

   a) *Claviceps perpuria*  b) *Claviceps plantaram*

   c) *Bacillus* spp  d) *Clostridium*

6. _______ is a substance added to foods to avoid undesirable drying of foods and to maintain the moisture level

   a) Antioxidant  b) Humectant

   c) Emulsifier  d) Starch

7. Humectants are

   a) Glycerin  b) Sorbitol

   c) Mannitol  d) Dextrose and corn syrup

   e) All

8. Water activity (aw) of fresh meats and vegetables is

   a) 0.95 to 1.00  b) 0.75 to 1.00

   c) 0.65 to 1.00  d) 0.7 to 1.00

9. Water activity (aw) of dried milk is

   a) 1.0  b) 0.2

   c) 0.5  d) 0.8

10. Water activity is

    a) RH / 100  b) RH×100

    c) 100/RH  d) 100×RH

11. Most bacteria require a water activity in the range of about

a) 0.65 to 1.00 b) 0.75 to 1.00

c) 0.90 to 1.00 d) 0.7 to 1.00

12. What are the important ingredients for decreasing Water activity in foods

a) Sugar b) Glycerol

c) Both a & b d) None

13. What is the range of water activity, where rate of lipid oxidation increases

a) ~ 0.2 to 0.3 b) ~ 0.35 to 0.4

c) ~ 0.4 to 0.5 d) ~ 0.65 to 0.85

14. What is the range of water activity, where onset of stickiness in foods occur

a) ~ 0.2 to 0.3 b) ~ 0.35 to 0.4

c) ~ 0.4 to 0.5 d) ~ 0.65 to 0.85

15. What is the range of water activity, where loss of crispness is takes place in foods

a) ~ 0.2 to 0.3 b) ~ 0.35 to 0.4

c) ~ 0.4 to 0.5 d) ~ 0.65 to 0.85

16. What is the range of water activity, where hardening of foods ( loss of moisture) takes place

a) ~ 0.2 b) ~ 0.3

c) ~ 0.4 d) ~ 0.5

17. What is the range of water activity, where microbial growth occurs

a) ~ 0.2 b) ~ 0.3

c) ~ 0.6 d) ~ 0.5

18. What is the range of water activity, where all the reactions (enzymes, Non- Enzymatic Browning (NEB) , lipid oxidation) takes place in foods

a) ~ 0.2 to 0.3 b) ~ 0.35 to 0.4

c) ~ 0.4 to 0.5 d) ~ 0.65 to 0.85

19. Most enzymatic reactions are slowed down at water activities

a) < 0.8 b) < 0.6

c) <0.7 d) <0.9

20. What is the optimum water activity of foods for Maillard reactions
    a) ~ 0.2 to 0.3 b) 0.6 to 0.7
    c) ~ 0.4 to 0.5 d) ~ 0.65 to 0.85
21. What is the optimum water activity of most of the spoilage bacteria
    a) 0.91 b) 0.88
    c) 0.80 d) 0.75
22. What is the optimum water activity of most of the spoilage yeasts
    a) 0.91 b) 0.88
    c) 0.80 d) 0.75
23. What is the optimum water activity of most of the spoilage molds
    a) 0.91 b) 0.88
    c) 0.80 d) 0.75
24. What is the optimum water activity of most of the Halophilic bacteria
    a) 0.91 b) 0.88
    c) 0.80 d) 0.75
25. What is the optimum water activity of most of the Xerophilic molds
    a) 0.91 b) 0.65
    c) 0.60 d) 0.75
26. What is the optimum water activity of most of the Osmophilic yeast
    a) 0.91 b) 0.65
    c) 0.60 d) 0.75
27. Factors used for food preservation are called as
    a) Preservatives b) Antioxidants
    c) Hurdles d) None
28. What are the hurdles used for food preservation by Low temperature
    a) Chilling b) Freezing
    c) Cryogenic freezing d) All
29. What are the hurdles used for food preservation by Reduced water activity
    a) Drying b) Curing
    c) Conserving d) All

30. What are the hurdles used for food preservation by Reduced redox potential
    a) Removal of oxygen b) Addition of ascorbate
    c) Both a & b d) None
31. Where Competitive flora is used as hurdle for food preservation
    a) Curing b) Fermentation
    c) Chilling d) Drying
32. Microbially derived hurdles for food preservation are
    a) Antibiotics b) Bacteriocins
    c) Protective cultures d) All
33. What is the pKa value of Acetic acid
    a) 4.75 b) 3.75
    c) 5.75 d) 6.75
34. Acetic acid is more effective against
    a) Yeast b) Bacteria
    c) Molds d) Both a & b
35. Water activity value of saturated salt solution is
    a) 0.65 b) 0.75
    c) 0.55 d) 0.25
36. pH of Low-acid canned foods (LACF) is
    a) 4.6 or higher b) 4.6 or lower
    c) 3.6 or lower d) 2.6 or lower
37. pII of Acidificd low-acid foods
    a) 4.6 or higher b) 4.6 or lower
    c) 3.6 or lower d) 2.6 or lower
38. _______ is a heat processing of vacuum-packed products at low temperatures for long times
    a) Vacuum drying b) Sous vide
    c) Vacuum packaging d) Both a & b
39. What is the pasteurization pressure in high pressure processing
    a) 6000 atmospheres b) 10000 atmospheres
    c) 12000 atmospheres d) 8000 atmospheres

40. What is the sterilization pressure in high pressure processing

a) 6000-8000 atmospheres b) 10000-12000 atmospheres

c) 12 000-15000 atmospheres d) 8000-15000 atmospheres

41. Irradiation treatment involves exposing packaged food to

a) Gamma rays b) Electron beams

c) X-rays d) All

42. What is the optimal radiation dose for extending the shelf-life of cooked and raw foods

a) 1–5 kGy b) 1–8 kGy

c) 1–10 kGy d) 1–100 kGy

43. What are the deterioration mechanisms are involved in soft fruits

a) Enzymic breakdown b) Mould growth

c) Moisture loss d) All

44. What are the deterioration mechanisms are involved in Frozen meat

a) Oxidation b) Freezer burn

c) Ice sublimation d) All

45. What are the deterioration mechanisms are involved in Snack foods

a) Moisture uptake b) Oxidation

c) Both a & b d) None

46. What are the deterioration mechanisms are involved in cakes

a) Moisture loss b) Starch changes

c) Microbial growth d) All

47. What are the deterioration mechanisms are involved in Breakfast cereals

a) Moisture migration b) Starch retrogradation

c) Oxidation d) All

48. What are the deterioration mechanisms are involved in Spices

a) Microbial growth b) Volatile loss

c) Chemical reactions d) All

49. What are the deterioration mechanisms are involved in Chocolate confectionery

a) Fat migration b) Oxidation

c) Moisture loss d) All

50. What are the deterioration mechanisms are involved in Fluid milk

a) Oxidation b) Hydrolytic reactions

c) Microbial growth d) All

51. What are the deterioration mechanisms are involved in Cheese

a) Oxidation b) Lactose crystallization

c) Microbial growth d) All

52. What are the deterioration mechanisms are involved in Fruit yoghurt

a) Syneresis b) Oxidation

c) Microbial growth d) All

53. Accelerated storage temperature for Cakes and pastry products

a) 30ºC/75% b) 27 ºC/50%

c) 17 ºC/50% d) 20 ºC/50%

54. Flat sour bacteria produces ____________ in canned foods

a) Gas b) Acid

c) Gas and Acid d) None

55. ________ is the time it takes to kill 90% of the microbial population

a) F-value b) Z-value

c) D-value d) Fo -value

56. __________ A whisky distilled from a fermented mash containing not less than 51% corn

a) Fenni b) Bourbon

c) Vodka d) Rum

57. Foods that are stored at temperature below their freezing point

a) Frozen foods b) Refrigerated foods

c) Fermented foods d) None

58. Foods that are stored at temperature above their freezing point

a) Frozen foods b) Refrigerated foods

c) Fermented foods d) None

59. A very vigorous fermentation with large quantities of gas is

a) Steady fermentation b) Fast fermentation

c) Stormy fermentation d) None

60. Blight disease caused by ______________ in potato and Tomato
    a) *Phytophthora infestans* b) Aspergillus *oryzae*
    c) *Bacillus* spp. d) *Mucor rouxii*
61. An enzyme amalyases are produced by
    a) *Rhizopus oligosporus* b) Aspergillus *oryzae*
    c) *Mucor rouxii* d) All
62. The organism involved in production of Temph
    a) *Rhizopus oligosporus* b) Aspergillus *oryzae*
    c) *Mucor rouxii* d) *Bacillus* spp.
63. The dominant spoilage flora of fish are
    a) *Pseudomanas* b) *Flavobacterium*
    c) *Micrococcus* d) All
64. Chemical indicators of quality of fish and sea foods are
    a) Indole b) Ammonia
    c) pH d) All
65. Riboflavin chelates
    a) Anions b) Cations
    c) Both a & b d) None
66. What are indicators of egg quality
    a) Formic acid b) Acetic
    c) Succinic acid d) All
67. Thermo stabilization temperature for the eggs is
    a) 54.4°C/15 min b) 52.4°C/15 min
    c) 50.4°C/15 min d) 54.4°C/10 min
68. What happens to the moisture content of food, If a food product with a water activity of 0.60 is placed in an environment with an ERH of 95%
    a) Decrease b) Increase
    c) No change d) None
69. Most commonly occurred bacteria in poultry meat and eggs is
    a) *Salmonella* b) *Clostridium*
    c) *Bacillus* d) *Pseudomonas*

70. What pH is used as a dividing line between medium acid and acid products.

a) 4.0 b) 3.0

c) 4.6 d) 3.7

71. The minimum growth temperature for most mesophilic organisms is

a) 15°C b) 10°C

c) 20°C d) 60°C

72. Storage of foods below 0°C but above freezing point is called as ________zone

a) Latent zone of chilling b) Latent zone of heating

c) Latent zone of Super heating d) None

73. Protease is an enzyme produced by

a) *Bacillus subtilis* b) *Mortierella semispora*

c) *S.cervisiae* d) a & b

74. Invertase is an enzyme produced by

a) *Bacillus subtilis* b) *Mortierella semispora*

c) *S. cervisiae* d) *Aspergillus niger*

75. Amylase is an enzyme produced by

a) *Bacillus subtilis* b) *Mortierella semispora*

c) *S.cervisiae* d) *Aspergillus niger*

76. Pectinases are enzymes produced by

a) *Aspergillus wentir* b) *Mortierella semispora*

c) *Byssochalamys fulva* d) a & c

77. Citric acid is produced by

a) *Bacillus subtilis* b) *Mortierella semispora*

c) *S.cervisiae* d) *Aspergillus niger*

78. Fumaric acid is produced by

a) *Rhizopus stolanifer* b) *Bacillus subtilis*

c) *S. cervisiae* d) *Bacillus subtilis*

79. Riboflavin (Vitamin $B_2$) is produced by

a) *Eremothecium ashbyl* b) *Streptomyces griseous*

c) *Cephalosporium acrimonium* d) *Pseudomonas denitrificans*

80. Penicillin is produced by

a) *Penicillium notatum* b) *P. chrysogenum*

c) *Cephalosporium acrimonium* d) a & b

81. Green rot of eggs is caused

a) *Penicillium notatum* b) *Pseudomonas fluorescens*

c) *Cephalosporium acrimonium* d) *P. chrysogenum*

82. _______ are the most heat resistant toxins of botulin and causes botulism in canned foods

a) Types A b) Types B

c) a & b d) None

83. ___________ Inoculation chambers are used for sterility testing.

a) Positive-pressure b) Negative-pressure

c) No change d) None

84. All glassware containing microbial cultures or otherwise contaminated by microorganisms should be sterilized by autoclaving

a) 121°C/40 min b) 120°C/40 min

c) 110°C/40 min d) 100°C/40 min

85. **ELISA** test is

a) Enzyme-linked immune syndrome assay

b) Enzyme-linked immunosorbent assay

c) Enzyme-linked immune assay

d) None

86. Thermophilic flat sours are

a) *G. stearothermophilus* b) *B. stearothermophilus*

c) *B.coagulans* d) b&c

87. Thermophilic non- $H_2S$ producing anaerobes are

a) *C. thermosaccharolyticum* b) *C. thermoaceticum*

c) *B. coagulans* d) a & b

88. Thermophilic $H_2S$ producing anaerobes are

a) *C. nigrificans* b) *C. thermoaceticum*

c) *C. thermosaccharolyticum* d) All

89. _________ are Gram-positive, aerobic or facultatively anaerobic, typically catalase-positive, rod-shaped, motile, endospore-forming bacteria.

a) Cocci  b) Bacillus
c) Aspergillus  d) Streptococci

90. Mycotoxins like Aflatoxin B1 and B2 are produced by

a) *A. flavus*  b) *A. parasiticus*
c) *A. nomius*  d) All

91. Sterigmatocystin is produced by

a) *A. versicolor*  b) *Emericella* spp.
c) Both a & b  d) None

92. Fumitremorgens is produced by

a) *A. versicolor*  b) *Emericella* spp.
c) *A. fumigatus*  d) *A. nomius*

93. ___________ has been isolated from foods and it produces Vero-cell active toxin

a) *B. thuringiensis*  b) *C. thermoaceticum*
c) *C. thermosaccharolyticum*  d) All

94. __________antigen site is used to differentiate *Salmonella typhi*

a) Somatic (O)  b) Flagellar (H)
c) Capsular (vi)  d) All

95. __________antigen site is used to differentiate *Salmonella Dublin*

a) Somatic (O)  b) Flagellar (H)
c) Capsular (vi)  d) All

96. __________antigen site is used to differentiate *Salmonella paratyphi*

a) Somatic (O)  b) Flagellar (H)
c) Capsular (vi)  d) All

97. Poisonous mushrooms are

a) *Amanitaphalloides*  b) *Amanita verna*
c) Both a & b  d) None

98. __________ *E .coli* do not produce entero toxins but causes diarrhea.

a) Entero pathogenic  b) Entero invasive
c) Entero haemorrhagic  d) Entero toxigenic

99. __________ *E. coli* do not produce entero toxins but causes ulceration and inflammation in the epithelial cells of the colon

a) Entero pathogenic b) Entero invasive

c) Entero haemorrhagic d) Entero toxigenic

100. __________ *E.coli* is associated with sporadic diarrhea out breaks.

a) Entero pathogenic b) Facultative entero pathogenic

c) Entero haemorrhagic d) Entero toxigenic

## Answers

| | | | | | | | | | |
|---|---|---|---|---|---|---|---|---|---|
| 1. e | 2. c | 3. d | 4. d | 5. b | 6. b | 7. e | 8. a | 9. b | 10. a |
| 11. c | 12. c | 13. a | 14. b | 15. c | 16. d | 17. c | 18. d | 19. a | 20. b |
| 21. a | 22. b | 23. c | 24. d | 25. b | 26. c | 27. c | 28. d | 29. d | 30. c |
| 31. b | 32. d | 33. a | 34. d | 35. b | 36. a | 37. b | 38. b | 39. a | 40. b |
| 41. d | 42. c | 43. d | 44. d | 45. c | 46. d | 47. d | 48. d | 49. d | 50. d |
| 51. d | 52. d | 53. a | 54. b | 55. c | 56. b | 57. a | 58. b | 59. c | 60. a |
| 61. d | 62. a | 63. d | 64. d | 65. b | 66. d | 67. a | 68. b | 69. a | 70. c |
| 71. b | 72. a | 73. d | 74. c | 75. d | 76. d | 77. d | 78. a | 79. a | 80. d |
| 81. b | 82. c | 83. a | 84. a | 85. b | 86. d | 87. d | 88. a | 89. b | 90. d |
| 91. c | 92. c | 93. a | 94. a | 95. b | 96. c | 97. c | 98. a | 99. b | 100. b |

1. Cultures are maintained by periodic transfer is called
   a) Sub culturing  b) At low temperature
   c) Lyophilisation  d) Storage in sterile soil
2. The principal constituent of cell walls of bacteria is
   a) Chitin  b) Cellulose
   c) Peptidoglycan  d) Lipoprotein
3. Volutin granules are reserve source of
   a) Phosphates  b) Fats
   c) Proteins  d) Carbohydrates
4. Microorganism that can tolerate high temperatures are called
   a) Mesophiles  b) Psychrophiles
   c) Thermophiles  d) Halophiles
5. Some bacteria that live in aquatic habitats form the structures that provide buoyancy
   a) Gas vacuoles  b) Episome
   c) PHB granules  d) Volutin granules
6. Teichoic acids are present in
   a) Gram positive bacteria  b) Gram negative bacteria
   c) Both  d) Moulds
7. The plasmid chromosome integrated into bacterial chromosome is called
   a) Polysome  b) Oxysome
   c) Peroxysome  d) Episome
8. The structures of microbes which damage mainly by ultra violet radiation is
   a) Peptidoglycan  b) DNA
   c) Cytoplasm  d) Proteins
9. Higher Protists include
   a) Algae  b) Fungi
   c) Protozoa  d) All
10. During the following phase death of the microorganisms is observed
   a) Lag phase  b) Log phase
   c) Stationary phase  d) Decline phase

11. The following are examples of food borne viruses
    a) Polio
    b) Hepatitis
    c) Gastero enteric viruses
    d) All
12. Complete oxidation of carbohydrates in aerobic respiration results in
    a) 38 ATP
    b) 32 ATP
    c) 42 ATP
    d) 20 ATP
13. The magnification of the electron microscope is
    a) 10000X
    b) 400000X
    c) 4000X
    d) 60000X
14. The following is the example for halophilic bacterium
    a) *Saccharomyces cerevisiae*
    b) *Halobacterium salinarum*
    c) *Bacillus megaterium*
    d) *Escherichia coli*
15. The organism responsible for a musty odor and taste in root beer
    a) *Micrococcus*
    b) *Leuconostoc*
    c) *Achromobacter*
    d) *Lactobacillus*
16. Hydrogen swells are favoured by
    a) Increasing acidities of foods
    b) Increasing temperatures of storage
    c) Imperfections in the tinning and lacquering of the interior of the can
    d) All of these
17. The acidity may be described as "clean" as produced by
    a) *Streptococcus lactis* & other lactics
    b) *Micrococcus*
    c) *Listeria monocytogenes*
    d) *Enterobacter aerogenes*
18. The frozen eggs are stored at
    a) 1 to 2° C
    b) -17.8 to -20.5° C
    c) -1 to -2° C
    d) 5 to 6° C
19. The process in which egg is held and rotated in front of a light to examine it for defects such as cracks, rots, molds, blood, developing embryo, crusted or sided yolk, weak white or large air cell is called as
    a) Chilling
    b) Candling
    c) Lightening
    d) Irradiation

20. The defect phosphorescence in meat caused by
    a) *Photobacterium* b) *Flavobacterium*
    c) *Lactobacillus* d) *Moraxella*
21. Bacteria which can survive in heat treatment are
    a) Halophiles b) Mesophiles
    c) Thermoduric bacteria d) Psychrophiles
22. Father of canning
    a) Spallanzani b) Nicolas Appert
    c) Louis Pasteur d) Robert Koch
23. TA spoilage caused by
    a) *Clostridium perfringens*
    b) Clostridium thermosaccharolyticum
    c) *Clostridium botulinum*
    d) *Clostridium sporogenes*
24. The following are examples for asexual spores
    a) Arthrospores b) Zygospores
    c) Ascospores d) Oospores
25. Source of Gamma rays is
    a) Cobalt 60 b) Cesium 137
    c) Both a & b d) Ca 20
26. The following chemical is not preservative
    a) Ozone b) Ethylene
    c) Propionates d) Benzoates
27. High speed centrifugation removes the spore by
    a) 25% b) 50%
    c) 70% d) 99%
28. The antioxidant used for the preservation of fish is
    a) Nordihydroguuaritic acid b) Ethyl gallate
    c) Ascorbic acid d) All
29. Radiation resistant microorganism is
    a) *Escherichia coli* b) *Micrococcus roseus*
    c) *Pseudomonas aeruginosa* d) *Bacillus subtilis*

30. Propionic acid forming bacteria grow actively in
    a) Bread b) Cheese
    c) Yogurt d) Milk
31. The smoking of foods have purpose like
    a) Adding desired flavors b) Preservation effect
    c) Both a & b d) Boiling effect
32. Smaller ice crystals are formed in
    a) Quick freezing b) Slow freezing
    c) Sharp freezing d) Chilling
33. Medium acid foods have pH
    a) Above 5.3 b) 4.5 to 5.3
    c) 3.7 to 4.5 d) less than 3.7
34. The temperature for preservation of frozen meats is
    a) -12.2 to -28.9°C b) -10 to -4.5°C
    c) 4 to 6°C d) 10-20°C
35. Green rot of eggs caused by
    a) *Alcaligenes viscolactis* b) *Pseudomonas flourescens*
    c) *Escherichia coli* d) *Pseudomonas aeruginosa*
36. Iced, Cut-up poultry often develops a slime that is accompanied by an odor described as
    a) Tainted b) Dishraggy
    c) Sour d) All
37. Chilling temperature required for commercial storage of eggs for 6 months or longer
    a) -1.7 to -0.55°C b) 0 °C
    c) 1-2 °C d) 0-4 °C
38. Rapid heating of cream is accomplished by injecting steam or by a combination of steam injection and evacuation in a process known as
    a) Vacreation b) Pasteurization
    c) Steam injection technique d) Ultrahigh pasteurization
39. The preservative added to Cheese and Cheese products is
    a) Sorbic acid b) Propionic acid
    c) Both a & b d) Acetic acid

40. Raw milk at temperatures above 50°C will spoil by

a) *Streptococcus lactis* b) *Streptococcus faecalis*

c) *Lactobacillus thermophilus* d) *Alcaligenes viscolactis*

41. Burnt or Caramel Flavor in milk produced by

a) *Aeromonas hydrophila*

b) *Streptococcus lactis* var. *maltigenes*

c) *Enterobacter oxytocum*

d) *Pseudomonas sapolactica*

42. The following organism is commonly called as bread mold

a) *Rhizopus stolonifer* b) *Mucor rouxii*

c) *Thamnidium elegans* d) *Aspergillus niger*

43. The following genus is an example for yeast

a) *Stemphylium* b) *Debaromyces*

c) *Pullularia* d) *Sporotrichum*

44. The yeast which form a pellicle on beers or wines is

a) *P\ichia membranaefaciens*

b) *Candida utilis*

c) *Saccharomyces cerevisiae*

d) *Escherichia coli*

45. Homofermentative lactic acid bacteria ferment sugars and produce

a) Mainly lactic acid and small amounts of acetic acid and $CO_2$

b) Volatilc products including alcohol and lactic acid

c) Only lactic acid

d) Pyruvic acid and lactic acid

46. The following Bacillus species grows at higher temperatures

a) *Bacillus subtilis* b) *Bacillus polymyxa*

c) *Bacillus stearothemophilus* d) *Bacillus macerans*

47. Sulfide stinker spoilage in canned foods caused by

a) *Clostridium botulinum* b) *Desulfatomaculum nigrificans*

c) *Erwinia caratovora* d) *Clostridium butyricum*

48. Thermoduric bacteria refers to

a) Bacteria which have an optimum temperature at least above 45°C

b) Bacteria which can survive pasteurization

c) Bacteria which cannot grow at higher temperatures

d) Bacteria which grow at chilling temperatures.

49. Moderate halophiles grow best in a medium with salt concentration of

a) 0.5 to 3.0% b) 3.0 to 15%

c) 15 to 30% d) 30 to 40%

50. Example for Osmophilic or Saccharophilic bacteria

a) *Leuconostoc mesenteroides* b) *Saccharomyces cerevisiae*

c) *Lactobacillus delbruckii* d) *Zygosaccharomyces rouxii*

51. Growth of one organism makes conditions favorable for growth of the second organism is

a) Synergism b) Antibiosis

c) Metabiosis d) Symbiosis

52. Example of False yeasts (Fungi Imperfectii)

a) *Candida* b) *Saccharomyces*

c) *Zygosaccharomyces* e) *Hansenula*

53. Pasteurization temperature was set based on *Coxiella burnetti* and this organism causes

a) Ropiness b) Q fever

c) Enteritis d) Nausea

54. The following species is example for mesophilic Clostridium

Clostridium butyricum Clostridium pasteurianum

Clostridium putrefaciens All of these

55. Musty flavours in butter caused by microorganisms belongs to

a) Fungi b) Actinomycetes

c) Viruses d) Bacteria

56. Extreme halophiles grows in the salt concentration of

a) 0.5 to 3.0 % b) 3.0 to 15%

c) 1.0 to 2.0% d) 15 to 30%

57. Surface Ropiness of milk caused by the microorganism
    a) Alcaligenes viscolactis
    b) Micrococcus freudenreichii
    c) Both a & b
    d) Enterobacter aerogenes
58. Extensive oxidation, usually following hydrolysis and the release of fatty acids can result in
    a) Oxidative rancidity
    b) Hydrolytic rancidity
    c) Ketonic rancidity
    d) Fungal oxidation
59. Water activity (aw) values for Chocloate, Confectionery and honey are
    a) Below 0.6
    b) 0.60-0.85
    c) 0.85-0.93
    d) 0.98 and above
60. False yeasts, which produce no ascospores or other sexual spores, belongs to
    a) Fungi Imperfectii
    b) Fungi perfectii
    c) Ascomycotina
    d) Basidiomycotina
61. Red, pink or yellow yeasts which may cause discolorations on foods is
    a) Trichosporon
    b) Rhodotorula
    c) Candida
    d) Torulopsis
62. Which of the following is an example for bacteria
    a) Kloeckera
    b) Alteromonas
    c) Kluyveromyces
    d) Stemphylium
63. The following microorganisms is an example for synergisms in milk
    a) Pseudomonas syncyanea and streptococcus lactis
    b) Leuconostoc mesenteroides and Lactobacillus plantarum
    c) Streptococcus thermophilus and Lactobacillus bulgaricus
    d) Micrococcus and Microbacterium
64. Hydrogen swells are favored by
    a) Increasing acidities of food
    b) Increasing temperature of storage
    c) Imperfections in the tinning and lacquering of the interior can
    d) All the above
65. The storage usually done in bins or boxes, for equalization of moisture or readdition of moisture to a desired level is
    a) Sweating
    b) Bag in box method
    c) Thawing
    d) Canning

66. The pH at which Sodium benzoate is most effective is

a) 5.0 to 6.0 b) 2.5 to 4.0

c) 7.0 to 8.0 c) 8.0 to 8.5

67. The antimicrobial substance present in onions and garlic is

a) Acrolein b) Butyl thiocyanate

c) Eugenol d) Cinnamic aldehyde

68. The antibiotics approved by FDA as dip for preserving poultry in the uncooked and dressed fowls is

a) Chloramphenicol

b) Neomycin

c) Streptomycin sulphate

d) Chlortetracycline and Oxytetracycline

69. The wave length of ultraviolet radiation which has germicidal effect

a) 254 nm b) 354 nm

c) 100 nm d) 554 nm

70. Red colour on bread caused by

a) Serratia marcescence b) Oidium aurantiacum

c) Monilia sitophila d) All

71. Antibiotic commonly used in bee keeping is

a) Neomycin b) Streptomycin

c) Both a & b d) Chloramphenicol

72. Curing agent for curing of meat is

a) Salt b) Sugar

c) Sodium nitrate d) All

73. Algae used for the production of single cell protein is

a) *Chlorella pyrenoidosa* b) *Spirulina maxima*

c) *Ṡcenedesnus acutus* d) All the above

74. Raw material used for the production of Bakers yeast is

a) Beet molasses b) Cane Molasses

c) Both a & b d) Corn steep Liquor

75. The term Single cell Protein was coined by

a) C.L. Wilson
b) Kinoshita
c) Emmerich
d) Alexander Flemming

76. The two processes are used commercially in the mass production of *Bacillus thuringiensis* is

a) Sub merged fermentation
b) Semi solid fermentation
c) Both a & b
d) Solid fermentation

77. The following method is long term preservation of microbes is

a) Agar slants
b) Storage at O°C
c) Lyophilization
d) Mineral oil storage

78. Enzyme cellulase is commercially produced from which organism

a) *Trichoderma viridae*
b) *Fusarium oxysporum*
c) *Rhizopus oryzae*
d) *Saccharomyces cerevisiae*

79. Use of Hallow fibres in the fermentors used for the production of

a) Monoclonal antibodies
b) Beer
c) Bakers yeast
d) Single cell protein

80. The bioreactor in which avoids bubbling and reduces the cause of foam formation

a) Air Lift Bioreactor
b) Novel See saw Bioreactor
c) Stirred tank Bioreactor
d) Cyclone Column Fermentor

81. The capacity of Laboratory fermentor is

a) 12-15 lit
b) 2000 gallons
c) 500 - 1000 lit
d) 10000 galllons

82. An example for antifoam agent is

a) Silicon Compounds
b) Corn oil
c) Soybean Oil
d) All of these

83. Commonly used plant material in Beer production

a) Hops
b) Cloves
c) Neem
d) Turmeric

84. The bacteria involved in probiotic product is

a) *Lactobacillus delbrueckii*
b) *Bacillus clausii*
c) *Bacillus pumilis*
d) All

85. The following microorganisms is used for the production of alcohol, when sulfite Waste liquor is used as a culture medium
    a) *Candida utilis* b) *Aspergillus niger*
    c) *Rhizopus stolonifer* d) *Saccharomyces cerevisiae*

86. Yogurt is commercially manufactured by the fermentation of milk with bacteria
    a) *Streptococcus thermophilus* c) *Lactobacillus bulgaricus*
    c) Both a & b d) None

87. The plant growth regulator responsible for production of lateral shoots is
    a) IAA b) Gibberellins
    c) Cytokinins d) Ethylene

88. The optimum temperature required for satisfactory production of vitamin $B_{12}$ is
    a) 80°F b) 98°F
    c) 100°F d) 30°F

89. Red rice produced by which organism
    a) Bacteria b) Yeast
    c) Virus d) Fungi

90. Fat soluble vitamins are
    a) Vitamin A b) Vitamin D
    c) Vitamin E d) All

91. An example for macrolide antibiotic
    a) Streptomycin b) Erythromycin
    c) Chloramphenicol d) Griseofulvin

92. Secondary metabolites are produced in which growth phase
    a) Lag phase b) Log phase
    c) Stationary Phase d) Decline Phase

93. The following technique is used in the purification of products
    a) Electrophoresis b) Chromatography
    c) Precipitation d) All

94. The microbial culture center "MTCC" located at
    a) Hyderabad b) Pune
    c) Chandigarh d) Kolkata

95. The type of fermentation in which more than one microorganism is employed

a) Batch fermentation b) Continuous fermentation

c) Dual or Multiple fermentation d) Fed batch fermentation

96. The straw coloured byproduct produced as a byproduct of Cheese making is

a) Liquor b) Whey

c) Yoghurt d) Butter milk

97. The potential raw material for the commercial production of Lactic acid is

a) Jerusalem artichokes b) Beet root

c) Rice starch c) Sugarcane molasses

98. The plant material used in the preparation of Poi is

a) Fleshy stems of taro plant b) Green olives

c) Pulp of Cucumber d) Soya beans

99. The enzyme rennet is produced by

a) Mucor pusillus b) Penicillium roqueforti

c) Rhizopus oligosporous d) Saccharomyces cerevisiae

100. The type of fermentation used for production of Bakers yeast is

a) Continuous fermentation b) Fed batch fermentation

c) Dual fermentation d) Batch fermentation

## Answers

| | | | | | | | | | |
|---|---|---|---|---|---|---|---|---|---|
| 1. a | 2. c | 3. a | 4. c | 5. a | 6. a | 7. d | 8. b | 9. d | 10. d |
| 11. a | 12. d | 13. b | 14. b | 15. c | 16. d | 17. a | 18. b | 19. b | 20. a |
| 21. c | 22. b | 23. b | 24. a | 25. c | 26. b | 27. d | 28. d | 29. b | 30. b |
| 31. c | 32. a | 33. b | 34. a | 35. b | 36. d | 37. a | 38. a | 39. c | 40. a |
| 41. b | 42. a | 43. b | 44. a | 45. a | 46. c | 47. b | 48. b | 49. b | 50. a |
| 51. c | 52. a | 53. b | 54. d | 55. b | 56. d | 57. c | 58. c | 59. a | 60. a |
| 61. b | 62. b | 63. a | 64. d | 65. a | 66. b | 67. a | 68. d | 69. a | 70. d |
| 71. c | 72. d | 73. d | 74. c | 75. a | 76. b | 77. c | 78. a | 79. a | 80. b |
| 81. a | 82. d | 83. a | 84. d | 85. a | 86. c | 87. c | 88. a | 89. b | 90. d |
| 91. b | 92. c | 93. d | 94. c | 95. c | 96. b | 97. a | 98. a | 99. a | 100. b |

1. In *Prokaryote* word *pro* means before and *karyon* means

   a) N*ut* b) K*ernel*

   c) S*hell* d) a & b

2. The use of microbes to reduce/degrade pollutants, industrial wastes and household garbage, a new area referred to as

   a) Bioremediation b) Biomagnifiacation

   c) Eutrification d) Purification

3. Which of the following invades the stomach and causes gastric ulcers ?

   a) Helminthes b) Helicobacter

   c) Protozoa d) Fungi

4. Monitoring, control and spread of diseases in communities is

   a) Epidemiology b) Endomology

   c) Edimology d) None

5. Pasteur suggested that mild heating at 62.8°C (145°F) for 30 min rather than boiling was enough to destroy the undesirable organisms in the year

   a) 1867 b) 1887

   c) 1897 d) 1987

6. *E.Coli*, the generic name of organism named after

   a) Theodor Escherich b) Joseph Escherich

   c) Christian Escherich d) Emil

7. Laminar flow hood works on the principle of application of

   a) Cloth

   b) High-efficiency particulate air filter

   c) Membrane filter

   d) All

8. In filters, HEPA means

   a) High-efficiency particulate

   b) High-efficiency particulate filter

   c) High-efficiency particulate air

   d) High-efficiency particulate atmosphere

9. Most of the organisms are killed at
   a) 100°C (i.e. 15lb/in$^2$) in15 min
   b) 121°C (i.e. 10 lb/in$^2$) in15 min
   c) 121°C (i.e. 15lb/in$^2$) in 10 min
   d) 121°C (i.e. 15lb/in$^2$) in15 min
10. Principle involved in sterilization in autoclave is
    a) Saturated steam under pressure
    b) Steam under vacuum pressure
    c) Heated air under pressure
    d) None
11. Saturated steam heats an object about ____times more efficiently than does not air at the same temperature.
    a) 150 b) 500
    c) 1500 d) 2500
12. Autoclaving is not recommended for
    a) Oils
    b) Powders
    c) Heat sensitive fluids and plastics
    d) All
13. Which of the following are often called as stirred tank reactors
    a) Kneaders
    b) Cementers
    c) Filters
    d) Fermenters
14. In food packaging AISI means Austeniti Stainless Steel
    a) American iron and steel institute
    b) Austeniti iron and Stainless Steel industry
    c) Austeniti industry on Stainless Steel and iron
    d) Australian iron and Stainless Steel industry
15. The type of steel recommended to use in food industries is
    a) AISI 304 b) AISI 316
    c) AISI 221 d) Both a & b

16. Germ theory of disease is given by
    a) Robert brown b) L. Pasteur
    c) Joseph Lister d) Robert Koch
17. Pure cultures of bacteria were 1st obtained by
    a) Robert Koch b) L. Pasteur
    c) Joseph Lister d) Robert brown
18. Mushroom is a
    a) Fungi b) Yeast
    c) Bacteria d) None
19. Yeast is a
    a) Fungi b) Mold
    c) Bacteria d) None
20. An antibiotic Streptomycin is discovered by
    a) Robert Koch b) Selman A Waksman
    c) A. Fleming d) All
21. In the presence of oxygen Facultative anaerobes employ
    a) Aerobic respiration b) Fermentation
    c) Both a & b d) None
22. In the absence of oxygen, Facultative anaerobes employ
    a) Aerobic respiration b) Fermentation
    c) Both a & b d) None
23. Fungus is the causative organism for disease in silkworm is first time recognized by
    a) *A.bassi* b) Leeuwenhoek
    c) Selman d) Joseph
24. In 1845 M J Berkeley had proved that Potato Blight of Ireland was caused by
    a) Fungus b) Mold
    c) Bacteria d) Virus
25. Pure culture technique for bacteria was developed by
    a) Pasteur b) Vent
    c) Koch d) Smith

26. Koch developed method for pure culture technique for bacteria is
    a) Streak plate method b) Pour plate method
    c) Rotating tube method d) Both a & b
27. For most bacteria optimum incubation temperature is
    a) 10-17ºC b) 20-27ºC
    c) 30-37ºC d) 40-47ºC
28. Nutrient medium for the growth of bacteria developed by
    a) Pasteur b) Vent
    c) Koch d) Smith
29. Microorganisms exist in two forms that is – a heat labile form (Vegetative) & heat resistant form (endospores), this statement is given by
    a) Arrhenius b) Luis
    c) John tyndall d) All
30. John tyndall developed a method of sterilization by _____ heating, called Tyndallisation.
    a) Continuous b) Discontinuous
    c) Batch d) All
31. Edward Jenner in 1798 used Cowpox Virus to immunize people from
    a) Small pox b) Cowpox
    c) Chickenpox d) None
32. TMV in crystalline form first time isolated by
    a) W.M. Stanley b) J.H. Northrup
    c) S.A. Waksman d) a & b
33. The bacteriophages, the viruses that kill the bacteria are first time discovered by
    a) Twort b) F.D. Herelle
    c) W.M. Stanley d) a & b
34. Endospores are generally formed in the stage of growth curve at the
    a) End of the active growth b) During stationary phase
    c) Lag d) Both a & b

35. Primary metabolites are generally formed in the stage of growth curve at the
    a) End of lag b) During log
    c) Decline d) Both a & b
36. Budding type reproduction is present in
    a) *Rhodopseudomonas* b) *Hyphomicrobium*
    c) Yeast d) All
37. Conidiospores or sporangiospores type reproduction is present in
    a) *Streptomyces* b) *Rhodopseudomonas*
    c) *Hyphomicrobium* d) Yeast
38. The time required for the cell number to double is
    a) Life time b) Generation time
    c) Growth time d) death time
39. Generation time of *E. coli* is
    a) 10 minutes b) 20 minutes
    c) 30 minutes d) 40 minutes
40. Generation time of *Treponema is*
    a) 33 sec b) 33 min
    c) 33 hours d) 33 days
41. The plotted slope of the line when log of cells versus time is called
    a) Death rate b) Growth rate
    c) Life rate d) Development time
42. In microbial growth curve 'Acclimatization phase' also called
    a) Lag phase b) Exponential phase
    c) Stationary phase d) Declined phase
43. In microbial growth curve primary metabolites are produced during
    a) Lag phase b) Exponential phase
    c) Stationary phase d) Declined phase
44. The device through which fresh medium is supplied and toxic products or metabolic wastes accumulated in the medium are removed to maintain a bacterial culture continuously is
    a) Chemostat b) Turbidostat
    c) Thermostat d) Both a & b

45. In microbial growth curve, secondary metabolites include toxins, antibiotics, sterols are formed in

a) Lag phase  b) Exponential phase
c) Stationary phase  d) Declined phase

46. Most microbes grow best at pH values around

a) 5.6  b) 6.0
c) 7.0  d) 8.5

47. Fruits will be spoiled by molds and yeasts, which can grow at value below

a) 3.5  b) 4.5
c) 5.5  d) 6.5

48. Most of the meats and sea foods have pH of

a) 4.6  b) 5.6
c) 6.6  d) 7.6

49. The oxidation reduction potential of a system is expressed by the symbol

a) OR  b) ORP
c) Eh  d) *M. volts*

50. The Eh values will be positive for the

a) Oxidized substances  b) Reduced substances
c) Neutral substances  d) All

51. The Eh values will be ____for the reduced conditions

a) Negative  b) Neutral
c) Positive  d) All

52. Aerobic bacteria such as *Bacillus* require oxidized conditions i.e. __________Eh.

a) Negative  b) Neutral
c) Positive  d) All

53. Which of the following aerobic microorganisms require positive Eh. Ex:,

a) *Pseudomonads*  b) *Acinetobacters*
c) *Bacillus*  d) All

54. Anaerobic micro organisms required negative Eh. Ex:

a) *Clostridium*  b) *Micrococcus*
c) *Bacillus*  d) *Acinetobacters*

55. Antimicrobial compound present in milk is

a) Lactenin b) Anticoliform factor

c) Lactoferrin d) All

56. Antimicrobial compound present in egg

a) Avidin b) Biotin

c) Lysozyme d) All

57. The lowest temperature at which a microorganism has been reported to grow is

a) -273°C b) -34°C

c) -4°C d) 5°C

58. The highest temperature at which a microorganism has been reported to grow is 90°C

a) 70°C b) 90°C

c) 121°C d) 150°C

59. Example for psychrotrophs

a) *Alcaligenes* b) *Seudomonas*

c) *Streptococcus* d) All

60. Example for thermophilic bacteria

a) *Bacillus* b) *Clostridium*

c) *Seudomonas* d) Both a & b

61. Predominant spoilage bacteria of most vegetables is

a) *Erwinia* b) *Pseudomonas*

c) Bacillus d) Both a & b

62. Predominant spoilage fungi of most vegetables is

a) *Botrytis* b) Bacillus

c) Yeast d) Mushroom

63. Predominant spoilage fungi of most fruits is

a) P*enicillium* b) R*hizopus*

c) A*spergillus* d) All

64. The temperatures used for canning high acid foods

a) 70°C b) 90°C

c) 100°C d) 121°C

65. The temperatures used for canning low acid foods

a) 70°C b) 90°C

c) 100°C d) 121°C

66. Fed batch fermentation also called as

a) Batch b) Continuous

c) Semi continuous d) Submerged

67. Most fermentation industries today use the _________culture methods.

a) Submerged b) Surface

c) Fed batch d) All

68. Citric acid fermentation process uses ___culture method

a) Submerged b) Surface

c) Fed batch d) All

69. Biogas production uses __________fermentation

a) Fed batch b) Submerged

c) Surface d) All

70. The fermentation process used for production of amylase using wheat bran is __fermentation.

a) Solid state b) Liquid state

c) Semisolid d) Amorphous

71. The fermentation process used for production of mushrooms ________ fermentation

a) Solid state b) Liquid state

c) Semisolid d) Amorphous

72. The Blue green algae are also being produced on large scale and used as biofertilizers in ______cultivation.

a) Rice b) Sorghum

c) Corn d) Finger millet

73. A symbiosis in which two phylogenetically unrelated microorganisms coexist overa prolonged period of time is

a) Antagonism b) Ammensalism

c) Parasitism d) Biocontrol

74. Antagonism (ammensalism) one microorganism produces a substance that is inhibitory to other microbial population results in a negative outcome for later one is

a) Parasitism b) Ammensalism

c) Antagonism d) b&c

75. The activities of various hyperparasites, i.e., those agents that parasitize plant pathogens, can result in

a) Parasitism b) Ammensalism

c) Antagonism d) Biocontrol

76. Biocontrol can occur when non-pathogens compete with pathogens for in and around the host plant.

a) Nutrients b) Habitat

c) Pathogenicity d) a & b

77. Ammonia produced during decompositions of proteins and amino acids at concentrations inhibitory to nitrite oxidizing populations of

a) Nitrobacter b) Nitrosomonas

c) Chlorella d) All

78. Bacteria grow fastest in the pH range of

a) 3.0 -5.0 b) 6.0 -8.0

c) 8.0-9.0 d) 9.0 -10.0

79. Yeasts grow fastest in the pH range

a) 3.0-5.0 b) 4.5-6.0

c) 6.0-7.0 d) 7.0-8.0

80. Filamentous fungi grow fastest in the pH range

a) 3.5-4.0 b) 3.0-5.0

c) 5.0-6.0 d) 6.0-7.0

81. Which of the following statement is correct

a) Active proteolytic bacteria, can grow in alkaline pH

b) Bacteria which are acid formers are favoured by moderate acidity

c) Molds can grow over a wide range of pH vales than the yeast and bacteria

d) All

82. The post – rigor mammalian muscle has longer storage life than fish because

a) $A_w$ b) $P^H$

c) Temperature d) Nutrient content

83. The pH of post – rigor mammalian muscle, round 5.6 and it is lower than that of fish___.

a) 6.2 - 6.5 b) 4.2-5.8.5

c) 5.2-7.5 d) 8.5-9.5

84. Yeast is a/an __________microorganism.

a) Aerobic b) Facultative

c) Anaerobic d) a & b

85. Bacteria may be

a) Aerobic b) Anaerobic

c) Facultative d) All

86. High O - R potential is not favorable to the ______________ organisms.

a) Aerobe b) Facultative

c) Anaerobe d) All

87. Low O-R potential favours the ____________organisms.

a) Anaerobic b) Facultative

c) Obligate d) a & b

88. Antinutritional factor present in Egg white is

a) Lysozyme b) Avidin

c) Mymosine d) Lactoferrin

89. Avidin ties up ______mating it unavailable to microorganisms and eliminating possible spoilage organisms.

a) Lysozyme b) Mymosine

c) Biotin d) All

90. Antimicrobial substance present in Cran berries

a) Benzoic acid b) Ferulic acid

c) Hydro cyanide d) Citric acid

91. Antimicrobial substance present in garlic, onion, leeks

a) Gingirol b) Allicin

c) Thymol d) Isoflavin

92. Antimicrobial substance present in allspice (pimento), cloves, cinnamon
    a) Isoflavin b) Thymol
    c) Eugenol d) Cinnamaldehyde
93. Antimicrobial substance present in thyme and oregano
    a) Linalool b) Thymol
    c) Oleuropein d) Isoethanol
94. Antimicrobial substance present in cinnamon and cassia
    a) Thymol b) Cinnamic aldehyde
    c) Isoflavin d) a & b
95. Antimicrobial substance present in green olives
    a) Linalool b) Ferulic acid
    c) Oleuropein d) Hydrophenol
96. Antimicrobial substance present in egg
    a) Ovotransferrin b) Avidin
    c) Ovolflaroprotein d) All
97. Ovoflavo protein and Avidin in egg white binds with the following vitamin and which restricting their availability
    a) Biotin b) Riboflavin
    c) Vitamin-B1 d) a & b
98. Milk has capacity to generate antimicrobials in the presence of
    a) Lactose b) Hydrogen peroxide
    c) Ozone d) Carbon dioxide
99. The milk enzyme lactoperoxidase will catalyse the oxidation of thiocyanate by $H_2O_2$ to produce ________________.
    a) Hypo chloride b) Hypo-thiocyanate.
    c) Ozone d) All
100. Lysozyme present in milk, egg is most active against __________ bacteria.
    a) Gram positive b) Gram negative
    c) Both a & b d) No effect

## Answers

| | | | | | | | | | |
|---|---|---|---|---|---|---|---|---|---|
| 1. d | 2. a | 3. b | 4. a | 5. a | 6. a | 7. b | 8. c | 9. d | 10. a |
| 11. d | 12. d | 13. d | 14. a | 15. d | 16. d | 17. c | 18. a | 19. a | 20. b |
| 21. a | 22. b | 23. a | 24. a | 25. c | 26. d | 27. c | 28. c | 29. c | 30. b |
| 31. a | 32. d | 33. d | 34. d | 35. d | 36. d | 37. a | 38. d | 39. d | 40. c |
| 41. b | 42. a | 43. b | 44. d | 45. c | 46. c | 47. a | 48. b | 49. c | 50. a |
| 51. a | 52. c | 53. d | 54. a | 55. d | 56. c | 57. b | 58. b | 59. d | 60. d |
| 61. d | 62. a | 63. d | 64. c | 65. d | 66. c | 67. a | 68. b | 69. a | 70. a |
| 71. a | 72. a | 73. c | 74. d | 75. d | 76. d | 77. a | 78. b | 79. b | 80. a |
| 81. d | 82. b | 83. a | 84. d | 85. d | 86. c | 87. d | 88. b | 89. c | 90. a |
| 91. b | 92. b | 93. b | 94. d | 95. c | 96. d | 97. d | 98. b | 99. b | 100. a |

# 7

# Food Chemistry and Nutrition

1. What is the percentage of water in an adult's body weight

   a) 50% b) 60%

   c) 90% d) 70%

2. __________ is a direct reactant in hydrolytic processes of many foods

   a) Water b) Moisture content

   c) Both a&b d) None

3. The bond angle of the isolated water molecule (vapor state) is

   a) 109° b) 104.5°

   c) 109°28' d) 103.6°

4. The dissociation energy of O-H hydrogen bond in water is about

   a) 25 kJ/mole b) 15 kJ/mole

   c) 5 kJ/mole d) 50 kJ/mole

5. What is the Moisture content in meat

   a) 55-60% b) 65-75%

   c) 80-90% d) 40-50%

6. What is the Moisture content in milk

   a) 77 % b) 67%

   c) 87% d) 97%

7. What is the Moisture content in fruits & vegetables

   a) 70-90% b) 60-80%

   c) 50-80% d) 90-95%

8. What is the Moisture content in bread

   a) 25% b) 15%

   c) 12% d) 35%

9. What is the Moisture content in Honey
   a) 12% b) 15%
   c) 20% d) 40%
10. What is the Moisture content in butter & margarine
    a) 16-18% b) 10-12%
    c) 12-15% d) 20-25%
11. What is the moisture content in Coffee beans, roasted
    a) 10% b) 12%
    c) 15% d) 5%
12. What is the moisture content in edible oil
    a) 2% b) 1.5%
    c) 0% d) None
13. What is the moisture content in granulated sugar
    a) 2% b) 1.5%
    c) 0% d) 2.3%
14. What is the moisture content in Watermelon (raw)
    a) 80-90% b) 91-92%
    c) 85-90% d) 97-99%
15. The dry matter that remains after moisture removal is commonly referred as
    a) Total solids d) Brix
    c) Baume d) None
16. For food manufacturer __________ is inexpensive filler.
    a) Water b) Salt
    c) Sugar d) Citric acid
17. Moisture is a quality factor in the preservation & stability of
    a) Dehydrated vegetables and fruits
    b) Dried milk & powdered eggs
    c) Spices & Seasonings
    d) All

18. Moisture data are used to express results of other analytical determinations based on
    a) Dry basis (db)  b) Wet basis (wb)
    c) Both a&b  d) None
19. Moisture is a quality factor in the following foods
    a) Jams and jellies to prevent sugar crystallization
    b) Sugar syrups
    c) processed cereal products
    d) All
20. The water retains its physical properties and thus acts as the dispersing agent for colloids and the solvent for salts is a
    a) Free water  b) Adsorbed water
    c) Water of hydration  d) All
21. The water is held tightly or is occluded in cell walls or protoplasm and is held tightly to proteins is a
    a) Free water  b) Adsorbed water
    c) Water of hydration  d) All
22. The water, which is bound chemically is
    a) Free water  b) Adsorbed water
    c) Water of hydration  d) All
23. What is the melting point of water at 101.3 kpa (1atm) is
    a) 100°C  b) 0°C
    c) 373.99°C  d) 0.01°C
24. What is the boiling point of water at 101.3 kpa (1atm) is
    a) 100°C  b) 0°C
    c) 373.99°C  d) 0.01°C
25. Triple point of water is
    a) 0.01°C & 4.589mmm of hg  b) 0.01°C & 611.73 Pa
    c) Both a&b  d) None
26. Maximum density of water is at
    a) 100°C  b) 0°C
    c) 4°C  d) 0.01°C

27. __________ is the water, which does not freeze at low temperature (-40°C)

 a) Bound water b) Free water

 c) Ice water d) All

28. What type of bound water is available for the growth of microorganisms, enzymatic activity (hydrolytic and oxidative) and non enzymic browning.

 a) Type IV water b) Type III water

 c) Type II water d) Type I water

29. Type I water is removed from foods by

 a) Dehydration b) Freezing

 c) Both a&b d) None

30. Water activity (aw) is

 a) P0/P = ERH/100 b) P/P0 = ERH/100

 c) P/P0 =100/ERH d) None

31. kinetics of phase transitions can be measured by

 a) Differential scanning calorimetry

 b) Spectroscopy

 c) Chromatography

 d) All

32. What is a brain center that controls activities such as maintenance of water balance, regulation of body temperature, and control of appetite.

 a) Neuron b) Hypothalamus

 c) Blood d) Oxygen

33. Recommended daily intake of water for adults is

 a) 1.0 to 1.5 ml/kcal expended b) 2.0 to 2.5 ml/kcal expended

 c) 1.5 ml/kcal expended d) None

34. Recommended daily intake of water for infants and athletes

 a) 1.0 to 1.5 ml/kcal expended b) 2.0 to 2.5 ml/kcal expended

 c) 1.5 ml/kcal expended d) None

35. The water that is typically contains 250 to 500 parts per million (ppm) of minerals

 a) Hard water b) Processed water

 c) Mineral water d) Distilled water

36. Mineral waters are high in
    a) Calcium b) Magnesium
    c) Potassium d) Sodium
37. Hard waters are high in
    a) Calcium b) Magnesium
    c) a&b d) Sodium
38. Soft water is characterized by high in
    a) Potassium b) Sodium
    c) a&b d) Calcium
39. Lowering of water activity of foods helps in
    a) Retards the growth of microorganisms
    b) Slows enzyme catalyzed reactions
    c) Retards non-enzymatic browning
    d) All
40. Decreased water activity of dried foods enhances
    a) Microbial growth d) Non Enzymatic browning
    c) Lipid autoxidation d) Catalyzed reactions
41. What is the water activity of intermediate moisture foods" (IMF).
    a) 0.4-0.5 b) 0.5-0.6
    c) 0.6-0.9 d) 0.91-0.99
42. What are the chemical substances added to increase the shelf life of reduced water activity of foods
    a) Preservatives b) Humectants
    c) Antibiotics d) All
43. The following one is a humectant
    a) Glycerol b) NaCl
    c) Sorbitol & Sucrose d) All
44. What is the water activity of Highly perishable (fresh) foods and canned fruits, vegetables, meat, fish, and milk
    a) 0.91-0.99 b) 1.00–0.95
    c) 0.85-0.9 d) 0.6-0.8

45. What is the water activity of Most fruit juice concentrates, sweetened condensed milk, chocolate syrup, maple and fruit syrups

a) 0.91-0.99 b) 1.00–0.95

c) 0.87–0.80 d) 0.6-0.8

46. Many yeasts( *Candida, Torulopsis, Hansenula)* grow best at water activity of

a) 0.91-0.99 b) 0.91–0.87

c) 0.87–0.80 d) 0.6-0.8

47. What is the water activity of Jam, marmalade, marzipan, glacé fruits, some marshmallows

a) 0.80–0.75 b) 0.91–0.87

c) 0.87–0.80 d) 0.6-0.8

48. What is the water activity of Pasta containing approximately 12% moisture

a) 0.7 b) 0.5

c) 0.80 d) 0.6

49. What is the water activity of spices containing approximately 10% moisture

a) 0.7 b) 0.5

c) 0.80 d) 0.6

50. What is the water activity of Whole egg powder containing approximately 5% moisture

a) 0.7 b) 0.5

c) 0.4 d) 0.6

51. What is the water activity of Cookies, crackers, bread crusts, etc. containing 3–5% moisture

a) 0.3 b) 0.5

c) 0.4 d) 0.6

52. What is the moisture content of Sorbitol at a water activity of 0.8

a) 332 b) 67

c) 20 d) 108

53. What is the moisture content of sodium chloride at a water activity of 0.8

a) 332 b) 67

c) 20 d) 108

54. What is the moisture content of Glycerol at a water activity of 0.8

a) 332 b) 67

c) 20 d) 108

55. Kinetics of phase transitions of foods can be measured by

a) Water activity

b) Differential scanning calorimetry

c) Spectrophotometer

d) Pcyanomterer

56. What is the viscosity of An amorphous solid

a) $10^{12}$ Pa sec. b) $10^{10}$ Pa sec.

c) $10^{6}$ Pa sec. d) $10^{5}$ Pa sec.

57. Collapse of food products are visibly evident during

a) Freezing b) Drying

c) a&b d) None

58. The kinetics of collapse is governed by

a) Water activity

b) Differential scanning calorimetry

c) Molecular mobility

d) Pcyanomterer

59. An example of glassy/rubbery behavioral characteristics of foods

a) Boiled sweets b) Dried foods

c) Frozen foods d) All

60. What is the shapc of Sorption isotherms

a) Hyperbola b) Cuboids

c) Sigmoidal d) Zig zig

61. Water absorbed at low water activity is referred as

a) Monolayer water

b) Brunauer, Emmett and Teller (BET) isotherm analysis

c) Comparable isotherm equations (Guggenheim–Anderson–de Boer)

d) All

62. What are building blocks for protein biosynthesis
    a) Amino acids b) Gloucose molecules
    c) Fatty acids d) All
63. The following components of foods contributes for flavor and color after thermal or enzymatic reactions
    a) Vitamins b) Proteins
    c) Minerals d) Short chain faty acids
64. How many amino acids are present in most of the proteins
    a) 50 b) 20
    c) 10 d) 15
65. Protein synthesis occurs in
    a) Mitochondria b) Nucleus
    c) Ribosomes. d) All
66. What is the nutritional energy value of proteins
    a) 17 kJ/g b) 4 kcal/g
    c) 9 kcal/g d) a&b
67. Important sources of protein are
    a) Grains & legumes b) Oil seeds
    c) Milk & meat d) All
68. An essential amino acid for infants
    a) Arginine b) Histidine
    c) Tryptophan d) Lysine
69. Proteins that are not enzymatically modified in cells are called
    a) Homoproteins b) Conjugated proteins
    c) Hetroproteins d) All
70. An example of conjugated proteins
    a) Ovalbumin b) k-casein
    c) Lipoproteins & metalloproteins d) All
71. An example of contractile proteins
    a) Actin b) Myosin
    c) Tubulin d) All

72. An example of transfer proteins
   a) Serum albumin  b) Transferrin
   c) Hemoglobin  d) All
73. An example of storage proteins
   a) Egg albumen  b) Seed proteins
   c) Tubers proteins  d) All
74. Naturally occurring amino acids are in
   a) L-form  b) R-Form
   c) Cis Form  d) D-form
75. When the valence of the carbon is satisfied by more than one similar atoms/ groups then the particular carbon atom is
   a) Symmetric carbon  b) Asymmetric carbon
   c) Functional carbon  d) All
76. When the valence of the carbon is satisfied by four different groups, then that particular carbon atom is
   a) Symmetric carbon  b) Asymmetric carbon
   c) Functional carbon  d) All
77. An example of amino acid containing Symmetric carbon
   a) Arginine  b) Histidine
   c) Glycine  d) Lysine
78. The following one is optically inactive amino acid
   a) Arginine  b) Histidine
   c) Glycine  d) Lysine
79. Imidazole group containing amino acid is
   a) Arginine  b) Histidine
   c) Glycine  d) Lysine
80. ε-amino group containing diamide amino acid is
   a) Arginine  b) Histidine
   c) Glycine  d) Lysine
81. Guanido group containing amino acid
   a) Arginine  b) Histidine
   c) Glycine  d) Lysine

82. Indole group containing amino acid
    a) Arginine b) Histidine
    c) Tryptophan d) Lysine
83. Hydrophobic phenol group containing amino acid
    a) Arginine b) Tyrosine
    c) Tryptophan d) Lysine
84. Aromatic hydrophobic neutral amino acid
    a) Phenylalanine b) Tyrosine
    c) Tryptophan d) Lysine
85. An example of sulphur containing amino acid.
    a) Methionine b) Cysteine
    c) a&b d) Lysine
86. An example of imino acid
    a) Phenylalanine b) Proline
    c) Tryptophan d) Lysine
87. Pyrrolidine ring containing amino acid
    a) Phenylalanine b) Proline
    c) Tryptophan d) Lysine
88. What is an indicator of connective tissue presence in comminuted meat products?
    a) Phenylalanine b) Proline
    c) Hydroxyproline d) Lysine
89. Processing of foods results in excess losses of
    a) Phenylalanine b) Proline
    c) Tryptophan d) Lysine
90. Which amino acid is essential in many biochemical processes as a methyl-donor.
    a) Methionine b) Cysteine
    c) Tryptophan d) Lysine
91. In an adults body water constitutes about
    a) 70% b) 80%
    c) 60% d) 90%

92. Life constituent that is involved in chemical reactions and hydrolytic processes

    a) Air b) Water

    c) Food d) Oxygen

93. An example of nucleoproteins

    a) Ribosomes b) Ovalbumin

    c) *β-caseins,* d) Hemoglobin

94. An example of metalloproteins

    a) Hemoglobin b) Myoglobin

    c) *β-caseins,* d) a&b

95. An example of contractile proteins

    a) Myosin b) Actin

    c) Tubulin d) All

96. An example of transfer proteins

    a) Serum albumin b) Transferrin

    c) Hemoglobin d) All

97. Protective proteins are

    a) Toxins b) Allergens

    c) a&b d) None

98. Imino acid containing amino acid is

    a) Phenylalanine b) Proline

    c) Tryptophan d) Lysine

99. Presence of connective tissue in comminuted meat products is identified by

    a) Phenylalanine b) Hydroxyproline

    c) Tryptophan d) Lysine

100. The following one is a highly reactive amino acid during processing

    a) Phenylalanine b) Hydroxyproline

    c) Tryptophan d) Lysine

## Answers

| | | | | | | | | | |
|---|---|---|---|---|---|---|---|---|---|
| 1. b | 2. c | 3. b | 4. a | 5. b | 6. c | 7. a | 8. d | 9. c | 10. a |
| 11. d | 12. c | 13. c | 14. b | 15. a | 16. a | 17. d | 18. a | 19. d | 20. a |
| 21. b | 22. c | 23. b | 24. a | 25. c | 26. c | 27. a | 28. b | 29. a | 30. b |
| 31. a | 32. b | 33. a | 34. b | 35. c | 36. d | 37. c | 38. c | 39. d | 40. c |
| 41. c | 42. b | 43. d | 44. b | 45. c | 46. b | 47. a | 48. b | 49. b | 50. c |
| 51. a | 52. b | 53. a | 54. d | 55. b | 56. a | 57. c | 58. c | 59. d | 60. c |
| 61. d | 62. a | 63. b | 64. b | 65. c | 66. d | 67. d | 68. b | 69. a | 70. d |
| 71. d | 72. d | 73. d | 74. a | 75. a | 76. b | 77. c | 78. c | 79. b | 80. d |
| 81. a | 82. c | 83. b | 84. a | 85. c | 86. b | 87. b | 88. c | 89. d | 90. a |
| 91. c | 92. b | 93. a | 94. d | 95. d | 96. d | 97. c | 98. b | 99. b | 100. d |

1. The Food Standards are meant for
   a) Protecting consumers health
   b) Ensuring fair practices in the food trade.
   c) Protecting producers
   d) Both a&b
2. Preferably the following number of samples drawn for the monitoring of food safety by the institutes or lab is
   a) 1 to 2 samples b) 5 to 8 samples
   c) 10-15 samples d) 1 sample only
3. The following one is right statement
   a) For number of packages < 100,minimum of 10 samples has to be sampled
   b) For number of samples > 100 in a batch, root of total number of samples need to be sampled
   c) For package weight less than 5 kg the entire contents can be taken as incremental sample.
   d) All
4. The refrigerated samples after sampling before reaching to analysis lab need to be cooled at
   a) 0-4°C b) 5-10°C
   c) <0°C d) 12°C
5. Evanescent ingredient in dried fruits is
   a) Vitamin C b) Sulphur dioxide
   c) Calcium d) $CO_2$
6. Water samples should be delivered for analysis within
   a) 48 hours b) 24 hours
   c) 36 hours d) 72 hours
7. In general as per FSSAI guidelines, the food sample should be analysed and reported within ________working days from the date of receipt of samples by the Public Analyst.
   a) 7-14 days b) >15 days
   c) 30 days d) 1 day

8. The following is a ideal characteristics of Sample container
   a) Container should be inert
   b) Container should be air tight for chemical analysis
   c) Sterile for microbiological analysis
   d) All

9. The 'portion of product' collected from a lot during the first stage of the sampling process is called
   a) Primary sample  b) Composite sample
   c) Final sample  d) None

10. The sample produced by carefully mixing the primary samples (items) from a lot of *pre-packaged* products; or by carefully mixing the primary samples (increments) from a *bulk* (*not* pre-packaged) lot.
   a) Primary sample  b) Composite sample
   c) Final sample  d) None

11. The following sample is obtained by method of reduction
   a) Primary sample  b) Composite sample
   c) Final sample  d) None

12. The sample in which the characteristics of the lot from which it is drawn are maintained is called
   a) Representative sample  b) Composite sample
   c) Final sample  d) None

13. In grains analysis Refractions means
   a) Foreign matter and other food grains
   b) Damaged and weevilled grains
   c) Broken and shriveled grains
   d) All

14. Karnal bunt in wheat is caused by
   a) *Neovossia* indica  b) *Claviceps pupurea*
   c) Both a&b  d) None

15. Ergot produces Ergotoxin and occurs in
   a) Wheat  b) Rye
   c) Millets  d) All

16. Heavier filth material separated from product by sedimentation based on
    a) Densities of filth
    b) Floating them in an oil
    c) Sieve analysis
    d) All
17. The following one is a right statement
    a) Hydrocyanic acid in Beans is determined by titrating with Silver nitrate in an ammonical medium in the presence of potassium iodide
    b) 1 ml 0.02 M Silver Nitrate = 1.08 mg of HCN
    c) The glucosides in Beans are hydrolysed and the liberated Hydrocyanic acid steam distilled before titeration
    d) All the above
18. The following one is a right statement
    a) The grounded food sample is neither too coarse nor too fine and passes through the 1.0 mm sieve before analysis of moisture
    b) Moisture analysis dishes are made of Aluminium or stainless steel approx 7.5 mm wide and 2.5 mm deep with tight fitting lids.
    c) The time for moisture analysis should be reckoned from the moment the oven attains 130°C after the dishes have been placed.
    d) All
19. Ash content in food samples measured by dry ashing method in muffle furnace at
    a) 130-150°C
    b) 550-600°C
    c) 100-105°C
    d) 600-650°C
20. The following one is a right statement
    a) FLAT (both ends of can concave)
    b) FLIPPER (a mechanical shock producing distortion in one end or both ends)
    c) SPRINGER (one end ofcan is distorted while other end is flat)
    d) All
21. Drained weight of less than 850 ml Canned fruits and vegetables is measured by using Sieve Diameter, Mesh opening and drained time respectively
    a) 20mm; 2.5 X 2.5mm and 2 minutes
    b) 30mm; 2.5 X 2.5mm and 2 minutes
    c) 20mm; 11.2 X 11.2mm and 2 minutes
    d) 20mm; 0.3 X 0.3 mm and 5 minutes

22. Drained weight of above 850 ml Canned fruits and vegetables is measured by using Sieve Diameter, Mesh opening and drained time respectively
    a) 20mm; 2.5 X 2.5mm and 2 minutes
    b) 30mm; 2.5 X 2.5mm and 2 minutes
    c) 20mm; 11.2 X 11.2mm and 2 minutes
    d) 20mm; 0.3 X 0.3 mm and 5 minutes
23. Drained weight of less than 850 ml tomatoes is measured by using Sieve Diameter, Mesh opening and drained time respectively
    a) 20mm; 2.5 X 2.5mm and 2 minutes
    b) 30mm; 2.5 X 2.5mm and 2 minutes
    c) 20mm; 11.2 X 11.2mm and 2 minutes
    d) 20mm; 0.3 X 0.3 mm and 5 minutes
24. Drained weight of Other Products in Sauce is measured by using Sieve Diameter, Mesh opening and drained time respectively
    a) 20mm; 2.5 X 2.5mm and 2 minutes
    b) 30mm; 2.5 X 2.5mm and 2 minutes
    c) 20mm; 11.2 X 11.2mm and 2 minutes
    d) 20mm; 0.3 X 0.3 mm and 5 minutes
25. Sodium Chloride (salt content) in brine is measured by titrating against
    a) Silver nitrate solution
    b) Sodium hydroxide
    c) Potassium chromate
    d) None
26. Titrable acidity can be expressed conveniently in gm acid per 100 gm or per 100 ml as appropriate, by using the factor 1 ml of 0.1 N NaOH equals
    a) Malic acid - 0.0067 gm
    b) Oxalic acid - 0.0045 gm
    c) Citric acid monohydrate - 0.0070 gm
    d) All
27. Titrable acidity can be expressed conveniently in gm acid per 100 gm or per 100 ml as appropriate, by using the factor 1 ml of 0.1 N NaOH equals
    a) Citric acid anhydrous - 0.0064 gm
    b) Tartaric acid - 0.0075 gm
    c) Lactic acid – 0.0090 gm
    d) All

28. Titrable acidity can be expressed conveniently in gm acid per 100 gm or per 100 ml as appropriate, by using the factor 1 ml of 0.1 N NaOH equals
    a) Acetic acid – 0.0060 gm
    b) Oleic acid – 0.00282 gm
    c) Citric acid monohydrate - 0.0070 gm
    d) All
29. Volatile acids in fruits and vegetables is determined by
    a) By distilling into Erlenmeyer flask
    b) Titrate rapidly with 0.1 N NaOH until pink colour persists 15 seconds.
    c) 1 ml of 1 N NaOH = 0.0060 gm acetic acid x 10 (10 gm test sample taken)
    d) All
30. The presence of added sucrose in fruits and vegetables is measured by
    a) Inversion by copper- reduction methods
    b) Fehling's method
    c) Both a and b
    d) None
31. Vitamin C (Ascorbic Acid) is measured in fruits and vegetables by
    a) 2, 6 -dichlorophenol indophenol is reduced to a colourless form by ascorbic acid
    b) Oxidation by Potassium dichromate in a sulphuric acid medium
    c) Phenol reduces to dehydrated ascorbate
    d) None
32. Fruit content in final products is measured by
    a) After addition of formaldehyde solution titrated against NaOH
    b) After addition of alkali titrated against formaldehyde solution
    c) Evaporation of water from the product
    d) By using Pearson equation
33. Presence of Caramel in Vinegar is measured by
    a) Mohar reaction
    b) Bindict reaction
    c) Fiehe's reaction
    d) All

34. Presence of Peroxidase enzyme in fruits and vegetables is determined by
    a) Reacting with Guaiacol solution and Hydrogen peroxide gives
    b) Reacting with Guaiacol solution and sodium hydroxide
    c) Titrating against Iodine solution
    d) None

35. The amino acids occurring in food and tissue proteins are optically active and posses __________ configuration
    a) D-Form b) L-form
    c) Cis form d) Trans form

36. a) Albumin proteins are soluble in water and coagulated by heat
    b) Globulins are insoluble in pure water but soluble in neutral NaCl solution
    c) Glutelins are insoluble in all neutral solvents but soluble in very dilute alkalis
    d) Prolamins soluble in 70% alcohol
    a) a&b correct b) Only b correct
    c) a,b&c correct d) All are correct

37. An example of Phosphoproteins
    a) Caseinogen (milk) b) Ovo vitelin
    c) Lipovitellins d) a&b

38. Two important proteins in milk are
    a) Casein b) Lactalbumin
    c) a&b d) Whey globulin and Zein

39. Gelatin is formed when__________ is heated with water
    a) Casein b) Gluten
    c) Collagen d) All

40. An example of natural winterized oil
    a) Corn oil b) Soybean oil
    c) Sun flower and Sesame oil d) All

41. Cotton seed oil is converted to salad oil by process called
    a) Refining b) Winterization
    c) Solvent extraction d) None

42. The following enzymes decreases bread staling
    a) Pentosanases b) Hemicellulases
    c) Naringinase d) Starchyase
43. The following enzymes reduces viscosity of coffee concentration
    a) Pentosanases b) Hemicellulases
    c) Naringinase d) Starchyase
44. The following enzyme debitter citrus products
    a) Pentosanases b) Hemicellulases
    c) Naringinase d) Starchyase
45. The following enzymes reduces flatulence due to starchyose and raffinose in leguminous products
    a) Pentosanases b) Hemicellulases
    c) Naringinase d) Starchyase
46. The following enzyme increases fermentable sugars in brewing process
    a) Pentosanases b) Pullulanse
    c) Naringinase d) Starchyase
47. The following enzyme facilitates filtration of barley wort in brewing process
    a) Pentosanases b) Pullulanse
    *c) α-Glucanases* d) Starchyase
48. The following enzyme removes glucose and oxygen from foods
    a) Glucose oxidase b) Catalase
    *c) α-Glucanases* d) Starchyase
49. The following enzyme reduces concentration of diacetyl in beer
    a) Glucose oxidase b) Diacetyl reductase
    *c) α-Glucanases* d) Starchyase
50. The following enzymes are used for coagulation of milk for cheese production
    a) Renin b) Microbial rennets
    c) a&b d) Pepsin
51. The following enzyme is a tenderize agent of meat
    a) Papain b) Ficin
    c) Bromelin d) All

52. The following enzyme destroy $H_2O_2$ used for food sterilization
    a) Glucose oxidase b) Catalase
    *c) α-Glucanases* d) Starchyase
53. Enzymes hydrolyzing cell walls are called
    a) Cellulases b) Catalase
    *c) α-Glucanases* d) Cytases
54. Coloring pigments containing phenolic groups includes
    a) Anthocyanins b) Leucoanthoxanthins
    c) Catechins d) All
55. The following coloring pigments contains Flavylium nucleus
    a) Anthocyanins b) Leucoanthoxanthins
    c) Catechins d) All
56. _________ is responsible for reddish water soluble colour of many fruits and vegetables
    a) Anthocyanins b) Leucoanthoxanthins
    c) Catechins d) All
57. An Intensely artificial sweetner obtained from Anthoxanthins is
    a) Naringin dihydrochalcone b) K-sulfimate
    c) Aspartame d) All
58. Astringency or puckeriness nature of some foods is due to
    a) Anthocyanins b) Leucoanthocyanins
    c) Catechins d) All
59. The basic building block of Leucoanthocyanins is
    a) Flavylium b) Phenol
    c) Flavan-3,4 diol d) Flavan-3ols
60. Catechins are poly hydroxyl_______
    a) Flavylium b) Phenol
    c) Flavan-3,4 diol d) Flavan-3ols
61. Betalins are red water soluble pigments and found in
    a) Beet root b) Cactus fruit
    c) Poke berries d) All

62. The pigment Juglone occurring in walnut is a typical

a) Betalins
b) Napthoquinone
c) Catechins
d) All

63. Polyphenolase contains Copper as a prosthetic group and the pH for activity is between

a) 5&7
b) 7&9
c) 1&3
d) 8&9

64. a) Maillard reaction involving interaction between reducing sugars, amino acids and proteins causes non enzymatic browning in foods

b) Reaction of oxidation products of ascorbic acid with proteins or amino acids causes non enzymatic browning in foods

c) Reaction of oxidation products of poly unsaturated fatty acids with amino acids and proteins causes non enzymatic browning in foods

d) Caramelization of sugars causes non enzymatic browning in foods

a) a Only correct
b) b&c correct
c) c Only wrong
d) All are correct

65. The reaction between $\alpha$-dicarbonyls and amino acids causes non enzymatic browning in foods and reaction is called

a) Strecker reaction
b) Hodge reaction
c) Bindict reaction
d) Haughs reaction

66. 100g of good quality wheat flour requires about__________ ml of water to obtain better quality final backed products

a) 50-54
b) 60-64
c) 70-74
d) 90 ml

67. The following can be used as a complete or partial substitute for cereals in the diet

a) Potato
b) Milk
c) Millets
d) None

68. Hard cheese without gas holes is called

a) Cheddar
b) Swiss
c) Brick
d) Primost

69. Hard cheese with gas holes is called

a) Cheddar
b) Swiss
c) Brick
d) Primost

70. Semi hard cheese ripened by bacteria is called

a) Cheddar
b) Roquefort
c) Brick
d) Primost

71. Semi hard cheese ripened by moulds is called

a) Cheddar
b) Roquefort
c) Brick
d) Primost

72. Semi hard cheese ripened by whey is called

a) Cheddar
b) Roquefort
c) Brick
d) Primost

73. Soft cheese ripened by bacteria is called

a) Limburger
b) Roquefort
c) Brick
d) Primost

74. Soft cheese ripened by moulds is called

a) Limburger
b) Roquefort
c) Camembert
d) Primost

75. The following alcohol is made commercially from glucose by hydrogenation

a) Methanol
b) Sorbitol
c) Propone
d) All

76. Scorodose is a Tetrasaccharide occurs in

a) Onions
b) Garlic
c) a&b
d) Legumes

77. Most of the enzymes are completely inactivated at

a) 40°C
b) 70°C
c) 37°C
d) None

78. Most of the enzymes act best at the temperature of

a) 40°C
b) 70°C
c) 37°C
d) None

79. Glucose on reduction gives

a) Methanol
b) Sorbitol
c) Propone
d) All

80. Fructose on reduction yields

a) Methanol b) Sorbitol

c) Mannitol d) b&c

81. Galactose on reduction yields

a) Alcohol b) Dulcitol

c) Mannitol d) a&b

82. ___________ is responsible for the stiffening of cooked rice on standing

a) Amylose b) Amylopectin

c) a&b d) Oryzein

83. a) Amylose gives blue color with iodine

b) Amylopectin insoluble and it gives a purple color with Iodine

a) a only correct b) b only correct

c) a& b correct d) b wrong

84. a) Specific dynamic action of proteins is highest and it is about 30%

b) Specific dynamic action of Carbohydrates is 6%

c) Specific dynamic action of Fats is 4%

a) a&b correct b) a wrong

c a,b&c wrong d) a,b&c correct

85. The limiting amino acids in different cereal proteins are

a) Lysine & Threonine b) Methionine & cystine

c) Arginine d) All

86. The limiting amino acids in Maize

a) Lysine & Threonine b) Methionine & cystine

c) Tryptophan d) a&c

87. The limiting amino acids in legume proteins are

a) Lysine & Threonine b) Methionine & cystine

c) Tryptophan d) b & c

88. Soybean proteins are rich sources of Lysine & Threonine but limiting in

a) Tryptophan b) Arginine

c) Methionine d) All

89. Sesame seed proteins are rich sources of Methionine but limiting in
    a) Tryptophan b) Arginine
    c) Lysine d) Threonine

90. Milk proteins are rich in all essential amino acids except__________ in which they are limiting to some extent
    a) Lysine & Threonine b) Methionine & cystine
    c) Tryptophan d) b & c

91. a) Complete proteins, example: egg
    b) Partially complete proteins, example: Wheat
    c) Incomplete protein: Gelatin or Zein
    a) a only correct b) b & c wrong
    c) a&c correct d) a, b & c correct

92. The ratio between the content of the most limiting amino acid in the test protein to the content of the same amino acid in egg protein expressed as a percentage is called
    a) Biological value b) Chemical score
    c) PER d) NPU

93. The chemical scores of gelatin and zein are
    a) 100 b) 65
    c) 0 d) 42

94. The chemical score of Egg protein is
    a) 100 b) 65
    c) 0 d) 42

95. In β-oxidation of stearic acid produces________ acetyle COA
    a) 9 b) 8
    c) 2 d) None

96. In β-oxidation of Palmitic acid produces________ acetyle COA
    a) 9 b) 8
    c) 2 d) None

97. In β-oxidation of Butyric acid produces________ acetyle COA
    a) 9 b) 8
    c) 2 d) None

98. Proline is present in all proteins while__________is present only in collagen

a) Methionine
b) Hydroxy proline
c) Lysine
d) Tryptophan

99. The following sugar does not occur in nature

a) Sucrose
b) Lactose
c) Maltose
d) Raffinose

100. Naturally occurring fatty acids in ____________configuration

a) Trans form
b) L-form
c) Cis form
d) D-from

## Answers

| | | | | | | | | | |
|---|---|---|---|---|---|---|---|---|---|
| 1. d | 2. b | 3. d | 4. a | 5. b | 6. b | 7. a | 8. d | 9. a | 10. b |
| 11. c | 12. a | 13. d | 14. a | 15. d | 16. a | 17. d | 18. d | 19. b | 20. d |
| 21. a | 22. b | 23. c | 24. d | 25. a | 26. d | 27. d | 28. d | 29. d | 30. c |
| 31. a | 32. a | 33. c | 34. a | 35. b | 36. d | 37. d | 38. c | 39. c | 40. d |
| 41. b | 42. a | 43. b | 44. c | 45. d | 46. b | 47. c | 48. a | 49. b | 50. c |
| 51. d | 52. b | 53. d | 54. d | 55. a | 56. a | 57. a | 58. b | 59. c | 60. d |
| 61. d | 62. b | 63. a | 64. d | 65. a | 66. b | 67. a | 68. a | 69. b | 70. c |
| 71. b | 72. d | 73. a | 74. c | 75. b | 76. c | 77. b | 78. c | 79. b | 80. d |
| 81. d | 82. a | 83. c | 84. d | 85. a | 86. d | 87. d | 88. c | 89. c | 90. b |
| 91. d | 92. b | 93. c | 94. a | 95. a | 96. b | 97. c | 98. b | 99. c | 100. c |

1. Milk sugar lactose is a disaccharide of
   a) Glucose + Glucose
   b) Glucose+ Galactose
   c) Glucose + fructose
   d) All
2. Raffinose is
   a) Glucose + Glucose+ Galactose
   b) Glucose + Fructose+ Galactose
   c) Glucose + Maltose+ Galactose
   d) Galactose + Fructose+ Galactose
3. The well known poly galactan is
   a) Gelatin
   b) Collagen
   c) Agar agar
   d) All
4. a) Cellulose; β-1-4 linkages
   b) Dextrans; 1-6 glycoside linkages
   c) Levans; 2-6 linkages between fructose units
   d) Hemi celluloses are polysaccharides containing pentoses, hexoses and uronic acid

   a) a&d correct
   b) a & c correct
   c) Only a correct
   d) All are correct
5. The following test is conducted to estimate the quantities of Oleic, Linoleic and Linolenic acids
   a) RM value
   b) Thiocyanogen value + Iodine value
   c) Peroxide value
   d) TBA test
6. When rancid fats react with TBA, the following a red colour compound is produced
   a) Ketone
   b) Malonaldehyde
   c) Peroxide
   d) None
7. Many refined oils undergo change in flavor before the onset of rancidity. This process is known as
   a) Reversion
   b) Hydroxylation
   c) Hydrogenation
   d) All

8. a) Fructose (fruit sugar)
   b) Maltose (malt sugar)
   c) Xylose (wood sugar)

   a) a & b correct  b) Only a is correct
   c) a & c correct  d) All are correct

9. The common amino sugars are

   a) Glucosamine  b) Galactosamine
   c) Glyceraldehydes  d) a & b

10. In following foods non enzymatic browning takes place

    a) Dates  b) Honey
    c) Chocolates  d) All

11. Cereal starches usually contain ________ % of amylase and ______% of amylopectin

    a) 75 & 25  b) 50 & 50
    c) 25 & 75  d) 40 & 60

12. a) Amylose contributes to the jelling characteristics of cooked and cooled starches
    b) Amylopectin is responsible for the thickened properties of starch preparations but it does not contribute to gel formation

    a) a&b correct  b) b only correct
    c) a only correct  d) None

13. The temperature at which the starch granules begin to swell rapidly and lose birefringence is called

    a) Jelling  b) Gelatinization
    c) Retrogradation  d) None

14. When the individual amylase molecules are not sufficiently enlarged to produce a gel, so that the growing micelles eventually causes precipitation this is known as

    a) Jelling  b) Gelatinization
    c) Retrogradation  d) None

15. The rate of Retrogradation influenced by

    a) Temperature  b) Size of starch granules
    c) Shape of starch granules  d) All

16. The water insoluble parent pectic substances that occurs in plants is known as

a) Premier pectin
b) Protopectin
c) Pectinic acid
d) Uronic acid

17. Most commonly found fatty acids in animal and vegetable fats

a) $C_{16}$
b) $C_{18}$
c) $C_4$
d) a & b

18. Most commonly encountered saturated fatty acids

a) Lauric and Myristic
b) Palmitic
c) Stearic
d) All

19 a) Oleic acid is found in almost all fats and Olive oils contains 75%

b) Linoleic acid present in high concentration in many fats and oils and sunflower contains 60-80%

c) Linolenic acid comprises 50-60% of the fatty acids in Linseed oil

d) Arachidonic acid ($C_{20}$) is found primarily in animal sources.

a) a, b & c correct
b) a & b correct
c) Only a is correct
d) All are correct

20. a) The specific heat of fat is useful in processing. It increases with increasing un saturation of fatty acids

b) Viscosity of fats increases with an increase in the average chain length of fatty acids in triglycerides and decreases with increasing unsatuartion

c) The refractive index of fats increases with the increasing chain length of fatty acids in triglycerides or with increasing unsatuartion

d) Tristearin can exit in 3 forms with melting point 53°C, 64.2°C and 71.7°C

a) a, b & c correct
b) a & b correct
c) Only a is correct
d) All are correct

21. The following law state that the intensity of a ray of monochromatic light decreases exponentially as the concentration of absorbing material increases.

a) Beer's law
b) Lambert's law
c) Boil's law
d) All

22. The following law state that the proportion of light absorbed by an absorbing substance is independent of the intensity of the incident light
    a) Beer's law  b) Lambert's law
    c) Boil's law  d) All
23. When the poly peptide contains about 100 amino acid units is called
    a) Protein  b) Peptidine
    c) Glutathione  d) Carnosine
24. The following is a metalo proteins
    a) Ferritin  b) Haemosiderin
    c) Casein  d) a&b
25. The most widely known Phosphoproteins
    a) Casein  b) Pepsin
    c) a & b  d) Glutathione
26. Conversion factor converting nitrogen to protein for wheat is
    a) 5.83  b) 6.25
    c) 5.95  d) 6.38
27. Conversion factor converting nitrogen to protein for Rice is
    a) 5.83  b) 6.25
    c) 5.95  d) 6.38
28. Conversion factor converting nitrogen to protein for pulses is
    a) 5.83  b) 6.25
    c) 5.95  d) 6.38
29. Conversion factor converting nitrogen to protein for maize is
    a) 5.83  b) 6.39
    c) 5.95  d) 6.38
30. Conversion factor converting nitrogen to protein for Soybean is
    a) 5.83  b) 6.25
    c) 5.41  d) 6.38
31. Conversion factor converting nitrogen to protein for milk and milk products is
    a) 5.83  b) 6.25
    c) 5.41  d) 6.38

32. Conversion factor converting nitrogen to protein for oils and fats is
    a) 5.83    b) 6.25
    c) 5.41    d) 6.38
33. Conversion factor converting nitrogen to protein for egg whole and egg powder is
    a) 5.83    b) 6.25
    c) 6.68    d) 6.38
34. Conversion factor converting nitrogen to protein for meat and meat products is
    a) 5.80    b) 6.25
    c) 6.68    d) 6.38
35. Conversion factor converting nitrogen to protein for coffee, tea and Cocoa is
    a) 5.80    b) 6.25
    c) 6.68    d) 6.38
36. Conversion factor converting nitrogen to protein for beer and wine is
    a) 5.80    b) 6.25
    c) 6.68    d) 6.38
37. a) 1 IU = 0.3 μg of Retinol or 0.6 μg of β-carotene
    b) 1IU of Vitamin-D = 0.025 mg of chole calciferol
    c) 1 IU of Vitamin-E = 1mg of α-tocopherol acetate
    a) a only correct    b) a&c correct
    c) b&c correct    d) All are correct
38. The principle sources of B-complex vitamins are
    a) Liver    b) Yeast
    c) Bran of cereals    d) All
39. Sulfur is a component of the following vitamins
    a) Thiamine    b) Biotin
    c) a & b    d) Vitamin C
40. a) Type IV water is available for growth of microorganisms and enzymatic and NEB
    b) Type III water can be easily removed to a moisture content of 12-25%

c) Type II water permits enzymatic and Non enzymatic browning

d) Type I water permits lipid rancidity

a) a, b & c correct b) a & b correct

c) Only a is correct d) All are correct

41. When phytol is removed from chlorophylls in presence of an enzyme called chlorophyllase produces a green colored compound called

a) Pheophytins b) Chlorophillides

c) Pheoporbides d) All

42. In presence of acid medium when magnesium removed from chlorophylls produces a dull olive brown compound called

a) Pheophytins b) Chlorophillides

c) Pheoporbides d) All

43. When phytol and magnesium removed from chlorophylls in presence of an enzyme called chlorophyllase and acid medium produces dull olive brown compound called

a) Pheophytins b) Chlorophillides

c) Pheoporbides d) All

44. The following Xanthones is a major yellow colour pigment in mango

a) Mangiferin b) Capsanthin

c) Bixin d) All

45. The nutritional value of food is__________ proportional to the crude fiber content

a) Directly b) Inversely

c) No effect d) None

46. a) The ash obtained can be used in determination of individual minerals in a given food sample

b) Alkalinity of soluble ash is a measure of alkali carbonate present in the material. It is calculated interms of $K_2O$, $K_2CO_3$

c) Acid insoluble ash is a measure of soil, sandy matter etc ., in substances like spices, talc and confectionery

d) Ash value figures help in grading of flours and sugars

a) a, b&c correct b) a&b correct

c) Only a is correct d) All are correct

47. The following organic acid is principle volatile acid in foods
    a) Citric acid b) Malic acid
    c) Acetic acid d) All
48. Formic acid presence in a food indicates____________ spoilage
    a) Bacteria b) Moulds
    c) Yeast d) All
49. Galacto-uronic acid presence in fruit products indicates_______spoilage
    a) Bacteria b) Moulds
    c) Yeast d) All
50. Karl fisher reagent composed of
    a) $I_2$, $SO_2$ b) Pyridine
    c) Methanol d) All
51. The following test is useful in distinguishing lactose from sucrose and other monosaccharides
    a) Fearson's test b) Bennedict test
    c) Barfoed test d) None
52. Sugar functions in foods
    a) Sweetening cum nutritional requirement
    b) Humectant, Plasticizer and Texturising agent
    c) Flavour producing agent
    d) All
53. The following fatty acid is most abundantly and widely distributed in almost every oil
    a) Butyric acid b) Palmitic acid
    c) Linoleic acid d) All
54. The following one is not a essential fatty acid
    a) Linoleic acid b) Linolenic acid
    c) Arachidonic acid d) Palmitic acid
55. The following test is used to detect oxidative spoilage of fats
    a) Hexa bromide test b) Halpen test
    c) Kries test d) Holde test

56. a) Halpen test for added cotton seed oil
    b) Hexa bromide test for added Linseed oil
    c) Holde Test for added mineral oil
    d) Bellier turbidity temperature test for groud nut oil
    e) Modified Badovin test for added sesame oil
    a) a, b&c correct    b) a&b correct
    c) Only a&e is correct    d) All are correct
57. Phloroglucinol used in the following test
    a) Hexa bromide test    b) Halpen test
    c) Kries test    d) Holde test
58. The following is a brown flavour
    a) Coffee    b) Malt
    c) Caramel    d) All
59. Lemon like flavor in lemon is due to
    a) Neral/geranial    b) Benzaldehyde
    c) Octen-3-ol    d) None
60. Basmathi rice flavor is due to
    a) 2-acetyl-1-pyrroline    b) Diacetyl 1-pyrroline
    c) 2-acetyl-3- methyl thiophene    d) 2-acetyl thiazole

## Answers

| | | | | | | | | | |
|---|---|---|---|---|---|---|---|---|---|
| 1. b | 2. b | 3. c | 4. d | 5. b | 6. b | 7. a | 8. d | 9. d | 10. d |
| 11. c | 12. a | 13. b | 14. c | 15. d | 16. b | 17. d | 18. d | 19. d | 20. d |
| 21. a | 22. b | 23. a | 24. d | 25. c | 26. a | 27. c | 28. b | 29. b | 30. c |
| 31. d | 32. d | 33. c | 34. a | 35. b | 36. b | 37. d | 38. d | 39. c | 40. d |
| 41. b | 42. a | 43. c | 44. a | 45. b | 46. d | 47. c | 48. a | 49. b | 50. d |
| 51. a | 52. d | 53. b | 54. d | 55. c | 56. d | 57. c | 58. d | 59. a | 60. a |

1. An example of methyl donor in many biochemical process
   a) Phenylalanine b) Hydroxyproline
   c) Methionine d) Lysine
2. Nutritionally tyrosine can be replaced by
   a) Phenylalanine b) Tryptophan
   c) Methionine d) Lysine
3. Bouillon" flavor of protein hydrolysates originates partly from a lactone derived from
   a) Phenylalanine b) Tryptophan
   c) Methionine d) Threonine
4. Tryptophan is a precursor in the biosynthesis of
   a) Vitamin C b) Nicotinic acid
   c) Vitamin E d) All
5. The amino acid involved in enzymatic oxidation of foods
   a) Phenylalanine b) Tryptophan
   c) Tyrosine d) Lysine
6. The biological valence (BV) of proteins is calculated by
   a) $\frac{\text{Urea-N (non-protein diet) +N balance}}{\text{N Intake}} \times 100$
   b) $\frac{\text{Protein content (Reference Protein)"Protein content (non-protein diet)}}{\text{Protein intake}} \times 100$
   c) $\frac{\text{Weight gain (g)}}{\text{Available protein (g)}}$
   d) $\frac{\text{Urea-N (non-protein diet) +N balance}}{\text{N Intake} \times 100}$
7. Net protein utilization" (NPU) is determined by
   a) $\frac{\text{Urea-N (non-protein diet) +N balance} \times 100}{\text{N Intake}}$
   b) $\frac{\text{Protein content (Reference Protein) - Protein content (non-protein diet)} \times 100}{\text{Protein intake}}$
   c) $\frac{\text{Protein content (Reference Protein) - Protein content (non-protein diet)}}{\text{Protein intake} \times 100}$

d) $\frac{\text{Urea-N (non-protein diet) +N balance}}{\text{N Intake} \times 100}$

8. The growth value (protein efficiency ratio= PER) of laboratory animals is calculated by

   a) $\frac{\text{Weight gain (g)}}{\text{Available protein (g)}}$

   b) $\frac{\text{Available protein (g)}}{\text{Weight gain (g)}}$

   c) $\frac{\text{Urea-N (non-protein diet) +N balance}}{\text{N Intake} \times 100}$

   d) All

9. Calculate the amount of calcium (mg/100 g) required to meet 50 and 100% of the DV

   a) 200 mg/100 g b) 220 mg/100 g

   c) 250 mg/100 g d) 200 mg/100 g

10. All food products require analysis as part

    a) Quality management program

    b) Product development process

    c) Production & Product in the market

    d) All

11. Proximate analysis of foods refers to determination of

    a) Moisture & Ash b) Protein & fat

    c) Carbohydrates d) All

12. The following one is not a major step in food analysis

    a) Select and prepare samples

    b) Perform the assay

    c) Calculate and interpret the results

    d) Boiling on a water bath

13. Blank test is a

    a) Same quantities of the same reagents are used

    b) Same procedure is repeated in every detail

    c) The substance being tested is omitted

    d) All

14. Following one is not a desiccant
    a) Anhydrous calcium chloride b) Sodium hydroxide
    c) Silica gel d) Phosphorus pentoxide
15. Any substance normally consumed as a constituent of food is a
    a) Chemical compound b) Nutrient
    c) Food additive d) Preservative
16. Carbohydrate polymers with ten or more monomeric units, which are not hydrolysed by the endogenous enzymes in the small intestine of humans are called as
    a) Polysaccharides b) Conjugated carbohydrates
    c) Dietary fiber d) All
17. Fatty acids with cis-cis methylene interrupted double bonds are called as
    a) Trans Fatty acids b) Polyunsaturated fatty acids
    c) Cholesterol d) Saturated fatty acids
18. A standardized statement or listing of the nutrient content of a food is called as
    a) Nutrition label b) Nutrient declaration
    c) Nutritional composition d) Nutritional claim
19. Nutritional Information or nutritional facts of the product shall be given on the label per
    a) Per 100 gm b) Per 100 ml
    c) Per serving d) All
20. As per FSSAI regulations, nutritional information may not be necessary for the following product
    a) Wheat b) Rice
    c) Table salt d) All
21. Sucrose is a
    a) Reducing sugar b) Non reducing sugar
    c) Monosaccharide d) None
22. In Phenol-Sulfuric acid method, the total carbohydrate are spectrophotometrically - measured at ______________ nm
    a) 490 nm b) 520 nm
    c) 410 nm d) 620 nm

23. ________________ is a reducing sugar

a) Glucose b) Fructose

c) Galactose d) All

24. The gravimetric determinations are entirely based upon the concept of

a) Weight b) Acid – base titrations

c) Photometry d) All

25. The number of Individual samples to be selected from a lot is

a) $n= C\sqrt{N}$ b) $n= N\sqrt{C}$

c) $n= \sqrt{NC}$ d) $n= c\sqrt{N}$

26. What are the major considerations for validation of any analytical method

a) Controls b) Standard reference materials

c) Check samples d) All

27. What is the mesh number of dried foods particles used for estimation of moisture, protein and mineral

a) 20-mesh b) 40 Mesh

c) 50 mesh d) 30 mesh

28. Ash content of foods represents

a) Vitamins content b) Minerals content

c) Fiber content d) Sugars content

29. Ash content of most fresh foods is

a) 2% b) 3%

c) 5% d) 11%

30. Wet ashing is carried out by

a) Acids b) Oxidizing agents

c) a&b d) None

31. What are the ideal characteristics of solvent used for fat extraction

a) High solventpower for lipids

b) Low or no solvent power for proteins

c) Low solvent power for carbohydrates

d) All

32. An example of Non solvent wet extraction method for fat is
    a) Goldfish method
    b) Soxhlet method
    c) Mojonnier method
    d) Babcock method
33. What are the major steps in kjeldahl method of protein estimation
    a) Digestion
    b) Neutralization & Distillation
    c) Titration
    d) All
34. Nitrogen to Protein Conversion Factors for milk is
    a) 6.25
    b) 6.38
    c) 5.17
    d) 5.33
35. In which method cupric ions are complexed with peptide bonds under alkaline conditions
    a) Kjeldahl method
    b) Lowry method
    c) Biuret method
    d) All
36. Dinitrosalicylic acid method is used for
    a) Reducing sugars
    b) Proteins
    c) Saturate fat
    d) All
37. In Somogyi–Nelson method, the total reducing sugars are spectrophotometrically - measured at __________nm
    a) 490 nm
    b) 520 nm
    c) 410 nm
    d) 620 nm
38. Specific gravity of carbohydrate solution is correlated with
    a) °Brix
    b) Baumé Modulus
    c) a&b
    d) None
39. What is a small volume of a concentrated standard that is added to the sample?
    a) Working standard
    b) Spike
    c) a&b
    d) None
40. Melting point of fat shortening is measured by
    a) Refractive index
    b) Clear point
    c) Flash point
    d) All
41. The temperature at which the oil or fat sample begins to smoke
    a) Smoke point
    b) Flash point
    c) Fire point
    d) Melting point

42. _____________ is a measure of the resistance of an oil to crystallization.
    a) Cloud point b) Cold test
    c) Fire point d) Melting point
43. _____________ is defined as the mg of KOH necessary to neutralize the free acids present in 1 g of fat or oil
    a) Iodine value b) Acid value
    c) Saponification Value d) None
44. ___________ is defined as the milliequivalents (mEq) of peroxide per kilogram of sample
    a) Iodine value b) Acid value
    c) Peroxide value d) None
45. The resistance of lipids to oxidation is known as
    a) Rancidity b) Oxidative stability
    c) Antioxidants d) b&c
46. Proteins can be selectively precipitated or solubilized by changing
    a) Ionic strength b) Buffer pH
    c) Dielectric constant d) All
47. The transmittance (T) of a solution in spectroscopy is measured by
    a) $T = p/p_0$ b) $T = P^0/p$
    c) $\% T = p/p_0 \times 100$ d) $\% T = P^0/p \times 100$
48. A basic spectrophotometer is composed of
    a) Light source & Monochromator
    b) Radiation detector & Read out device
    c) Sample/reference holder
    d) All
49. The most commonly encountered detectors in spectroscopy are
    a) Phototube b) Photomultiplier tube
    c) Photodiode d) All
50. In paper and thin-layer chromatography, components of a mixture are characterized by
    a) Distance moved by component/ Distance moved by solvent
    b) Distance moved by solvent / Distance moved by component
    c) a&b
    d) None

51. Food safety analysis must be conducted on all products to assess
    a) Physical hazards b) Chemical hazards
    c) Biological hazards d) All
52. Food Business Operator (FBO) should establish quality standards to meet
    a) Consumer needs b) Market needs
    c) Legal requirements d) a & c
53. In a food industry product specifications specifies the standards for
    a) Processes
    b) Raw materials & packaging materials
    c) Acceptance tests & finished products
    d) All
54. Physical Systems Hazard Control include
    a) Identification of physical hazards
    b) Development and maintaining flow diagrams
    c) Document systems for control for all hazards
    d) All
55. What are the least traceable unit of distribution
    a) Date b) Shift
    c) Batch No d) All
56. Characteristics of quality are
    a) Sensory b) Hidden
    c) Quantitative d) All
57. Sensory attributes are
    a) Apperance & Color b) Flavour
    c) Taste d) All
58. ____________ is a prime sensory characteristic which also determines the flavour, texture nutritive value and wholesomeness of food
    a) Appearance b) Taste
    c) Aroma d) None
59. Fracturability encompasses
    a) Crumbliness & Crispiness b) Crunchiness
    c) Brittleness d) All

60. Graininess is perceived in fruit juices like
    a) Apple b) Guava
    c) Sapota d) b & c
61. Flavour is combination of
    a) Texture & aroma b) Taste & Aroma
    c) Mouth feeling & aroma d) None
62. The following one is not a hidden attributes of foods
    a) Taste b) Nutritive value
    c) Toxic compounds d) Nutrient content
63. Higher the Shrinkage ratio of foods is ___________ will be the unit cost
    a) Lower b) Greater
    c) No change d) None
64. An example of subjective method for determination of quality of food by an industry is
    a) Sensory method b) Physical methods
    c) Chemical method d) Microbial method
65. The following one is not an objective method for determination of quality of food by an industry is
    a) Chemical method b) Physical methods
    c) Sensory method d) Microbial method
66. Sensory techniques are mainly used in
    a) Quality control b) Product development
    c) a&b d) None
67. In Munscll color system, the wavelength of "hue" is
    a) 400 to 500 nm (blue) b) 500 to 600nm (green and yellow)
    c) 600 to 800 nm (red) d) None
68. In Munsell color system, the wavelength of "value" is
    a) 400 to 500 nm (blue) b) 500 to 600nm (green and yellow)
    c) 600 to 800 nm (red) d) None
69. In Munsell color system, the wavelength of "chroma" is
    a) 400 to 500 nm (blue) b) 500 to 600nm (green and yellow)
    c) 600 to 800 nm (red) d) None

70. General appearance characteristics of carbonated beverages is
    a) Degree of hue b) Degree of effervescence
    c) Degree of sweetness d) All
71. The amount of volatiles that escape from a food product is affected by
    a) Temperature b) Nature of compounds
    c) Condition of a surface d) All
72. What is the viscosity of water
    a) 100cP b) 10 cP
    c) 1 cP d) 0 cP
73. An example of Tactile feel properties,
    a) Grainy b) Gritty
    c) Crystalline & Flaky d) All
74. Tip of the tongue (Anterior of the tongue) tastes?
    a) Salt b) Sour
    c) Sweet d) Bitter
75. Lateral portions (first edges) of the tongue tastes?
    a) Salt b) Sour
    c) Sweet d) Bitter
76. Lateral portions (Second edges) of the tongue tastes?
    a) Salt b) Sour
    c) Sweet d) Bitter
77. Back of the tongue (Posterior)) of the tongue tastes?
    a) Salt b) Sour
    c) Sweet d) Bitter
78. The taste Umami is stimulated by
    a) Monosodium L-glutamate (MSG,E621),
    b) Disodium 5'-inosinate+Disodium 5'-guanylate (I+G, E631+E627),
    c) Yeast extracts & HVP
    d) All
79. The burning sensation in the mouth caused by
    a) Capsaicin b) Thymol
    c) Curcumin d) None

80. The cool sensation in the mouth or nose produced by
    a) Capsaicin b) Menthol
    c) Thymol d) All
81. In Hunter L, a, b system, vertical co-ordinate (L) = 0 indicates_______ color
    a) Black b) White
    c) Blue d) Yellow
82. In Hunter L, a, b system, vertical co-ordinate (L) = 100 indicates __________ color
    a) Black b) White
    c) Blue d) Yellow
83. In Hunter L, a, b system, horizontal co-ordinate "–a" indicates ________ colour
    a) Black b) Green
    c) Blue d) Yellow
84. In Hunter L, a, b system, horizontal co-ordinate "+a" indicates ________ colour
    a) Black b) Green
    c) Red d) Yellow
85. In Hunter L, a, b system, horizontal co-ordinate "–b " indicates ________ colour
    a) Blue b) Green
    c) Red d) Yellow
86. In Hunter L, a, b system, horizontal co-ordinate " +b " indicates ________ colour
    a) Blue b) Green
    c) Red d) Yellow
87. In which system colour of a product is defined by three co-ordinates.
    a) Hunter L, a, b system b) Munsell system
    c) Lovibond meter d) All
88. Pungency is the essential characteristic of
    a) Spices b) Cruciferae vegetables
    c) Both a&b d) None

89. Common physical tests used for determination of color of food products
    a) Colour difference meter b) Munsell colour system
    c) Spectrophotometry d) All
90. In which method, color of food product is measured by light reflectance at differentwavelengths
    a) Colour difference meter b) Munsell colour system
    c) Spectrophotometry d) Hunter color system
91. Common physical tests used for determination of viscosity of food and beverage products
    a) Ostwald Viscometer b) Rotating Spindle
    c) Falling weight d) All
92. __________measured by weight loss due to evaporation
    a) Ash b) Moisture
    c) Crude fiber d) Fat
93. What are the major factors influencing quality of food
    a) Genetic b) Harvest
    c) Pre harvest& Post harvest d) All
94. What are the pre-harvesting factors influencing quality of food
    a) Temperature & Light b) Rains & Wind
    c) Humidity d) All
95. Fruit size get effected by deficiency of __________nutrient during cultivation
    a) Phosphorus b) Nitrogen
    c) Potassium d) None
96. Green colour of green leafy vegetables is maintained by
    a) Auxins b) Gibberlic acid
    c) Cytokinin d) Ethylene
97. An example of ripening hormone
    a) Auxins b) Gibberlic acid
    c) Cytokinin d) Ethylene
98. ______________ is used as sprout inhibitor in onions bulbs
    a) Malic acid b) Malic hydrazide
    c) Malate hydrate d) Ethylene

99. Higher temperature after post harvesting causes

| a) Off flavour | b) Weight loss |
|---|---|
| c) Wilting | d) All |

100. Low RH after post harvesting of fruits and vegetables causes

| a) Microbial growth | b) Weight loss |
|---|---|
| c) Wilting | d) b&c |

## Answers

| | | | | | | | | | |
|---|---|---|---|---|---|---|---|---|---|
| 1. c | 2. a | 3. d | 4. b | 5. c | 6. a | 7. b | 8. a | 9. b | 10. d |
| 11. d | 12. d | 13. d | 14. b | 15. b | 16. c | 17. b | 18. b | 19. d | 20. d |
| 21. b | 22. a | 23. d | 24. a | 25. a | 26. d | 27. a | 28. b | 29. c | 30. c |
| 31. d | 32. d | 33. d | 34. b | 35. c | 36. a | 37. b | 38. c | 39. b | 40. b |
| 41. a | 42. b | 43. b | 44. c | 45. b | 46. d | 47. a | 48. d | 49. d | 50. a |
| 51. d | 52. d | 53. d | 54. d | 55. d | 56. d | 57. d | 58. a | 59. d | 60. d |
| 61. b | 62. a | 63. b | 64. a | 65. c | 66. c | 67. a | 68. b | 69. c | 70. b |
| 71. d | 72. c | 73. d | 74. c | 75. a | 76. b | 77. d | 78. d | 79. a | 80. b |
| 81. a | 82. b | 83. b | 84. c | 85. a | 86. d | 87. a | 88. c | 89. d | 90. c |
| 91. d | 92. b | 93. d | 94. d | 95. b | 96. c | 97. d | 98. b | 99. d | 100. d |

1. The measure of chemical stability, microbial activity and flow characteristics of water in foods is called
   a) Moisture estimation b) Water activity
   c) Water Analysis d) Free Water
2. The loss of specific properties of the native protein due to change in secondary or tertiary structure of the protein is called
   a) Coagulation b) Emulsification
   c) Denaturation d) Germination
3. If a starch product is subjected to dry heat, carbohydrate compounds called __________ are formed
   a) Dextrins b) Carbons
   c) Amylopectin d) Gel
4. The products of photosynthetic process that takes place in green plants by the reaction of carbon dioxide and water in the presence of sunlight
   a) Carbohydrates b) Proteins
   c) Fats d) Vitamins
5. Oligosaccharides have ______ number of sugar units joined together by glycosidic bond
   a) 1 b) 2
   c) 3-10 d) More than 10
6. An example of analogus carbohydrate is
   a) Pectin b) Inulin
   c) Lignin d) Saponin
7. An example of fibrous protein is
   a) Myoglobin b) Collagen
   c) Hemoglobin d) Cellulose
8. The irreversible loss of crystalline regions in starch granules that occur upon heating in presence of water is referred to as
   a) Retrogradation b) Dextrinisation
   c) Gelatinisation d) Texturization
9. This process increases the digestibility, palatability and nutritive value of grains
   a) Decortication b) Germination
   c) Fermentation d) Extrusion

10. Most of the glucose form liver and muscles is changed into glycogen by the process of

    a) Glycogenesis b) Glycogenolysis

    c) HMP shunt pathway d) TCA cycle

11. When soluble fiber is fermented, ________ are produced

    a) Long chain fatty acids b) Medium chain fatty acids

    c) Essential chain fatty acids d) Short chain fatty acids

12. One of the following is an example of Disaccharide

    a) Lactose b) Maltose

    c) Fructose d) Both a&b

13. Breaking of one or more peptide bonds in a protein to form smaller polypeptide chains is called

    a) Denaturation b) Salting - in

    c) Hydrolysis d) Oxidation

14. Minerals constitute about __ % of the total body weight.

    a) 2 b) 4

    c) 6 d) 8

15. Each gram of dietary protein provides about 4 calories, a gram of fat provides about ___ calories.

    a) 5 b) 9

    c) 11 d) 7

16. The vitamin essential to the absorption and utilization of calcium and phosphorus is vitamin ____.

    a) C b) A

    c) D d) E

17. The vitamin essential for the formation of photopigments of the retina is vitamin ____.

    a) E b) A

    c) D d) C

18. The term metabolism refers to
    a) Anabolic reactions
    b) Catabolic reactions
    c) Oxidation & reduction
    d) All chemical reactions of the body
19. The breakdown of complex organic molecules into smaller ones is known as
    a) Anabolism b) Catabolism
    c) Metabolism d) Embolism
20. A chemical reaction that requires energy is
    a) Anabolism b) Catabolism
    c) Metabolism d) Oxidation
21. Each gram of a carbohydrate produces about
    a) 2.0 kilocalories b) 4.0 kilocalories
    c) 6.0 kilocalories d) 8.0 kilocalories
22. Glucose is stored in the liver and muscles as
    a) Starch b) Cellulose
    c) Fat d) Glycogen
23. The rate of glucose transport across cell membranes is greatly increased by
    a) Insulin b) Glucagon
    c) Thyroxine d) ADH
24. The net gain of ATP for each molecule of glucose which undergoes glycolysis is
    a) 8 b) 10
    c) 2 d) 6
25. Where in a cell does the Krebs cycle occur?
    a) Mitochondria b) Cytoplasm
    c) Ribosomes d) Golgi apparatus
26. Before entering the Krebs cycle, pyruvic acid has to be changed and combined with
    a) ATP b) CoA
    c) NADH d) FADH

27. The largest amount of ATP molecules are produced during the
    a) Krebs Cycle b) Glycolysis
    c) Electron transport chain d) Transition reation
28. Glycogen breakdown occurs when
    a) Blood levels of glucose drop b) Glucagon is released
    c) Blood levels of glucose are high d) Both a&b
29. The formation of glucose from proteins and fats is called
    a) Glycogenolysis b) Lipolysis
    c) Gluconeogenesis d) Lipogenesis
30. Ketone bodies are the result of catabolism of
    a) Glycerol b) Fatty acids
    c) Amino acids d) Glucose
31. Before entering the Krebs cycle, fatty acids are converted into molecules of
    a) Fructose-6-phosphate b) Glyceraldehyde-3 phosphate
    c) Pyruvic acid d) Acetyl CoA
32. Excess carbohydrates are synthesized into
    a) Amino acids b) Proteins
    c) Ketones d) Triglycerides
33. Protein synthesis occurs in
    a) Ribosomes b) Mitochondria
    c) Lysosomes d) Chromosomes
34. Which of the following vitamins can be made by the body?
    a) Vitamin C&B b) Vitamin D&K
    c) Vitamin E&B d) Vitamin K&C
35. Which of the following vitamins is water-soluble?
    a) Vitamin B b) Vitamin D
    c) Vitamin E d) Vitamin K
36. Which of the following factors affect the metabolic rate?
    a) Age & Sex b) Exercise
    c) Hormones d) All

37. The transfer of heat from a warmer object to a cooler one without physical contact is called

a) Conduction b) Evaporation

c) Convection d) Radiation

38. The thermostat of the human body is located in the

a) Thalamus b) Hypothalamus

c) Cortex d) Medulla oblongata

39. Coenzyme A is a derivative of

a) Vitamin A b) Vitamin B

c) Vitamin C d) Vitamin D

40. How many molecules of ATP are generated during the complete oxidation of a glucose molecule?

a) 30 to 32 b) 32 to 34

c) 34 to 36 d) 36 to 38

41. The process of deamination refers to:

a) Removal of a nitrogen from an amino acid

b) Anabolism of amino acids into proteins

c) Increase in basal metabolic rate that comes from an excess of proteins

d) Anabolism of lipids from excess amino acids

42. Energy derived from nutrients is measured in

a) Ergs b) Kcalories

c) Newtons d) BTU's

43. Which of the following does NOT describe minerals?

a) They are inorganic

b) They are needed in small quantities

c) They can be synthesized by the body

d) Levels are controlled by hormones

44. A food scientist has a sample of a plant oil and wants to determine if the oil contains saturated or unsaturated fatty acids. Which of the following spectroscopic techniques would be most useful for this purpose?

a) Ultraviolet spectroscopy b) Visible spectroscopy

c) Infrared spectroscopy d) Mass spectroscopy

45. Which of the following molecules can have both *cis* and *trans* geometrical isomers

a) $CH_3(CH_2)_2CH_3$ b) $CH_2CHCH_2CH_3$
c) $CH_3CCCH_3$ d) $CH_3CHCHCH_3$

46. Which of the following substances is likely to exhibit the highest boiling point due to intermolecular forces

a) $CH_3OH$ b) $CH_3SH$
c) $CH_3F$ d) $CH_2O$

47. Which of the following quantities contains the greatest number of moles?

a) 30g of $N_2$ b) 35g of $NH_3$
c) 60g of NaCl d) 75g of CaO

48. A food that has a low glycemic index is

a) Honey b) Pasta
c) Instant Noodles d) Margarine

49. The products of photosynthetic process that takes place in green plants by the reaction of carbon dioxide and water in the presence of sunlight

a) Carbohydrates b) Proteins
c) Fats d) Vitamins

50. Which of the following is a banned food colour

a) Tartrazine & Ponceau 4R b) Quinoline & sunset yellow
c) Carmoisine & Allura red d) All

51. The disaccharides consist of two molecules of monosaccharide units joined by _____ bonds

a) Peptide bonds b) Covalent bonds
c) Glycosidic bonds d) Ionic bond

52. A substance used in doughs and batters that causes a foaming action which lightens and softens the finished product

a) Leavening agent b) Emulsifying agent
c) Thickening agent d) Acidulant

53. The edible part of plants or analogous carbohydrates that are resistant to digestion and absorption in the human small intestine with complete or partial fermentation in the large intestine.

a) Starch b) Fiber
c) Protein d) Vitamins

54. Radical scavenging antioxidants break free radical chain reaction by donating ________ to free radicals

a) Oxygen  b) Carbondioxide
c) Nitrogen  d) Hydrogen

55. Formation of melanoidins by heating sugar to high temperatures in low water conditions is called

a) Oxidation  b) Crystallization
c) Caramelization  d) Dehydration

56. At physiological pH, basic amino acids will be ____________ charged.

a) Negatively  b) Positively
c) Neutrally  d) Equally

57. Fats can exist in different crystalline forms, and this phenomenon is known as

a) Halogenation  b) Winterisation
c) Polymorphism  d) Saponification

58. A substance used to help retain moisture in foods

a) Humectant  b) Surfactant
c) Acidulant  d) Oxidant

59. The mineral which acts as antioxidant is

a) Magnesium  b) Selenium
c) Molybdinum  d) Potassium

60. As the number of double bonds in fats *increases*, the melting point

a) Increases  b) Decreases
c) Doesn't change  d) Fluctuates

61. Green flavor of peas, pepper, beans comes from

a) Terpinoids  b) Esters
c) Aldehydes  d) Ketones

62. The two fatty acids designated as essential fatty acids are

a) Oleic & Arachadonic  b) Linoleic acid and Alpha-linolenic
c) Palmitic & Butyric  d) All

63. ________ is referred to as the sweetest of all nutritive sugars

a) Fructose  b) Lactose
c) Glucose  d) Galactose

64. The millard reaction causes loss of the amino acids
    a) Lysine  b) Arginine
    c) Tryptophan  d) All
65. Top of Form 'Table sugar' is which type of sugar? Bottom of Form
    a) Fructose  b) Galactose
    c) Glucose  d) Sucrose
66. Top of Form Regular coffee, tea, and cola all contain which stimulant?
    a) Caffeine  b) Sucrose
    c) Theobromine  d) Theophylline
67. Top of Form Sodium bicarbonate is commonly used in cooking as
    a) Alum  b) Baking Powder
    c) Baking Soda  d) Cream of tartar
68. Top of Form tomato gets its red color from
    a) β-Carotene  b) Lycopene
    c) Limonene  d) Betalin
69. Top of Form Leafy vegatable get most of their green color from
    a) Carotenes  b) Chlorophyll
    c) Mitochondria  d) Xanthophyll
70. Top of Form The bubbles in champagne and soda are
    a) $CO_2$  b) Hydrogen
    c) Nitrogen  d) Oxygen
71. Top of Form Tshe type of alcohol in beverages such as wine, beer, and vodka is
    a) Ethyl alcohol  b) Isopropyl alcohol
    c) Methyl alcohol  d) Propyl alcohol,
72. Which of the following contains the most potential energy?
    a) A glow stick  b) A gallon of gasoline
    c) A mole of candy bar  d) A hot pack
73. Which of the following would be most likely to be a liquid at room temperature?
    a) Animal fat  b) Unsaturated fat
    c) Saturated fat  d) All

74. Aspirin blocks the activity of an enzyme that makes prostaglandins, mediators of fever, pain, and inflammation. Aspirin is ______.

a) Antibody
b) Steroid
c) Enzyme inhibitor
d) Restriction enzyme

75. Which statement best describes the chemistry of soaps?

a) They contain only a charged group
b) They are highly polar, charged compounds
c) They contain only a nonpolar group
d) They contain both a nonpolar group and a polar group

76. Which of the following is *not* true about acid-base indicators?

a) They act as sensors of H+ by changing color.
b) They account for the fact that roses are red and violets are blue
c) They are found in pH paper
d) They form the basis of the Scott test for cocaine

77. Water boils at 100°C; methanol, at 65°C; ammonia, at -33°C. The relative strengths of the intermolecular forces for these molecules are as follows:

a) water > methanol > ammonia
b) water > ammonia > methanol
c) ammonia > methanol > water
d) methanol > water > ammonia

78. 100 mL of 0.1 M NaCl is added to 0.4 L water. What is the concentration of NaCl?

a) 0.02 M
b) 2 M
c) 0.2 M
d) 0.06 M

79. Glucose ($C_6H_{12}O_6$; molar mass = 180 g/mol) undergoes combustion in the body to release 2540 kJ of heat per mole. The products are $CO_2$ and H2O. How much is heat is released if 900 grams of glucose undergo this reaction?

a) 1270 kJ
b) 12700 kJ
c) 127 kJ
d) 127000 kJ

80. A closed vessel contains 104 g of $C_2H_2$ and 300 g of O2 that react to form carbon dioxide and water. 236 kJ of heat are released per mole of $C_2H_2$. How much heat is released in this reaction?

a) 8850 kJ
b) 855 kJ
c) 885 kJ
d) 8550 kJ

81. Among $CH_4$, HCN, and CaO, in terms of expected boiling point rank the highest

a) $CH_4$ b) HCN

c) CaO d) All have same boiling point

## Answers

| | | | | | | | | | |
|---|---|---|---|---|---|---|---|---|---|
| 1. b | 2. c | 3. a | 4. a | 5. c | 6. c | 7. b | 8. c | 9. c | 10. a |
| 11. d | 12. d | 13. c | 14. b | 15. b | 16. c | 17. b | 18. d | 19. c | 20. a |
| 21. b | 22. d | 23. a | 24. c | 25. a | 26. b | 27. c | 28. d | 29. c | 30. b |
| 31. d | 32. d | 33. a | 34. b | 35. a | 36. d | 37. d | 38. b | 39. b | 40. d |
| 41. a | 42. b | 43. c | 44. c | 45. d | 46. a | 47. b | 48. b | 49. a | 50. d |
| 51. c | 52. a | 53. b | 54. d | 55. c | 56. b | 57. c | 58. a | 59. b | 60. b |
| 61. b | 62. b | 63. a | 64. d | 65. d | 66. a | 67. c | 68. b | 69. b | 70. a |
| 71. a | 72. c | 73. b | 74. c | 75. d | 76. d | 77. a | 78. a | 79. b | 80. c |
| 81. c | | | | | | | | | |

1. Polysaccharides contain hundreds or thousands of __________ units.
   a) Protein  b) Carbohydrate
   c) Fat  d) Saccharide
2. A flavoring agent called _____ sugar is produced by the hydrolysis of sucrose under acidic conditions
   a) Convert  b) Demerara
   c) Invert  d) Beet
3. Two monosaccharides can be linked together through a glycosidic linkage to form a
   a) Disaccharide  b) Oligosaccharide
   c) Polysaccharide  d) All the above
4. The simplest carbohydrate which exists in two isomeric forms that are mirror images of each other
   a) Glucose  b) Glyceraldehyde
   c) Glycogen  d) Glycerine
5. Which among the following have an OH group on the carbon next to the carbonyl group react with a basic solution of Cu2+ (Benedict's reagent) to form a red-orange precipitate of copper(I) oxide ($Cu_2O$).
   a) Sucrose and Lactose  b) Lactose and Maltose
   c) Sucrose and Fructose  d) Aldehydes and ketones
6. Despite their high molecular weights, the presence of large numbers of OH groups make the monosaccharides much more _____ soluble than most molecules of similar MW.
   a) Water  b) Fat
   c) Solvent  d) Lipids
7. A dextrorotatory (+) substance rotates polarized light to
   a) Right  b) Left
   c) Both Right and Left  d) Centre
8. Molecules which are _______ of each other have exactly the same physical properties (melting point, boiling point, index of refraction, etc.) but not their interaction with polarized light.
   a) Ploymers  b) Diamers
   c) Enantiomers  d) All the above

9. Starches are usually insoluble in water because of the high molecular weight, but they can form thick colloidal suspensions with

   a) Water b) Fat

   c) Solvent d) Lipids

10. Amylopectin consists of long chains of glucose (up to 105 molecules) connected by $\alpha(1 \rightarrow 4)$ glycosidic linkages, with $\alpha(1 \rightarrow 6)$ branches every ____ glucose units along the chain.

    a) 10 to 14 b) 14 to 23

    c) 24 to 30 d) 31 to 50

11. Glycogen is abundant in the liver and muscles; on hydrolysis it forms_____, which maintains normal blood sugar level and provides energy.

    a) Sucrose b) Fructose

    c) Glucose d) Lactose

12. Cellulose is an important structural________, and is the single most abundant organic compound on earth.

    a) Monosaccharide b) Disaccharide

    c) Oligosaccharide d) Polysaccharide

13. Aspartame is about 160 times sweeter than sugar; it is composed of the amino acids aspartic acid and______, neither of which has a sweet taste.

    a) Phenylalanine b) Valine

    c) Cystine d) Methionine

14. The Maillard Reaction occurs between foods containing carbohydrates and proteins under______, which it results in development of a brown color.

    a) Homogenizing b) Heating

    c) Evaporation d) Cooling

15. The three most important disaccharides are sucrose, lactose and _______.

    a) Fructose b) Maltose

    c) Galactose d) Glucose

16. A protein is a linear sequence of amino acids linked together by _________ bonds.

    a) Covalent b) Ionic

    c) Peptide d) Sulphide

17. Phenylalanine has a bulky non-polar ________ ring as part of its side chain
   a) Phenyl
   b) Chlorine
   c) Alanine
   d) Benzene
18. Lysine contains an _______ group that is an unusually strong nucleophile.
   a) Amine
   b) Carbonyl
   c) Hydroxyl
   d) All the above
19. In carbohydrates a special functional groups are present in the given following
   a) Alcohol & Carboxyl groups
   b) Aldehyde & Ketone groups
   c) Hydroxyl groups & Hydrogen groups
   d) Carboxyl groups & Others
20. The most common way to denature proteins is through
   a) Heating
   b) Warming
   c) Cold treatment
   d) Blending
21. Any factor or combination of factors that increase the forces forcing a protein apart or reduces those holding them together may cause
   a) Homogenization
   b) Denaturation
   c) Maturation
   d) Saturation
22. Polyphenoloxidase activity is a plant defense mechanism designed to turn natural plant ______ into antimicrobial products in the case of tissue injury.
   a) Proteins
   b) Vitamins
   c) Sterols
   d) Phenols
23. Which among the following catalyzes the formation of lipid peroxides from polyunsaturated fatty acids.
   a) Lipoxygenase
   b) Amylase
   c) Amylose
   d) Protease
24. Amylase enzymes are used to produce _______ and glucose from suspensions of starch.
   a) Maltins
   b) Proteins
   c) Dextrins
   d) Xanthans

25. The twenty common amino acids are often referred to using _____ letter abbreviations.

a) 1 b) 2

c) 3 d) 4

26. For amino acids, the R-group is often called the "______ chain" or "variant group."

a) Side b) Front

c) Back d) Top

27. Polar neutral amino acids have polar (hydrophilic) side-chains and their predominant forms have uncharged side-chains at physiological _____

a) Temperature b) pH

c) Texture d) All the above

28. The end of the peptide structural formula that has a quaternary ammonium group is called the ______ terminus

a) X b) E

c) M d) N

29. Primary structure of a protein is the linear sequence of amino acids connected by ______ bonds.

a) Ionic b) Peptide

c) Glycosidic d) Covalent

30. A key feature of secondary protein structure is that it only involves ______ bonding between peptide groups within an individual peptide chain.

a) Oxygen b) Nitrogen

c) Carbon d) Hydrogen

31. A _____ bridge is an attractive force between the positive formal charge on polar basic amino acid residue and a negative formal charge on a polar acidic residue

a) Sweet b) Glucose

c) Salt d) Chloride

32. Dipole-dipole attractive forces can occur between ____ side-chains and/or peptide groups.

a) Polar b) Nonpolar

c) Molar d) Neutral

33. Which proteins function as enzymes, chemical signaling compounds, transporters of other compounds, and antibodies.

a) Primary b) Secondary

c) Teritary d) Globular

34. Human albumin is the most abundant protein in human

a) Muscle b) Blood plasma

c) Skeleton d) Cartilage

35. Antibodies, also referred to as immunoglobulins (Ig), are _____ proteins.

a) Primary b) Secondary

c) Teritary d) Globular

36. Fibrous proteins play important roles in providing _______ rigidity and in contractile movement.

a) Muscular b) Structural

c) Cultural d) Integral

37. Which is the most abundant protein in the body. Its function is to provide structural rigidity and stiffness.

a) Blood b) Collagen

c) Hair d) Cholesterol

38. Scurvy is caused by deficiency of which of these?

a) Vitamin A b) Vitamin D

c) Vitamin K d) Vitamin C

39. _______ shape is maintained by the attractive forces involved in secondary, tertiary, and quaternary structures.

a) Carbohydrate b) Protein

c) Fat d) Mineral

40. The substances that increase the rates of chemical reactions

a) Reagents b) Enzymes

c) Catalysts d) Microbes

41. Which of the following are the sources of fats?

a) Oil, Ghee, Butter b) Eggs

c) Rice, wheat, potatoes d) Iron and calcium

42. Examples of enzymes with relative specificity include_____, which catalyze the hydrolysis of various proteins.

a) Amlyases
b) Proteases
c) Lipases
d) Oxygenases

43. A non-protein compound that must be permanently or temporarily bound to an enzyme in order for the enzyme to function.

a) Refractor
b) Catalyst
c) Pro-factor
d) Co-factor

44. Unlike temperature, pH, and denaturing agents, which affect all types of enzymes, _______ will only affect specific enzymes.

a) Prohibitors
b) Inhibitors
c) Contractors
d) Co-factors

45. Fermentation is the degradation of complex organic substances into simpler ones by the activity of living cells through the agency of

a) Acid
b) Alkali
c) Enzymes (Zymase)
d) Oxidizing substances

46. Phenylketonuria (PKU) is an ______ related disease that can be controlled by a special diet.

a) Fat
b) Vitamin
c) Enzyme
d) Mineral

47. Fatty acids usually have an even number of carbon atoms because they are built from ____ carbon molecules.

a) 1
b) 2
c) 3
d) 4

48. Beeswax is an ester formed by combining ____ acid with a 30 carbon alcohol

a) Palmitic
b) Linoleic
c) Acetic
d) Oleic

49. The component of food which is the major energy supplier is

a) Cellulose
b) Protein
c) Carbohydrate
d) Fiber

50. Unsaturated triglycerides are often further subcategorized as either monounsaturated or______.

a) Polyunsaturated b) Saturated

c) Oligosacturated d) Disaturated

51. The indigestible material in food is known as

a) Fiber b) Minerals

c) Vitamins d) Fats

52. Which of the following are essential for body growth and formation of new cells?

a) Carbohydrates b) Protein

c) Minerals d) Vitamins

53. Vegetable oils and fish oils have a lower degree of ______ than animal fats, and are therefore liquid at room temperature

a) Maturation b) Liquefaction

c) Saturation d) Purifaction

54. Since animal fats have a relatively high degree of saturation, they are ______ at room temperature

a) Liquid b) Semi solid

c) Solid d) Greasy

55. Phospholipids get their name from the presence of ______ ion

a) Potassium b) Phosphate

c) Polar d) Photon

56. Glycerophospholipids are made by combining glycerol, two fatty acids, a phosphate group and ______.

a) Alcohol b) Glycerine

c) Alkali d) Acid

57. Sphingophospholipids are made by combining sphingosine, a fatty acid, a ______ group, and an alcohol.

a) Potassium b) Phosphate

c) Polar d) Photon

58. The steroid found most often in humans and other animals is

a) Phytosterol b) Plasmasterol

c) Cholesterol d) Ergosterol

59. Glycocholate, taurocholate, and other bile salts are released from the gallbladder into the small intestine, where they aid digestion by forming _____ with dietary lipids.

a) Fractions	b) Sols
c) Gels	d) Emulsions

60. Low HDL and high LDL levels in the blood are warning signs of____, the buildup of cholesterol-containing deposits in arteries.

a) Fever	b) Diabetes
c) Cancer	d) Atherosclerosis

61. Arachidonic acid, eicosapentaenoic acid, or dihomo-gamma-linolenic acid undergo reactions that transforms them into the various classes of eicosanoids such as

a) Prostaglandins	b) Thromboxanes
c) Leukotrienes, and prostocyclin	d) All the above

62. When pure, the fats are colourless, virtually odourless and possess an extremely_____.

a) Bland taste	b) Pungent aroma
c) Acidic taste	d) Tart Taste

63. The melting point of fats depends on the chain length of the constituent fatty acid and the degree of

a) Saturation	b) Unsaturation
c) Maturation	d) Solution

64. Fats containing saturated fatty acids from C 4 to C 8 are liquid at room temperature, but those containing C 10 or higher saturated fatty acids are solid and their _____ points increase with increasing chain length.

a) Smoke	b) Cloud
c) Melting	d) Heat

65. Greater the degree of unsaturation (or higher the number of double bonds) of the constituent fatty acid, the lower is the ______ point of the fat.

a) Smoke	b) Cloud
c) Melting	d) Heat

66. The specific gravity of the fats is less than 1 (about 0.86) and, therefore, they _____ on water surface.

a) Mix	b) Emulsify
c) Float	d) Bloat

67. The presence of double bond (s) in the unsaturated fatty acid part of the fat molecule produces _____ isomerism.

a) Polymerase b) Concrete

c) Photometric d) Geometric

68. An emulsifying agent helps in the production of a finely divided suspension of a fat in an _____ medium.

a) Alkaline b) Acidic

c) Aqueous d) Agar

69. The force with which the surface molecules are held together is called the _____ tension.

a) Molecule b) High

c) Surface d) Forced

70. When liquid fat is poured on water, it spreads uniformly over the surface of water in the form of a unimolecular layer and reduces the ___ of water.

a) Gravity b) pH

c) Taste d) Surface tension

71. The fats are hydrolyzed by the enzymes lipases to yield fatty acids and ___.

a) Sterol b) Alcohol

c) Glycerol d) All the above

72. The hydrolysis of fats by alkali is called

a) Ionization b) Emulsion

c) Hydrogenation d) Saponification

73. Calcium soaps are used industrially as _____.

a) Shampoos b) Cosmetics

c) Lubricating greases d) Cleaning agents

74. Zinc soaps are employed in the manufacture of talcum powder and other cosmetics, while lead and magnesium soaps are used in paints industry to hasten the process of

a) Drying b) Frying

c) Cooling d) Spreading

75. Rancidity is caused by the growth of microorganisms which secrete enzymes like

a) Lipases b) Proteases

c) Oxidases d) Amylases

76. Unsaturated fatty acids, either free or combined in lipids, react with gaseous hydrogen to yield

a) Solid fats b) Cream

c) Butter d) Ghee

77. The phenomenon in which unsaturated fatty acids and their esters can take up halogens like Br2 and I2 at their double bond at room temperature in acetic acid or methanol solution is called

a) Saponification b) Hydrogenation

c) Chlorination d) Halogenation

78. The oxidation reactions have been extensively used in establishing the position of double bond(s) in the fatty acid chain, which help in knowing ______ structure.

a) Carbohydrate b) Protein

c) Lipid d) Nutrient

79. Oxidative rancidity is observed more frequently in animal fats than in vegetable fats, due to the absence of

a) Natural antioxidants b) Vitamins

c) Proto-oxidants d) Minerals

80. Fats, when heated in the presence of a dehydrating agent, NaHSO4 or KHSO4 produce an unsaturated aldehyde called acrolein from the glycerol moiety. This is known as

a) Quality test b) Saturation test

c) Acrolein test d) All the above

81. Number of milligrams of KOH required to neutralize the free fatty acids present in 1 gm of fat.

a) Iodine value b) Fat value

c) Acid value d) Koettstorfer number

82. Number of milligrams of KOH required to saponify 1 gm of fat.

a) Koettstorfer number b) Saponification number

c) Iodine number d) Potassium number

83. Number of grams of iodine absorbed by 100 g of fat.
    a) Koettstorfer number b) Saponification number
    c) Iodine number d) Both a&c
84. Number of millilitres of 0.1N KOH required to neutralize the insoluble fatty acids
    a) Koettstorfer number b) Saponification number
    c) Iodine number d) Polenske number
85. Number of millilitres of 0.1N KOH required to neutralize the soluble, volatile fatty acids derived from 5 g of fat.
    a) Saponification number b) Reichert-Meissl number
    c) Polenske number d) Iodine number
86. Number of milligrams of KOH required to neutralize the acetic acid obtained by saponification of 1 gm of fat after it has been acetylated
    a) Reichert-Meissl number b) Polenske number
    c) Iodine number d) Acetyl number
87. While sugars (carbohydrates) and proteins provide 4 Calories (kilocalories) of energy per gram, fat of any type provides _____ Calories
    a) 6 b) 7
    c) 8 d) 9
88. Fish oils are unsaturated and have ______ chain length fatty acids.
    a) Short b) Medium
    c) Low d) Long
89. Linolenic acid is classified as ______ fatty acid.
    a) Omega - 3 b) Omega - 6
    c) Both A and B d) None of the above
90. The following vitamin helps to release energy from foods, promotes normal appetite, and is important in maintaining proper nervous system function.
    a) Vitamin B1 b) Vitamin C
    c) Vitamin E d) Vitamin K
91. The following vitamin helps to release energy from foods, promotes good vision, and healthy skin. It also helps to convert the amino acid tryptophan (which makes up protein) into niacin
    a) Vitamin B2 b) Vitamin C
    c) Vitamin E d) Vitamin K

92. Ultraviolet light is known to destroy the following vitamin, which is why most milk is packaged in opaque containers instead of clear.

a) Riboflavin b) Vitamin C

c) Vitamin E d) Vitamin K

93. Pellagra is a disease state that occurs as a result of severe ____ deficiency.

a) Thiamin b) Riboflavin

c) Niacin d) Pyridoxin

94. Vitamin B6, otherwise known as _____, aids in protein metabolism and red blood cell formation.

a) Pyridoxine b) Pyridoxal

c) Pyridoxamine d) All the above

95. Folate, also known as folic acid or folacin, aids in protein metabolism, promoting red blood cell formation, and lowering the risk for ____ birth defects.

a) Vision related b) Spinal cord

c) Genetic d) Neural tube

96. Vitamin B12, also known as _____, aids in the building of genetic material, production of normal red blood cells, and maintenance of the nervous system.

a) Cobalt b) Pyridoxin

c) Cobalamin d) Riboflavin

97. Which among thc following is involved in energy production, and aids in the formation of hormones and the metabolism of fats, proteins, and carbohydrates from food.

a) Pantothenic acid b) Hydrocloric acid

c) Ascorbic acid d) Citric acid

98. Which among the following vitamin aids in wound healing, bone and tooth formation, improving immune system function, increasing absorption and utilization of iron and act as antioxidant.

a) Vitamin A b) Vitamin D

c) Vitamin C d) Vitamin E

99. The fat-soluble vitamins are stored in the body for long periods of time and generally pose a greater risk for _____ when consumed in excess

a) Toxicity b) Carcinogenicity

c) Teratogenicity d) All the above

100. Night blindness and very dry, rough skin may indicate a lack of vitamin

a) A b) D

c) E d) K

101. Vitamin ___ plays a critical role in the body's use of calcium and phosphorous.

a) A b) D

c) E d) K

102. About 60 percent of vitamin E in the diet comes from

a) Vegetable oil b) Cereals

c) Animal food d) Pulses

103. Which vitamin is naturally produced by the bacteria in the intestines, and plays an essential role in normal blood clotting, promoting bone health, and helping to produce proteins for blood, bones, and kidneys

a) A b) D

c) E d) K

104. Based on, _____ vitamins are classified as either fat-soluble (lipid soluble) or water-soluble.

a) Stability b) Solubility

c) Reproducability d) All the above

105. Total calcium in the human body is 1 to 1.5kg, out of which ______ % is seen in bones.

a) 50 b) 75

c) 88 d) 99

106. Factors responsible for increase in calcium absorption include

a) Vitamin D b) Parathyroid hormone

c) Acidity and amino acids d) All the above

107. Bone and teeth contains bulk quantity of

a) Potassium b) Iron

c) Zinc d) Calcium

108. Active form of calcium is

a) Calcitriol
b) Calcium oxide
c) Calcium phosphate
d) Calcium carbide

109. Calcitriol increase reabsorption of calcium and _____ by renal tubules.

a) Potassium
b) Iron
c) Zinc
d) Phosphorus

110. Plasma calcium level less than 8mg/dl is called

a) Hypercalcemia
b) Opticlacemia
c) Hypocalcemia
d) Pseudoclacemia

111. Bone posses _____ % of total phosphorus and muscle contains 10% of total phosphorus.

a) 50
b) 70
c) 80
d) 90

112. Total body content of iron is 3 to 5gm out of which 75% is recorded in ___ and rest of them are recorded from liver, spleen, bone marrow and muscle.

a) Blood
b) Lymph
c) Plasma
d) Lungs

113. The transport form of iron in blood is

a) Transportin
b) Transferin
c) Myoglobin
d) Albumin

114. Enzymes such as cytochrome oxidase, tyrosinase, lysyl oxidase, allanine synthase, monoamine oxidase, superoxide dismutase and phenol oxidase contains

a) Iron
b) Mangesium
c) Zinc
d) Copper

115. The abnormal metabolism of copper leads to and

a) Menke's kidney hair syndrome
b) Wilson's disease
c) Both A & B
d) None

116. In the human total body the content of Zinc is 2gm, out of which 60 % is encountered in

a) Liver
b) Kidney
c) Skeletal muscle
d) Skin

117. More than 300 enzymes in human body are ______dependent

a) Iron b) Copper

c) Selenium d) Zinc

118. Fluoride is well known for their protective effect on.

a) Rabies b) Scabies

c) Caries d) Diabetes

119. The safe limit of fluorine is about _____ PPM in water.

a) 1 b) 10

c) 15 d) 20

120. Glutathione Peroxidase and 5-de-iodinase are dependent on

a) Magnesium b) Chromium

c) Selenium d) Potassium

121. Selenium toxicity could be termed as

a) Selenosis b) Melanosis

c) Scerosis d) Fluorosis

122. The major source of Manganese are

a) Oils b) Fats

c) Nuts d) Fruits

123. These are solid crystalline, chemical elements that cannot be decomposed and synthesized by ordinary chemical reactions.

a) Vitamins b) Enzymes

c) Minerals d) Nutrients

124. A chemical that acts as a messenger transmitting a signal from one cell to another

a) Chromosome b) Hormone

c) Robosome d) All the above

125. Hormones that bind to the surfaces of cells communicate with intracellular metabolic processes through intermediary molecules called ____ messengers

a) First b) Second

c) Primary d) Teritary

126. A polypeptide hormone synthesized in the pancreas by β-cells

a) Insulin
b) Inulin
c) Ptyalin
d) Choline

127. Glucagon, a peptide hormone synthesized and secreted from the α-cells of the islets of Langerhans of pancreas, raises ___ levels.

a) Blood lipid
b) Blood pressure
c) Blood glucose
d) Bood urea

128. An enzyme of the hydrolyase class that catalyzes the hydrolysis of 1,4-α- glycosidic linkages in polysaccharides.

a) Ω - Amylase
b) β-Amylase
c) Σ - Amylase
d) α-Amylase

129. Which among the following catalyses the reversible interconversion of lactate and pyruvate.

a) Lactate Dehydrogenase
b) Pyruvate dehydrogenase
c) Lactose intolerence
d) All the above

130. Sugars having ____ group are called aldoses and sugars with keto group are called ketoses.

a) Aldehyde
b) Alkanes
c) Acidosis
d) All the above

131. Compounds having same structural formula but differing in spatial configuration are known as

a) Positional Isomers
b) Stereo Isomers
c) Structural Isomers
d) Functional Isomers

132. When a beam of plane polarized light is passed through a solution of carbohydrates, which rotates the light right is known as

a) Levorotatory
b) Hexarotatory
c) Dextrorotatory
d) All the above

133. Equimolecular mixture of optical isomers has no net rotation and hence it is called as

a) Optically active mixture
b) Enantiomers
c) Racemic mixture
d) Both a&c

134. When sugars are different from one another, only in configuration with regard to the single carbon atom, other than the reference carbon atom, they are called

a) Diastereomers
b) Enantiomers
c) Mesomers
d) Epimers

135. In mild alkaline solutions, carbohydrates containing free sugar group (aldehyde or keto group) will tautomerise to form

a) Alkadienes
b) Enediols
c) Alkanols
d) Enediones

136. Maltose consists of two ____________, which act as reducing disaccharide

a) α-D-fructose molecules
b) α-D-glucose molecules
c) α-D-galactose molecules
d) α-D-lactose molecules

137. Amylose molecules consist typically of 200 to 20,000 _____ units which form a helix as a result of the bond angles between the glucose units.

a) Fructose
b) Maltose
c) Glucose
d) Lactose

138. Glucose is stored as glycogen in animal tissues by the process of

a) Glycolysis
b) Gluconeogenesis
c) Glucogenesis
d) Glycogenesis

139. Majority of the monosaccharides found in the human body are of

a) D-type
b) L-type
c) DL-types
d) None of the above

140. In glycolysis pathway glucose is converted to pyruvate under aerobic condition or ____ under anaerobic condition

a) Acetate
b) Lactate
c) Citrate
d) Malate

141. Under aerobic conditions the end product of glycolysis is ______ acid.

a) Pyruvic
b) Latic
c) Citric
d) Acetic

142. The goal of glycolysis, glycogenolysis, and the citric acid cycle is to conserve energy as ATP from the catabolism of

a) Carbohydrates
b) Proteins
c) Fats
d) Aminoacids

143. In glycogenolysis, _____ stored in the liver and muscles, is converted to glucose

a) Fat
b) Protein
c) Cholesterol
d) Glycogen

144. Gluconeogenesis is a metabolic pathway that results in the generation of glucose from _______ sources.

a) Non-carbohydrate
b) Non-protein
c) Non-fat
d) All the above

145. Glucose is the major breakdown product of _____ metabolism.

a) Cellular
b) Microbial
c) Molecular
d) Biochemical

146. Lactose intolerance is a common digestive problem where the body is unable to digest lactose, a type of sugar mainly found in

a) Fruits
b) Vegetables
c) Milk and dairy foods
d) Cereals

147. Carbohydrates have a ____ sparing effect which means that when adequate carbohydrates are taken in diet the amino acids in diet are spared from energy metabolism and used for synthesis of proteins in the body.

a) Lipid
b) Protein
c) Vitamin
d) Mineral

148. High biological value proteins and are termed as ______ class proteins.

a) First
b) Second
c) Third
d) Poor

149. Which among the following proteins are superior biologically as they contain all essential amino acids

a) Pulses
b) Tubers
c) Legumes
d) Animal foods

150. A diet containing right amount of energy, carbohydrates, proteins, fats, fiber, vitamins, minerals and water to fulfill requirement of body is called

a) Nutrition
b) Perfect Diet
c) Balanced diet
d) Food pyramid

## Answers

| | | | | | | | | | |
|---|---|---|---|---|---|---|---|---|---|
| 1. b | 2. c | 3. a | 4. b | 5. d | 6. a | 7. a | 8. c | 9. a | 10. c |
| 11. c | 12. d | 13. a | 14. b | 15. b | 16. c | 17. d | 18. a | 19. b | 20. a |
| 21. b | 22. d | 23. a | 24. c | 25. c | 26. a | 27. b | 28. d | 29. b | 30. d |
| 31. c | 32. | 33. d | 34. b | 35. d | 36. b | 37. b | 38. d | 39. b | 40. c |
| 41. a | 42. b | 43. d | 44. b | 45. c | 46. c | 47. b | 48. a | 49. c | 50. a |
| 51. a | 52. b | 53. c | 54. c | 55. b | 56. a | 57. b | 58. c | 59. d | 60. d |
| 61. d | 62. a | 63. b | 64. c | 65. c | 66. c | 67. d | 68. c | 69. c | 70. d |
| 71. c | 72. d | 73. c | 74. a | 75. a | 76. a | 77. d | 78. c | 79. a | 80. c |
| 81. c | 82. b | 83. d | 84. d | 85. b | 86. d | 87. d | 88. d | 89. a | 90. a |
| 91. a | 92. a | 93. c | 94. d | 95. d | 96. c | 97. a | 98. c | 99. a | 100. a |
| 101. D | 102. a | 103. d | 104. b | 105. d | 106. d | 107. d | 108. a | 109. d | 110. c |
| 111. c | 112. a | 113. b | 114. d | 115. c | 116. c | 117. d | 118. c | 119. a | 120. c |
| 121. a | 122. c | 123. c | 124. b | 125. b | 126. a | 127. c | 128. d | 129. a | 130. a |
| 131. b | 132. c | 133. c | 134. d | 135. b | 136. b | 137. c | 138. d | 139. a | 140. b |
| 141. a | 142. c | 143. d | 144. a | 145. a | 146. c | 147. b | 148. a | 149. d | 150. c |

1. A recommended intake value based experimentally determined approximations of nutrient intake healthy people that are assumed to be adequate is
   a) Zero Intake  b) Minimum Intake
   c) Average Intake  d) Maximum Intake
2. A quantitative estimate of a nutrient intake that is used as a reference value for planning diets for healthy people is
   a) Body mass index  b) Basic metabolic rate
   c) Daily recommended intake  d) Dietary Recommended Intake
3. Useful fortification vehicles for Vitamin A fortification are
   a) Sugar  b) Margarine
   c) Cereal flours  d) All
4. Chemical form of iodine that suitable for use as food fortificant is
   a) Iodate  b) Iodide
   c) Iodous  d) a&b
5. Iodate and iodide are usually added to foods as
   a) Potassium salt  b) Magnesium salt
   c) Manganese salt  d) iron salt
6. Universal salt iodization (USI) is the iodization of all salt for consumption of
   a) Human  b) Livestock
   c) Plants  d) a&b
7. Chemical form of iodine solution used in the wet iodization method is
   a) Potassium iodate  b) Potassium iodide powder
   c) a&b  d) None of the above
8. Chemical form of iodine solution used in the dry iodization method is
   a) Potassium iodate  b) Potassium iodide powder
   c) a&b  d) None of the above
9. Type of packaging that help to prevent iodine losses is
   a) HDPE bags laminated with LDPE
   b) PP bags laminated with PVC
   c) PVC bags laminated with PP
   d) PVC bags laminated with LDPE

10. Zinc compound that suitable for use as food fortificant is
    a) Sulfate
    b) Chloride
    c) Gluconate
    d) All
11. Zinc absorption from food is dependent on
    a) Cadmium
    b) Amount of zinc consumed
    c) Ratio of phytate to zinc in the meal
    d) All
12. Fortificant form of thiamine ($B_1$) is
    a) Thiamine hydrochloride
    b) Thiamine mononitrate
    c) Thymol
    d) a & b
13. Fortificant form of riboflavin is
    a) Sodium salt of riboflavin 22 - phosphate
    b) Potassium salt of riboflavin 22 - phosphate
    c) Sodium salt of riboflavin 52 - phosphate
    d) Potassium salt of riboflavin 52 - phosphate
14. Fortificant form of Niacin is
    a) Nicotinic Acid
    b) Niacinamide
    c) Nisin
    d) a & b
15. Fortificant form of Pyridoxine
    a) Pyridoxine hydrazide
    b) Pyridoxine hydrochloride
    c) Pyridine
    d) a & b
16. Fortificant form of Folic acid (B9)
    a) Pteroyl monoglutamic acid
    b) Pteroyl diglutamic acid
    c) Pteroyl triglutamic acid
    d) Pteroyl tetraglutamic acid
17. Characteristic of pure form of Vitamin B12 is soluble in
    a) Fats
    b) N-hexane
    c) Acetone
    d) Water

18. Folic acid ($B_9$) is soluble in

a) Dilute acid
b) Dilute alkali
c) Strong acid
d) a & b

19. Characteristics of Pyridoxine ($B_6$)

a) Water soluble
b) Stable in oxygen and heat
c) Sensitive to UV light
d) All

20. Niacin is soluble in

a) Alkali
b) Fats
c) Water
d) a&c

21. Riboflavin is unstable in

a) Low temperatures
b) Light
c) High temperatures
d) b & c

22. Thiamine is stable to oxygen in the absence of

a) Moisture
b) Light
c) Fat
d) a&b

23. The amount of niacin added to wheat flour typically ranges from _________ mg/kg.

a) 5 to 10 mg/kg
b) 10 to 200 mg/kg
c) 15 to 70 mg/kg
d) 55 to 90 mg/kg

24. The amount of vitamin $B_1$ added to foods typically ranges from ________mg/kg.

a) 1.5 to 11 mg/kg
b) 5.5 to 55 mg/kg
c) 2.5 to 25 mg/kg
d) 7.5 to 66 mg/kg

25. The amount of vitamin $B_{12}$ added to foods typically ranges from ________ mg/kg.

a) 1.1 to 2.1 mg/kg
b) 1.3 to 4 mg/kg
c) 2.3 to 8 mg/kg
d) 5.3 to 10 mg/kg

26. Wheat flour was first fortified with_______ in the United Kingdom in 1943 in order to restore the calcium lost during milling.

a) Vitamin D
b) Calcium
c) Phosphorus
d) Gluten

27. It is compulsory to add ____________ mg calcium carbonate/kg to white and brown (but not wholegrain) flours milled in the United Kingdom.
    - a) 4–6 mg
    - b) 40–60 mg
    - c) 40–560 mg
    - d) 940–1560 mg
28. For food fortification purposes, the _________ are generally considered to be the most suitable source of selenium.
    - a) Phosphorus salts
    - b) Calcium salts
    - c) Sodium salts
    - d) Fluoride salts
29. The most commonly used fluoride compound for large scale water fortification is
    - a) Tri-fluoro-silicate acid
    - b) Tetra-fluoro-silicate acid
    - c) Penta-fluoro-silicate acid
    - d) Hexa-fluoro-silicate acid
30. Sodium selenate is an ingredient in some sports drinks the function of Sodium selenate is activation of ________ enzymes.
    - a) Lipase
    - b) Protease
    - c) Amylase
    - d) Antioxidant
31. The vitamin essential for absorption of calcium is
    - a) Vitamin A
    - b) Vitamin B
    - c) Vitamin C
    - d) Vitamin D
32. 1$\mu$g calciferol equals to __________ IU of vitamin D
    - a) 10
    - b) 20
    - c) 40
    - d) 60
33. Niacin (vitamin B3) is unique in that it can be synthesized from _________ amino acid.
    - a) Glycine
    - b) Valine
    - c) Proline
    - d) Tryptophan
34. 1 mg niacin can be generated from approximately _______ mg tryptophan.
    - a) 10
    - b) 20
    - c) 40
    - d) 60
35. For which of the following foods niacin fortification is profitable?
    - a) Peanut flour
    - b) Green pea flour
    - c) Maize flour
    - d) All

36. The level that will provide the greatest number of at-risk individual with an adequate micronutrient intake without causing an unacceptable risk of excess intake in the whole population is
    a) Feasible Fortification Level (FFL)
    b) Target Fortification Level (TFL)
    c) Minimum Fortification Level (mFL)
    d) Maximum Fortification Level (TFL)
37. The average micronutrient concentration of a fortified food product measured at the factory is
    a) Feasible Fortification Level (FFL)
    b) Target Fortification Level (TFL)
    c) Minimum Fortification Level (mFL)
    d) Maximum Fortification Level (TFL)
38. The lower limit of the micronutrient content to be achieved by the fortification process is
    a) Feasible Fortification Level (FFL)
    b) Target Fortification Level (TFL)
    c) Minimum Fortification Level (mFL)
    d) Maximum Fortification Level (MFL)
39. The upper limit of the micronutrient content to be achieved by the fortification process is
    a) Feasible Fortification Level (FFL)
    b) Target Fortification Level (TFL)
    c) Minimum Fortification Level (mFL)
    d) Maximum Fortification Level (MFL)
40. The minimum micronutrient content of a fortified food as defined in regulations and standards is
    a) Legal Minimum Level b) Target Fortification Level
    c) Minimum Fortification Level d) Maximum Fortification Level
41. Maximum micronutrient content that a fortified food can present as it is established in food law whose purpose is to minimize the risk of excess intake of certain micronutrients is
    a) Feasible Fortification Level b) Target Fortification Level
    c) Maximum Fortification Level d) Maximum Tolerable Level

42. Recommended Nutrient Intakes (RNIs) of Vitamin A for pregnant women
    a) 700 µg b) 800 µg
    c) 900 µg d) 1000 µg
43. Recommended Nutrient Intakes (RNIs) of Vitamin A for lactating women
    a) 850 µg b) 750 µg
    c) 650 µg d) 550 µg
44. Recommended Nutrient Intakes (RNIs) of Niacin for 19–50 years female
    a) 13 mg b) 14 mg
    c) 15 mg d) 16 mg
45. Recommended Nutrient Intakes (RNIs) of Niacin for pregnant women
    a) 18 mg b) 19 mg
    c) 20 mg d) 21 mg
46. Recommended Nutrient Intakes (RNIs) of Folate for lactating women
    a) 200 µg b) 500 µg
    c) 700 µg d) 900 µg
47. Recommended Nutrient Intakes (RNIs) of Folate for 19–50 years female
    a) 200 µg b) 400 µg
    c) 600 µg d) 800 µg
48. Recommended Nutrient Intakes (RNIs) of Folate for pregnant women
    a) 400 µg b) 600 µg
    c) 800 µg d) 900 µg
49. Recommended Nutrient Intakes (RNIs) of Iodine for lactating women
    a) 100 µg b) 200 µg
    c) 300 µg d) 400 µg
50. Recommended Nutrient Intakes (RNIs) of Iodine for pregnant women
    a) 100 b) 200 µg
    c) 300 d) 400
51. Recommended Nutrient Intakes (RNIs) of Vitamin $B_6$ for lactating women
    a) 1 mg b) 2 mg
    c) 3 mg d) 4 mg

52. Recommended Nutrient Intakes (RNIs) of Riboflavin for 19–50 years female

a) 0.1 mg  b) 1.0 mg
c) 1.1 mg  d) 10.1 mg

53. Recommended Nutrient Intakes (RNIs) of Riboflavin for 19–50 years male

a) 1.1 mg  b) 1.2 mg
c) 1.3 mg  d) 1.4 mg

54. Recommended Nutrient Intakes (RNIs) of Zinc for pregnant women

a) 4.2 mg  b) 4.3 mg
c) 4.4 mg  d) 4.5 mg

55. Recommended Nutrient Intakes (RNIs) of Zinc for lactating women

a) 5.7 mg  b) 5.8 mg
c) 5.9 mg  d) 6.0 mg

56. Recommended Nutrient Intakes (RNIs) of Zinc for 19–50 years female

a) 2 mg  b) 3 mg
c) 5 mg  d) 9 mg

57. Recommended Nutrient Intakes (RNIs) of Zinc for 4–6 years children

a) 1.6 mg  b) 2.5 mg
c) 5.5 mg  d) 2.9 mg

58. Recommended Nutrient Intakes (RNIs) of Selenium for 4–6 years children

a) 11  b) 22 μg
c) 31  d) 51

59. Recommended Nutrient Intakes (RNIs) of Vitamin C for 19–50 years male

a) 45 mg  b) 60 mg
c) 75 mg  d) 40 mg

60. Recommended Nutrient Intakes (RNIs) of Vitamin C for 19–50 years female

a) 30 mg  b) 45 mg
c) 50 mg  d) 70 mg

61. Recommended Nutrient Intakes (RNIs) of Vitamin $B_6$ for 4–6 years children

a) 0.01 mg  b) 0.1 mg
c) 0.6 mg  d) 1.2 mg

62. Recommended Nutrient Intakes (RNIs) of Calcium for 19–50 years male
    a) 1000 mg b) 500 mg
    c) 250 mg d) 125 mg
63. Recommended Nutrient Intakes (RNIs) of Vitamin $B_{12}$ for 4–6 years children
    a) 2.5 µg b) 5.6 µg
    c) 1.9 µg d) 1.2 µg
64. Recommended Nutrient Intakes (RNIs) of Vitamin A for 19–50 years male
    a) 600 µg b) 200 µg
    c) 50 µg d) 80 µg
65. Recommended Nutrient Intakes (RNIs) of Riboflavin for 4–6 years children
    a) 2.0 mg b) 1.2 mg
    c) 0.6 mg d) 1.01 mg
66. Recommended Nutrient Intakes (RNIs) of Thiamine for 4–6 years children
    a) 6.0 mg b) 3.0 mg
    c) 0.6 mg d) 0.3 mg
67. Recommended Nutrient Intakes (RNIs) of Vitamin C for 4–6 years children
    a) 10 mg b) 30 mg
    c) 50 mg d) 90 mg
68. Recommended Nutrient Intakes (RNIs) of Vitamin E for 1–3 years children
    a) 1 mg b) 2 mg
    c) 4 mg d) 5 mg
69. Recommended Nutrient Intakes (RNIs) of Vitamin D for 1–3 years children
    a) 8 µg b) 6 µg
    c) 7 µg d) 5 µg
70. Recommended Nutrient Intakes (RNIs) of Vitamin A for 1–3 years children
    a) 100 µg b) 400 µg
    c) 200 µg d) 500 µg

71. Visible characteristic of Folic acid ($B_9$) is _______ colour.
    a) Yellow
    b) Orange
    c) Yellow-orange
    d) Bluish green
72. Form of zinc used for food fortifications is
    a) Zinc chloride
    b) Zinc oxide
    c) Zinc carbide
    d) Zinc bromide
73. 1 μg Retinol equivalent is equal to _______ IU vitamin A.
    a) 1.11
    b) 2.22
    c) 3.33
    d) 4.44
74. 1 μg calciferol is equal to _____ IU vitamin D.
    a) 10
    b) 20
    c) 30
    d) 40
75. Thiamine is unstable in the presence of
    a) Sulfites
    b) Neutral pH
    c) Alkaline pH
    d) All
76. Characteristics of Folic acid ($B_9$)
    a) Moderately stable to heat
    b) Stable in solution at neutral pH
    c) Unstable in UV light
    d) All
77. Niacin is very stable to
    a) Oxygen
    b) Heat
    c) Light
    d) All
78. Characteristics of Riboflavin
    a) Stable in white bread
    b) Very unstable in light.
    c) Rapid loss from milk on exposure to light
    d) All
79. Which of the following is stable in white bread
    a) Thiamin
    b) Riboflavin
    c) Niacin
    d) b & c
80. Sodium selenate is an ingredient added in some sports drinks with an amount of
    a) 10 μg/l
    b) 20 μg/l
    c) 50 μg/l
    d) 60 μg/l

81. Universal fortification is equivalent to

a) Targeted fortification
b) Market-driven fortification
c) Mass fortification
d) Others

82. Form of iodine for food fortification is potassium iodate. Percentage of the iodine in potassium iodate is

a) 20%
b) 40%
c) 60%
d) 80%

83. Source of the vitamin B12

a) Meat
b) Milk
c) Eggs
d) All

84. The capacity of an intervention such as fortification to achieve the desired impact under ideal circumstances is refers to

a) Effectiveness
b) Value
c) Efficacy
d) Worth

85. Addition of essential nutrients to foods in which nutrients are unavoidably lost during processing (such as milling), storage or handling is called

a) Fortification
b) Enrichment
c) Value addition
d) Restoration

86. Suggested fortified food for people living in higher latitude regions during the winter months is

a) B-complex vitamins enriched wheat flour
b) Iron enriched wheat flour
c) Vitamin D fortified milk
d) a & b

87. Example for voluntary fortification foods

a) Margarine
b) Salt
c) Meat
d) both a & b

88. Source of phytase enzyme

a) Whole rye
b) Whole wheat
c) Whole rice
d) a & b

89. The reason why margarines and oils are the ideal foods for vitamin A fortification is
    - a) Vitamin A is oil-soluble
    - b) Oil protects the vitamin A from oxidation
    - c) Hydrogenated oils can stored for longer periods
    - d) All
90. The dry forms of retinol esters for vitamin A fortification usually are
    - a) Starch coated
    - b) Gum-coated
    - c) Gelatin coated
    - d) All
91. Ferrous sulfate fortified flour may prone to
    - a) Fermentation
    - b) Lump formation
    - c) Rancidity
    - d) All
92. Main purpose of addition of Ascorbic acid to flours is to
    - a) Enhance iron absorption
    - b) Act as a raising agent
    - c) Bleaching agent
    - d) Act as acidulant
93. Keshan disease is caused due to deficiency of
    - a) Boron
    - b) Selenium
    - c) Cobalt
    - d) Iron
94. Worldwide, the three most common forms of Micronutrient malnutrition is/are
    - a) Iodine deficiency
    - b) Vitamin A deficiency
    - c) Iron deficiency
    - d) All
95. Which of the following is involved in antioxidant activity?
    - a) Iron
    - b) Copper
    - c) Iodine
    - d) Selenium
96. Disease caused due to deficiency of riboflavin is
    - a) Ariboflavinosis
    - b) Cheilosis
    - c) Edema
    - d) All
97. Disease caused due to deficiency of biotin vitamin is
    - a) Dermatitis
    - b) Acidosis
    - c) Alopecia
    - d) All

98. Disease caused due to deficiency of pyridoxine vitamin is
    a) Dermatitis  b) Glossitis
    c) Sideroblastic anemia  d) All
99. Megaloblastic anemia is caused due to deficiency of
    a) Cobalamin  b) Folate ($B_9$)
    c) Pyridoxine  d) a & b

10.0 Recommended food for the people with sickle cell anemia disease
    a) Calcium rich foods  b) Vitamin D fortified foods
    c) Vitamin A rich foods  d) a & b

## Answers

| | | | | | | | | | | | | | | | | | | | |
|---|---|---|---|---|---|---|---|---|---|---|---|---|---|---|---|---|---|---|---|
| 1. | c | 2. | d | 3. | d | 4. | d | 5. | a | 6. | d | 7. | a | 8. | c | 9. | a | 10. | d |
| 11. | d | 12. | d | 13. | c | 14. | d | 15. | b | 16. | a | 17. | d | 18. | d | 19. | d | 20. | d |
| 21. | d | 22. | d | 23. | c | 24. | a | 25. | b | 26. | b | 27. | d | 28. | c | 29. | d | 30. | d |
| 31. | d | 32. | c | 33. | d | 34. | d | 35. | c | 36. | a | 37. | b | 38. | c | 39. | d | 40. | a |
| 41. | d | 42. | b | 43. | a | 44. | b | 45. | a | 46. | b | 47. | b | 48. | b | 49. | b | 50. | b |
| 51. | b | 52. | c | 53. | c | 54. | a | 55. | b | 56. | b | 57. | d | 58. | b | 59. | a | 60. | b |
| 61. | c | 62. | a | 63. | d | 64. | a | 65. | c | 66. | c | 67. | b | 68. | d | 69. | d | 70. | b |
| 71. | c | 72. | b | 73. | c | 74. | d | 75. | d | 76. | d | 77. | d | 78. | d | 79. | d | 80. | a |
| 81. | c | 82. | c | 83. | d | 84. | c | 85. | d | 86. | c | 87. | d | 88. | d | 89. | d | 90. | d |
| 91. | c | 92. | b | 93. | b | 94. | d | 95. | d | 96. | d | 97. | d | 98. | d | 99. | d | 100. | d |

1. One of the following is used to test the protein quality
   a) PV  b) TBN
   c) PER  d) Iodine value
2. The deficiency of thiamine results in
   a) Beri -Beri  b) Night blindness
   c) Scurvy  d) Pellagra
3. A procedure that usually replaces some of the kidney's normal functions
   a) Haemolysis  b) Dialysis
   c) Transfusion  d) Transplantation
4. One gram of fat gives
   a) 8 Kcal  b) 4 Kcal
   c) 9 Kcal  d) 6 Kcal
5. The energy value of foods is measured by using
   a) PH meter  b) Bomb calorimeter
   c) Spectrophotometer  d) All
6. Mineral which is structural constituent of Haem is
   a) Calcium  b) Phosphorous
   c) Cobalt  d) Iron
7. Egg protein is referred as
   a) Reference protein  b) Preference protein
   c) Non Amino acid Protein  d) Biological protein
8. Type I diabetes is also known as
   a) Non insulin dependent diabetes
   b) Insulin dependent diabetes
   c) Adult onset diabetes
   d) Both a & c
9. Development of off flavours in fat is called
   a) Hydrogenation  b) Oxidation
   c) Rancidity  d) Purification
10. Nutrient required for growth and maintenance of body
    a) Iron  b) Fats
    c) Carbohydrate  d) Protein

11. The enzyme in TCA cycle are located in
    a) Cytosol b) Mitochondria
    c) Endoplasmic reticulum d) Golgi apparatus
12. Polyphenol oxidase requires the following as metallic activator
    a) Mn b) Cu
    c) Zn d) K
13. In gel filtration technique the following proteins are separated which of the following is eluted first
    a) Haemoglobin (m.wt 64,000) b) Myoglobin (m.wt 16800)
    c) Serum albumin (m.wt 68,500) d) All the above
14. An example of sulphur containing amino acid is
    a) Cysteine b) Methionine
    c) Cystine d) All the above
15. Synthesis of DNA from RNA is catalyzed by a RNA dependent DNA polymerase called
    a) Transcription b) Replication
    c) Trasamination d) Reverse transcription
16. In krebs cycle FADH2 Is formed during convertion of
    a) Succinate to Fumerate b) Fumerate to Malate
    c) Malate to Oxalate d) Isocitrate to Oxalosuccinate
17. Which of the following is a sangers reagent
    a) 1 Fluro 2,4 dinitrobenzene b) Phenyl isothiocynate
    c) 4-Hydroxy proline d) None of these
18. The enzyme used in the food industry to convert starch to glucose is
    a) Renin b) Glucose isomerase
    *c) α-amylase* d) Proteases
19. The nitrogen base not present in DNA structure
    a) Adenine b) Guanine
    c) Cytosine d) Uracil
20. The back bone of nucleic acid structure is constructed by
    a) Peptide bond b) Glycosidic bond
    c) Phosphodiester bidges d) All of them

21. The following is a non protein amino acid
    a) Ornithine  b) Homocysterine
    c) Histamine  d) All of these
22. Biologically active form of vitamin D is
    a) Cholecalciferol  b) Ergocalciferol
    c) Dehydrocholestrol  d) Calcitrol
23. Milk is the richest source of following mineral
    a) Ca  b) Mg
    c) Ni  d) Fe
24. End product of glycolysis is
    a) Pyruvic acid  b) Acetyl coA
    c) CO2 &H2O  d) None
25. In double reciprocal plot of enzyme catalyzed reaction the slope is denoted by
    a) $\frac{1}{v\max}$  b) $\frac{1}{Km}$
    c) $\frac{v\max}{Km}$  d) $\frac{Km}{v\max}$
26. Anabolic pathway are generally
    a) Exergonic  b) Endergonic
    c) Both  d) Isothermal
27. Km is the substrate concentration when the reaction velocity is
    a) Maximum  b) Half of maximum
    c) Minimum  d) Half of minimum
28. The anticodon complementary to the initiation codon is
    a) TAU  b) UAC
    c) UGA  d) None of the above
29. In proteins the secondary structure is maintained by
    a) Covalent bond  b) Hydrogen bonds
    c) Peptide bonds  d) Ionic bond
30. The most predominant and versatile element of life
    a) Nitrogen  b) Oxygen
    c) Carbon  d) Hydrogen

31. The enzyme that hydrolyze α-1,4 linkages is
    a) Amylo α-1,4 glucosidase  b) α-amylase
    c) β-amylase  d) None
32. Which of the following is widely used in confectionary industry
    a) Fructose  b) Sucrose
    c) Glucose  d) Invert sugar
33. Freezing point of milk is
    a) -0.53 to-0.55°C  b) -0.63 to-0.65°C
    c) -0.73 to-0.75°C  d) -0.83 to-0.85°C
34. Gas dispersed in liquid is known as
    a) Emulsion  b) Colloids
    c) Smoke  d) Foam
35. An example of solid sol is
    a) Ruby glass  b) Alloy
    c) Both a&b  d) None of the above
36. Brownian movement is
    a) An optical property  b) Kinetic property
    c) Electrical property  d) None
37. An solid substance dispersed in a liquid is known as
    a) Foam  b) Emulsion
    c) Sol  d) Smoke
38. Lipid oxidation in fatty acids is retarded by the addition of
    a) Acids  b) Bases
    c) Antioxidants  d) Stabilizers
39. Protective action of different colloids is known as
    a) Silver number  b) Gold number
    c) Iodine number  d) Peroxide number
40. Which of the following function as antioxidant
    a) Ascorbic acid  b) Carotenoids
    c) BHA  d) All the above
41. When heated in water starch granules undergo a process called as
    a) Hydrogenation  b) Precipitation
    c) Gelatinization  d) Solubilization

42. An example of Humectant is
    a) Glycerol b) Sorbital
    c) Sucrose d) All the above
43. One of the following contains high sucrose
    a) Carrot b) Onion
    c) Beet root d) Tomato
44. Major phospholipids of egg yolk are
    a) Phosphotidyl choline b) Phosphotidyl ethanol amine
    c) Lecithin d) Both a&c
45. Dispersion medium is also called as
    a) Discontinuous medium b) Continuous medium
    c) Constant phase d) Mobile phase
46. Normal starch contains ________________% of amylase
    a) 20-30% b) 30-40%
    c) 50-60% d) 60-70%
47. In specific density of cow's milk ranges from
    a) 1.029-1.039 b) 1.345-1.346
    c) 0.125-0.213 d) 1.030-1.304
48. Stiff greases is an example of______________emulsion
    a) Oil in water b) Water in oil
    c) Both a&b d) All the above
49. Palm oil and coconut oil are rich in
    a) Myristic acid b) Stearic acid
    c) Lauric acid d) All the above
50. An example of saturated fatty acid is
    a) Oleic acid b) Stearic acid
    c) Palmitic acid d) All the above
51. Which of the following flavor compound is derived from shikimic acid pathway
    a) Eugenol b) Cinnamyl alcohol
    c) Vanilin d) All the above

52. Mineral associated with the structure of myoglobin is
    a) Mg    b) Cu
    c) Iron    d) Selenium
53. The principal allomerization of myoglobin is
    a) 10 Methoxy lactone    b) 10-Hydroxy chlorophyll
    c) Both a&b    d) None
54. Mean particle size in case of lakes is
    a) <10 μm    b) <1μm
    c) <0.1 μm    d) None of the above
55. Mineral that causes black discoloration in canned foods
    a) Fe    b) Cu
    c) S    d) All the above
56. The pungent principal present in chilli is
    a) Cinnamaldehyde    b) Capsanthin
    c) Lentin    d) Allicin
57. Which of the following are phenolic compounds
    a) Hydroxycinnamic acids    b) Flavan -3-ols
    c) Anthocyanidins    d) All the above
58. Betalanines exhibit maximum light absorbance at
    a) 502-507 nm    b) 408-410 nm
    c) 537-538 nm    d) None
59. Pure dye content of commercially available colorant is
    a) 65-75%    b) 75-85%
    c) 85-95%    d) 90-93%
60. Classic salty taste is represented by
    a) NaCl    b) LiCl
    c) KCL    d) $CaCl_2$
61. Magnesium free derivatives of chlorophyll is known as
    a) Phyllins    b) Phoephytins
    c) Chlorophyllides    d) Rhodine
62. Anthocyanins show their greatest tinctorial strength at PH
    a) 1    b) 2
    c) 3    d) 4

63. The compounds that are structurally similar to anthocyanidins
    a) Proanthocyanidins  b) Quinones
    c) Xanthins  d) None
64. The Federal Food ,Drug and cosmetics act was passed on
    a) 1938  b) 1960
    c) 1906  d) All the above
65. ———————————— is the only indigoid type Dye
    a) FD&C Blue No.2  b) FD&C Red No.3
    c) BOTH A&B  d) None
66. Synthetic forms of vitamin-A that are widely used in food fortification
    a) Retinyl palmitate  b) Retinal acetate
    c) Retinal  d) Both a&b
67. An example for non nutritive ,nontoxic mineral
    a) Al  b) B
    c) Zn  d) All the above
68. An example for thickener and stabilizer
    a) CMC  b) Guar and locust bean gums
    c) Xanthans  d) All the above
69. Predominant form of anthocyanin that exists in an aqueous medium of pH 4-5 is
    a) Quinonoid  b) Flavylium
    c) Carbinol base  d) Chromenol
70. EEC number of anthocyanin juice concentrate
    a) E 163  b) E 162
    c) E 153  d) E 140
71. ISO 9001 refers to
    a) Quality assurance model basically used by companies that design, produce, inspect and installation
    b) Quality assurance model basically used by companies that produce, inspect and installation but not design.
    c) Quality assurance model basically used by companies must inspect and test the items
    d) None of the above

72. Sucrose is a
    a) Reducing sugar b) Non reducing sugar
    c) Monosaccharide d) None of these
73. In Phenol-Sulfuric acid method, the total carbohydrate are spectrophotometrically - measured at ________________nm
    a) 490 nm b) 520 nm
    c) 410 nm d) 620 nm
74. ________________is a reducing sugar
    a) Glucose b) Fructose
    c) Galactose d) All of these
75. The gravimetric determinations are entirely based upon the concept of
    a) Weight b) Acid – base titrations
    c) Photometry d) All of these
76. The number of Individual samples to be selected from a lot is
    a) n= C"N b) n= N"C
    c) n= "NC d) n= c/"N
77. What are the major considerations for validation of any analytical method
    a) Controls b) Standard reference materials
    c) Check samples d) All of these
78. What is the mesh number of dried foods particles used for estimation of moisture, protein and mineral
    a) 20 mesh b) 40 mesh
    c) 50 mesh d) 30 mesh
79. Ash content of foods represents
    a) Vitamins content b) Minerals content
    c) Fiber content d) Sugars content
80. Ash content of most fresh foods is
    a) 2% b) 3%
    c) 5% d) 11%
81. Wet ashing is carried out by
    a) Acids b) Oxidizing agents
    c) Both a&b d) None of these

82. What are the ideal characteristics of solvent used for fat extraction
    a) High solvent power for lipids
    b) Low or no solvent power for proteins
    c) Low solvent power for carbohydrates
    d) All of these
83. An example of Non solvent wet extraction method for fat is
    a) Goldfish method b) Soxhlet method
    c) Mojonnier method d) Babcock method
84. What are the major steps in kjeldahl method of protein estimation
    a) Digestion b) Neutralization & Distillation
    c) Titration d) All of these
85. Nitrogen to Protein Conversion Factors for milk is
    a) 6.25 b) 6.38
    c) 5.17 d) 5.33
86. In which method cupric ions are complexed with peptide bonds under alkaline conditions
    a) Kjeldahl Method b) Lowry method
    c) Biuret method d) All of these
87. Dinitrosalicylic acid method is used for
    a) Reducing sugars b) Proteins
    c) Saturate fat d) All of these
88. In Somogyi–Nelson method, the total reducing sugars are spectrophotometrically - measured at ________________nm
    a) 490 nm b) 520 nm
    c) 410 nm d) 620 nm
89. Specific gravity of carbohydrate solution is correlated with
    a) °Brix b) Baumé Modulus
    c) Both a&b d) None of these
90. What is a small volume of a concentrated standard that is added to the sample?
    a) Working standard b) Spike
    c) Both a&b d) None of these

91. Melting point of fat shortening is measured by
    a) Refractive index
    b) Clear point
    c) Flash point
    d) All of these
92. The temperature at which the oil or fat sample begins to smoke
    a) Smoke point
    b) Flash point
    c) Fire point
    d) Melting point
93. _______________ is a measure of the resistance of oil to crystallization.
    a) Cloud point
    b) Cold test
    c) Fire point
    d) Melting point
94. ______________is defined as the mg of KOH necessary to neutralize thefree acids present in 1 g of fat or oil
    a) Iodine value
    b) Acid value
    c) Saponification Value
    d) None of these
95. ______________is defined as the milliequivalents (mEq) of peroxide per kilogram of sample
    a) Iodine value
    b) Acid value
    c) Peroxide value
    d) None of these
96. The resistance of lipids to oxidation is known as
    a) Rancidity
    b) Oxidative stability
    c) Antioxidants
    d) Both b&c
97. Proteins can be selectively precipitated or solubilized by changing
    a) Ionic strength
    b) Buffer pH
    c) Dielectric constant
    d) All of these
98. The transmittance (T) of a solution in spectroscopy is measured by
    a) $T = p/p_0$
    b) $T = P^0/p$
    c) $\% T = p/p_0 \times 100$
    d) $\% T = P^0/p \times 100$
99. The most commonly encountered detectors in spectroscopy are
    a) Phototube
    b) Photomultiplier tube
    c) Photodiode
    d) All

100. In paper and thin-layer chromatography, components of a mixture are characterized by

   a) Distance moved by component / Distance moved by solvent

   b) Distance moved by solvent / Distance moved by component

   c) Ratio of solvent/solution

   d) All

## Answers

| | | | | | | | | | |
|---|---|---|---|---|---|---|---|---|---|
| 1. c | 2. a | 3. b | 4. c | 5. b | 6. d | 7. a | 8. b | 9. c | 10. d |
| 11. b | 12. b | 13. c | 14. d | 15. d | 16. a | 17. a | 18. c | 19. d | 20. d |
| 21. a | 22. d | 23. a | 24. a | 25. d | 26. b | 27. b | 28. c | 29. b | 30. b |
| 31. a | 32. d | 33. a | 34. d | 35. c | 36. a | 37. c | 38. c | 39. b | 40. d |
| 41. c | 42. d | 43. c | 44. d | 45. b | 46. a | 47. a | 48. b | 49. d | 50. d |
| 51. d | 52. c | 53. a | 54. b | 55. b | 56. b | 57. d | 58. c | 59. d | 60. a |
| 61. b | 62. a | 63. a | 64. b | 65. a | 66. c | 67. d | 68. d | 69. c | 70. a |
| 71. a | 72. b | 73. a | 74. d | 75. a | 76. a | 77. d | 78. a | 79. b | 80. c |
| 81. c | 82. d | 83. d | 84. d | 85. b | 86. c | 87. a | 88. b | 89. c | 90. b |
| 91. b | 92. a | 93. b | 94. b | 95. c | 96. b | 97. d | 98. a | 99. d | 100. a |

# 8

# Nutraceuticals and Functional Foods

1. The word 'Nutraceutical' is coined by
   a) Batson b) Waksman
   c) DeFelice d) Vent
2. Nutraceuticals can
   a) Prevent diseases b) Cause diseases
   c) Cure diseases d) a&c
3. Functional foods
   a) Improve health or performance
   b) Reduce the risk of chronic disease
   c) Provide health benefits beyond basic nutrition
   d) All
4. Japan is the second largest market in the world for nutraceutical products after the___
   a) China b) United States
   c) Korea d) South Africa
5. Naturally occurring chemical compounds contained in, or derived from, a plant, animal or marine source, that exert the desired health/wellness benefit are called
   a) Neutral compounds b) Active compounds
   c) Bioactive compounds d) All
6. Standardized and characterized preparations, fractions or extracts of agri-commodities of varying purity are called
   a) Functional ingredients b) Industrial ingredients
   c) Traditional Food Ingredients d) All
7. Conventional foods that are manufactured by the traditional food processing industry called
   a) Traditional Processed Foods b) Traditional Whole Foods
   c) Designer Foods d) Novel Foods

8. Foods that have been modified by genetic manipulation come under
   a) Traditional Processed Foods  b) Traditional Whole Foods
   c) Designer Foods  d) Novel Foods
9. Foods that naturally contain or enriched with cancer-preventing substances such as phyto-chemicals are called
   a) Traditional Processed Foods  b) Traditional Whole Foods
   c) Designer Foods  d) Novel Foods
10. Conventional foods that have been grown by agricultural producers for generations are
    a) Traditional Processed Foods  b) Traditional Whole Foods
    c) Designer Foods  d) Novel Foods
11. Fruit juice fortified with vitamin C is a
    a) Fortified product  b) Enriched product
    c) Altered product  d) Enhanced commodity
12. Eggs with increased omega-3 content achieved by altered chicken feed is a
    a) Fortified product  b) Enriched product
    c) Altered product  d) Enhanced commodity
13. Meat or ice cream with fibers as fat releasers is a
    a) Fortifiedproduct  b) Enriched product
    c) Altered product  d) Enhanced commodity
14. Margarine with plant sterol ester, probiotics, prebiotics is a
    a) Fortified product  b) Enriched product
    c) Altered product  d) Enhanced commodity
15. A food in which one of the components has been naturally enhanced through special growing conditions, new feed composition, genetic manipulation is called
    a) Fortified product  b) Enriched product
    c) Altered product  d) Enhanced commodity
16. A food from which a deleterious component has been removed, reduced or replaced with another substance with beneficial is called
    a) Fortified product  b) Enriched product
    c) Altered product  d) Enhanced commodity

17. A food with added new nutrients or components not normally found in a particular food is a

a) Fortified product　　b) Enriched product

c) Altered product　　d) Enhanced commodity

18. Non-digestible food ingredients that beneficially affect the host by stimulating the growth and/or activity of one or a limited number of bacteria in the colon is called

a) Probiotics　　b) Prebiotics

c) Symbiotics　　d) Synbiotics

19. Live micro-organisms which, when administered in adequate amounts, confer a health benefit on the host is called

a) Probiotics　　b) Prebiotics

c) Synbiotics　　d) All

20. Food ingredients or dietary supplements combining probiotics and prebiotics in a form of synergism are called

a) Probiotics　　b) Symbiotics

c) Synbiotics　　d) All

21. Prebiotics enhance the growth and survival of the probiotic cultures by influencing the growth and metabolites of

a) Probiotic　　b) Starter

c) Human beings　　d) Both a&b

22. Which of the following is a prebiotic component

a) Inulin　　b) Isomalto-oligosaccharides

c) Polydextrose　　d) All

23. Which of the following is a prebiotic component

a) Lactose　　b) Lactulose

c) Lactic acid　　d) Galactose

24. Which of the following is not a prebiotic component

a) Inulin　　b) Starch

c) Resistant starch　　d) All

25. Which of the following can be considered as a obesity controlling agents

a) Monosaccharides　　b) Oligosaccharides

c) Polysaccharides　　d) All

26. Following is a prebiotic component
    a) Starch b) Cellulose
    c) Chitin d) Fructo-oligosaccharide
27. An agent that increases the effectiveness of another agent when combined with it is called
    a) Synergists b) Symbiotics
    c) Sour foods d) Sweet foods
28. Which of the following is a functional food?
    a) Food for specified health use b) Food with omega-3 and soy
    c) Food with calcium and inulin d) All
29. The functional drinks those with intention of eye health are prepared with
    a) Omega-3 b) Lutein
    c) Inulin d) Calcium
30. The functional drinks those with intention of bone health are prepared with combination of
    a) Calcium and Inulin b) Inulin and soy
    c) Lutein and soy d) Calcium and omega-3
31. The functional drinks those of cholesterol lowering drinks are prepared with combination of
    a) Omega-3 and soy b) Lutein and soy
    c) Calcium and soy d) Inulin and soy
32. Fiber present in cereals is
    a) Water soluble b) Oil soluble
    c) Acetone soluble d) Hexane soluble
33. Which of the following can be suggested to fulfill the prebiotic concept?
    a) Beta-glucan b) Arabinoxylan
    c) Oligosaccharides d) All
34. Material used for encapsulation of probiotics ?
    a) Starch b) Lipid
    c) Protein d) Vitamins
35. Which of the following is a prebiotic?
    a) Galacto-oligosaccharides b) Fructo oligosaccharides
    c) Resistant starch d) All

36. Plant components that have health-promoting properties are called
    - a) Phytols
    - b) Phyto-chemicals
    - c) Phyto-nutrients
    - d) Both b&c
37. Special dietary food intended for use solely under medical supervision to meet nutritional requirements in specific medical conditions are called
    - a) Phyto-chemicals
    - b) Medical Foods
    - c) Phyto-nutrients
    - d) GM Foods
38. Chemicals in their natural from isolated from medicinal plants withpotential benefits in human health and disease prevention are called
    - a) Phyto-chemicals
    - b) Medical Foods
    - c) Phyto-nutrients
    - d) Phyto-pharmaceuticals
39. A product that is intended to supplement the diet that bears or contains on or more dietary ingredients called as
    - a) Diet formula
    - b) Dietary foods
    - c) Dietary Supplements
    - d) Dietary mixes
40. Anti-malarial agent present in _____ plant.
    - a) Egg plant
    - b) Jasmine
    - c) Eucalyptus
    - d) Cinchona
41. The plant's compound can be used for treatment of heart ailments is
    - a) Datura
    - b) Deadly night shade
    - c) Hibiscus
    - d) a&b
42. The plant's compound can be as pain killer
    - a) Ginger
    - b) Opium poppy
    - c) Fenugreek
    - d) Neem
43. Anti-malarial agent present in *Cinchona* plant is
    - a) Coumarin
    - b) Morphine
    - c) Quinine
    - d) Scopoletin
44. Which of the following compound can be used of heart ailments
    - a) Quinine
    - b) Tropane alkaloid
    - c) Morphine
    - d) Ergot alkaloid
45. Which of the following compound acts as Abortifacient?
    - a) Quinine
    - b) Tropane alkaloid
    - c) Morphine
    - d) Ergot alkaloid

46. Which of the following compound has hypoglycemic effect?
    a) Coumarin b) Scopoletin
    c) Fenugreekine d) All
47. Antidepressant, pain killer agent present in Opium poppy is?
    a) Scopoletin b) Morphine
    c) Coumarin d) trigonelline
48. Hypoglycemic agent present in Fenugreek?
    a) Fenugreekine b) Scopoletin
    c) Coumarin d) All
49. α-carotene is an
    a) Antioxidant b) Anti carcinogenic agent
    c) Blood pressure reducing agent d) a&b
50. Trigonelline is a hypoglycemic agent present in
    a) Tomatoes b) Fenugreek
    c) Potato d) Cucumber
51. Carotenoid present in carrots
    a) α-carotene b) β-carotene
    c) γ-cryptoxanthin d) Lutein
52. Antioxidant and anticancer agent present in Oranges & tangerines
    a) α-carotene b) β-carotene
    c) γ-cryptoxanthin d) Lutein
53. Zeaxanthin is present in
    a) Corn b) Avocado
    c) Oranges d) a&b
54. Lycopene is present in
    a) Tomatoes b) Pink grapefruit
    c) Watermelon d) All
55. Lutein is present in
    a) Kale b) Spinach
    c) Watercress d) All
56. The pigment lycopene is present in
    a) Guava b) Tomatoes
    c) Papaya d) All

57. The pigment present in parsley is
   a) α-carotene b) β-carotene
   c) γ-cryptoxanthin d) Lutein
58. Which of the following is a correct match of carotenoid and its medicinal property
   a) β-carotene - Antioxidant, anticarcinogenic
   b) α-carotene - Antioxidant
   c) γ-cryptoxanthin - Antioxidant, anticancer
   d) All
59. Which of the following is a correct match of non-carotenoid terpenoid and its medicinal property
   a) Perillyl alcohol - Anticancer
   b) Saponins - Reduces cholesterol levels in blood
   c) Terpenol -Anticancer
   d) All
60. Perillyl alcohol is a non-carotenoid terpenoid, present in
   a) Cherries b) Mints
   c) Peas d) a&b
61. Saponins is a non-carotenoid terpenoid, widely present in
   a) Fruits b) Vegetables
   c) Oil seeds d) Pulses
62. Terpenol is a non-carotenoid terpenoid, present in
   a) Carrots b) Tomatoes
   c) Corn d) Avocado
63. Terpene limonoids is a non-carotenoid terpenoid, present in
   a) Peels of citrus fruits b) Membranes of citrus fruits
   c) Legumes d) a&b
64. Which of the following is a flavonoid polyphenolic compound?
   a) Anthocyanins b) Isothiocyanates
   c) Phenethyl isothiocyanate d) Indoles
65. Which of the following is not a flavonoid polyphenolic compound?
   a) Flavonone b) Catechin
   c) Betacyanin d) Lignan

66. Which of the following is a non-flavonoid polyphenolic compound?
   a) Curcumin  b) Resveratrol
   c) Lignan  d) All
67. Which of the following is a glucosinolate compound?
   a) Indoles  b) Sulforaphene
   c) Thiosulfonates  d) All
68. Which of the following is not a glucosinolate compound?
   a) Isothiocyanates  b) Phytosterol
   c) Phenethyl isothiocyanate  d) Menthol
69. Terpenes include
   a) Menthol  b) Borneol
   c) Santonin  d) All
70. Menthol is a
   a) Monoterpene  b) Sesquiterpene
   c) Diterpene  d) Triterpene
71. Correct match about terpenes
   a) Borneol-Monoterpene  b) Santonin-Sesquiterpene
   c) Gossypol-Sesquiterpene  d) All
72. Terpenes are derived biosynthetically from units of
   a) Isocyanide  b) Isoprene
   c) Isoamyl  d) Isobutyl
73. Molecular formula of isoprene
   a) $C_5H_8$  b) $C_6H_8$
   c) $C_7H_8$  d) $C_8H_8$
74. Number of Isoprene units present in hemiterpene
   a) 1  b) 2
   c) 3  d) 4
75. Number of Isoprene units present in monoterpene
   a) 1  b) 2
   c) 3  d) 4
76. Number of Isoprene units present in sesquiterpene
   a) 1  b) 2
   c) 3  d) 4

77. Number of Isoprene units present in diterpene
    a) 1 b) 2
    c) 3 d) 4
78. Number of Isoprene units present in triterpene
    a) 5 b) 6
    c) 7 d) 8
79. Number of Isoprene units present in tetraterpene
    a) 5 b) 6
    c) 7 d) 8
80. Flavor compound present in hops
    a) Myrcene b) β-pinene
    c) β–caryophyllene d) All
81. Isoflavone is present in
    a) Soybean b) Citrus fruits
    c) Grapefruit d) Asparagus
82. Rutin is a nutraceutical that protects us against
    a) Cancer b) Cardio vascular ailments
    c) Cataracts d) Diabetics
83. Rutin is present in
    a) Asparagus b) Buckwheat
    c) Citrus fruits d) All
84. Hesperetin is present in
    a) Asparagus b) Buckwheat
    c) Citrus fruits d) Soybean
85. Food source of Quercitin is
    a) Red onions b) Red grapes
    c) Buckwheat d) All
86. Quercitin is present in
    a) Green tea b) Apple skins
    c) Custard apple leaves d) All
87. Silymarin is present in
    a) Artichoke b) Milk thistle
    c) Cranberries d) a&b

88. Which of the following phytochemical is present in tea
    a) Tannin b) Catechin
    c) Quercitin d) All
89. Ellagic acid is present in
    a) Strawberries b) Raspberries
    c) Blueberries d) a&b
90. Medicinal property of Ellagic acid is
    a) Hypoglycemic b) Antineoplastic
    c) Anti-malarial d) Anticancer
91. Resveratrol is present in
    a) Grapes b) Curcuma
    c) Tomatoes d) Pine apple
92. Isothiocyanates are present in
    a) Radish b) Horseradish
    c) Mustard d) All
93. Thiosulfonates are present in
    a) Garlic b) Onion
    c) Broccoli d) a&b
94. Barbaloin is a phytochemical present in
    a) Barbados cherry b) Blue berry
    c) Aloe d) Pine oil
95. The test used measure the antioxidant activity of a foods and beverages is
    a) ORAC test b) TAC Assay
    c) DPPH d) All
96. Expand ORAC
    a) Oxygen range absorbance capacity
    b) Oxygen radical absorbance control
    c) Oxygen reduced absorbance capacity
    d) Oxygen radical absorbance capacity

97. DPPH
    a) 1,1-Diphenyl-2-picryl-hydrazyl
    b) 1,2-Diphenyl-2-picryl-hydrazyl
    c) 1,1-Diphenyl-3-picryl-hydrazyl
    d) 1,1-Diphenyl-4-picryl-hydrazyl
98. TAC
    a) Total activity capacity
    b) Total antioxidant capacity
    c) Telometric activity control
    d) Total antioxidant control
99. TRAP
    a) Tin radical absorbance protocal
    b) Telometric radicle amplification protocol
    c) Telometric report amplification protocol
    d) Telometric repeat amplification protocol
100. ABTS
    a) 2,2'-azino-bis (3-methylbenzoline-6-sulphonic acid)
    b) 2,2'-azino-bis (3-ethylbenzoline-6-sulphonic acid)
    c) 2,2'-azino-bis (3-butylbenzoline-6-sulphonic acid)
    d) 2,2'-azino-bis (3-propylbenzoline-6-sulphonic acid)

## Answers

| | | | | | | | | | |
|---|---|---|---|---|---|---|---|---|---|
| 1. c | 2. d | 3. d | 4. b | 5. c | 6. d | 7. a | 8. d | 9. c | 10. b |
| 11. a | 12. d | 13. c | 14. b | 15. d | 16. c | 17. b | 18. d | 19. a | 20. c |
| 21. d | 22. d | 23. b | 24. b | 25. b | 26. d | 27. a | 28. d | 29. b | 30. a |
| 31. a | 32. a | 33. d | 34. a | 35. d | 36. d | 37. b | 38. d | 39. c | 40. d |
| 41. d | 42. b | 43. c | 44. b | 45. d | 46. d | 47. b | 48. d | 49. d | 50. b |
| 51. a | 52. c | 53. d | 54. d | 55. d | 56. d | 57. d | 58. d | 59. d | 60. d |
| 61. d | 62. a | 63. d | 64. a | 65. d | 66. d | 67. d | 68. d | 69. d | 70. a |
| 71. d | 72. b | 73. a | 74. a | 75. b | 76. c | 77. d | 78. b | 79. d | 80. d |
| 81. a | 82. b | 83. d | 84. c | 85. d | 86. d | 87. d | 88. d | 89. d | 90. d |
| 91. a | 92. d | 93. d | 94. c | 95. d | 96. d | 97. a | 98. b | 99. d | 100. b |

1. A species that possesses one or several unpaired electrons in the outer shell, and is thus very reactive is called
   a) Free radical b) Proton
   c) Cyanide d) Bioactive compound
2. A substance that offers protection against the harmful effect of free radicals and other reactive oxygen species is
   a) Radicle b) Antioxidant
   c) Proton d) Cyanide
3. A plant-derived substance that is nutritionally, medicinally or physiologically highly active is called
   a) Phytochemical b) Bimolecular
   c) Photochemical d) Physical
4. Free radicals are
   a) Highly reactive and unstable
   b) Generated by normal metabolism
   c) Contain unpaired valence electrons
   d) All
5. Which of the following is a reactive oxygen species
   a) Superoxide anion b) Hydroxyl radical
   c) Singlet oxygen d) All
6. Which of the following is a free radical
   a) Hydrogen peroxide b) Nitric oxide
   c) Peroxynitrite d) All
7. An overproduction of radicals may be triggered by
   a) Radiation b) Pollutants
   c) Cigarette smoke d) All
8. Free radical can be formed by
   a) Alcohol b) Immune system
   c) Stress d) All
9. Lipids may be especially prone to damage due to a process known as lipid__________.
   a) Peroxidation b) Rancidity
   c) Flavor reversion d) All

10. Free radicals are react with cellular components such as
    a) Lipids b) Proteins
    c) Nucleic acids d) All
11. Free radicals are involved in
    a) Ageing process b) Heart diseases
    c) Rheumatism d) All
12. Intake of _______it is believed to decrease the risk of diseases like cancer, cardiovascular disease and Alzheimer's disease
    a) Carbohydrates b) Proteins
    c) Lipids d) Antioxidants
13. Colour of carotenoids
    a) Yellow b) Orange
    c) Orange red d) All
14. Carotenoids are present in
    a) Fruits b) Flowers
    c) Green vegetables d) All
15. Carotenoids are
    a) Monoterpenoids b) Diterpenoids
    c) Sesquiterpenoids d) Tetraterpenoids
16. Which of the following carotene is in red of red-orange colour
    a) α- carotene b) β-carotene
    c) Lycopene d) All
17. Which of the following carotene is in yellow colour?
    a) Lutein b) Violaxanthin
    c) Zeaxanthin d) All
18. In spinach and other vegetables, bioavailability of lutein is ____times higher than that of β-carotene.
    a) 1-2 b) 5-9
    c) 9-10 d) 10-12
19. Which of the following is a flavonoid?
    a) Flavones b) Anthocyanins
    c) Catechins d) All

20. Which of the following is not a flavonoid?
    a) Flavans b) Proanthocyanidins
    c) Isoflavones d) Resin
21. Flavonoids occur mainly as
    a) Glucosides b) Ketones
    c) Aldehydes d) Amines
22. The antioxidative activity of Flavonoids depends on
    a) Chemical structure b) Number of hydroxyl groups
    c) Number of ketol groups d) a & b
23. Flavonoid present in vegetables
    a) Luteolin b) Apigenin
    c) Flavonols d) All
24. Flavonoid present in vegetables
    a) Quercetin b) Kaempferol
    c) Myricetin d) All
25. Good dietary source of flavonoids
    a) Kale b) Onions
    c) Broccoli d) All
26. Percentage of aqueous acetone used for the extraction, isolation and characterization and identification of of phytochemicals or bioactive compounds is
    a) 50% b) 60%
    c) 70% d) 80%
27. Tomato is rich in
    a) Flavonoids b) Flavones
    c) Cinnamic acid derivatives d) All
28. The compound present in tomoto is
    a) Lycopene b) β-carotene
    c) Oxalic acid d) All
29. The main polyphenols found in eggplant
    a) Chlorogenic acid b) Caffeic acid
    c) Pcoumaric acid d) All

30. Eggplant is deficient in
    a) Provitamin A
    b) Vitamin E
    c) Dietary fiber
    d) Both a & b
31. Chilli pepper contains
    a) Luteolin
    b) Myricetin
    c) Capsaicinoids
    d) All
32. The antioxidative capability of the compounds is generally evaluated by
    a) CAP-e assay
    b) CUPRAC
    c) FRAP
    d) All
33. Expand CAP-e assay
    a) Cellulose-based Antioxidant Protection on Erythrocytes
    b) Callus-based Antioxidant Protection in Erythrocytes
    c) Cell-based Antioxidant Protection in Erythrocytes
    d) Cell-based Antioxidant Propagation in Erythrocytes
34. Expand CUPRAC
    a) Cupric reducing antioxidant capacity
    b) Cuprous reducing antioxidant capacity
    c) Cuprous reducing antioxidant captivity
    d) Cuprous reducing antioxidant capacity
35. Abbreviate FRAP
    a) Ferrous reducing/antioxidant power assay
    b) Ferric reducing/antioxidant power assay
    c) Ferric reducing antioxidant protection
    d) Ferrous reducing antioxidant protection
36. The antioxidant activity of phenolic compounds is done by
    a) Scavenging free radicals
    b) Donating hydrogen atoms or electron
    c) Chelating metal cations
    d) All
37. The structure of phenolic compounds is a key determinant of their radical scavenging and metal chelating activity and this is referred to as
    a) SAR
    b) PAR
    c) CAR
    d) DAR

38. Expand the word 'SAR' related to radical scavenging mechanism of phenolic compounds
    a) Structural autolysis rate
    b) Structure autolysis relationships
    c) Structure activity relationships
    d) Structure activity rate
39. The highest quantity of fructans is found in
    a) Garlic b) Artichoke
    c) Shallots d) All
40. Fructans percentage present in fresh vegetables is
    a) 1.0 to 1.4 b) 1.2 to 11.4
    c) 1.2 to 15.4 d) 1.2 to 17.4
41. Isoflavones are plant derived phytoestrogens and belong to a class of compounds known as
    a) Flavonoids b) Flavours
    c) Aroma compounds d) None
42. Isoflavones are predominantly found in and
    a) Soybeans b) Fermented soy foods
    c) Non-fermented soy foods d) a&b
43. Isoflavones are biologically inactive glucoside conjugates comprising ____% of the total isoflavonen concentration.
    a) 60 to 75% b) 70 to 85%
    c) 80 to 95% d) 90 to 95%
44. Isoflavones are present mostly as $\beta$-glucoside conjugates which includes
    a) Daidzin b) Genistin
    c) Glycitin d) All
45. Isoflavones are produced almost exclusively by the members of
    a) Glycitin derivatives b) Malonyl derivatives
    c) Acetyl derivatives d) All
46. The beany flavor of soybean is primarily the result of
    a) Polyphenoloxidase b) Lipoxygenase
    c) Protease d) Diastase

47. The concept that promotes the consumption of all the components of the grains, in the same proportions as in the native grain is

a) Complete food concept b) Whole grain concept

c) Wholesome food concept d) None

48. Antioxidant property of wheat is due to __________that present in arabinoxylans in the aleurone layer of wheat.

a) Formic acid b) Ferulic acid

c) Ketones d) Aldehydes

49. Part of wheat that contains tocopherol is

a) Epidermis b) Endosperm

c) Aleurone layer d) Germ

50. Part of wheat that is used as a valuable ingredient and a good way of fortifying food products

a) Testa b) Endosperm

c) Aleurone layer d) Germ

51. Body heat is produced by burning of ____in the mitochondria (energy factories) in your cells.

a) Carbohydrates b) Proteins

c) Fats d) Vitanins

52. Dietary ingredients suggested to promote optimal metabolism and thermogenesis

a) Green Tea (EGCG) b) Alpha-Lipoic Acid

c) D-Ribose d) All

53. Dietary ingredients suggested to promote optimal metabolism and thermogenesis

a) Caffeine b) Chromium

c) B Vitamin Complex d) All

54. Dietary ingredients suggested to promote optimal metabolism and thermogenesis

a) Acetyl-L-Carnitine b) Cinnamon

c) Pipcrinc d) All

55. Dietary ingredient is not suggested to promote optimal metabolism and thermogenesis

a) Capsaicin b) Coenzyme Q10

c) Piperine d) Menthol

56. Isoflavones are produced almost exclusively by the members of

a) Fabaceae b) Malvaceae

c) Solanaceae d) Cruciferae

57. Epigallocatechin gallate (EGCG) is present in

a) Tea b) Ginger

c) Garlic d) Pepper

58. Capsaicin is the primary pungent principle in.

a) Black pepper b) Red hot peppers

c) Cinnamon d) Garlic

59. Capsaicin can directly influence the activity of_____channel proteins in the cell membrane and thereby bring about changes in cellular metabolism.

a) Calcium b) Potassium

c) Sodium d) Manganese

60. Top functional foods are

a) Fruits & vegetables b) Fish/Fish Oils

c) Dairy products d) All

61. The type of carotene that helps in prevention of the oxidation of LDL cholesterol

a) Beta- carotene b) Delta- carotene

c) Zeta- carotene d) Alpha- carotene

62. Foods with a highest ORAC value is

a) Kale b) Onion

c) Egg plant d) Broccoli flowers

63. Which of the following food has high ORAC value compared to others?

a) Beets b) Brussels sprout

c) Spinach d) Red bell Pepper

64. Correct statement about carotene

a) Carotene can be used in cakes

b) Carotene has EU number as E160

c) Carotene also present in cow milk

d) All

65. CAP-e assay is used for evaluation of potential of

a) Free radicals b) Minerals

c) Antioxidant d) Oils

66. Which of the following is acts as Omega-3 fatty acid?

a) Oleic acid b) Linolenic acid

c) Alphalinolenic acid d) None

67. Nature's richest source of alphalinolenic acid (which is an Omega-3 fatty acid) is

a) Flaxseed b) Cranberry seed

c) Apple d) a&b

68. Alpha-linolenic acid is

a) L-amino acid b) D- amino acid

c) Essential fatty acid d) Nonessential fatty acid

69. Coenzyme Q10 (Co-Q10) is

a) Powerful antioxidant

b) Essential for body cells, tissues, and organs

c) Needed for energy production

d) All

70. Catechin is a

a) Polyphenol b) Enzyme

c) Protein d) Phospholipid

71. Amino acid present in L-Glutathione protein

a) Glutamic acid b) Cysteine

c) Glycine d) All

72. Function of L-Glutathione is

a) Neutralization of free radicals

b) Regulation of Vitamins C&E

c) Cellular metabolism

d) All

73. Which of the following is a polyphenol?

a) Ferulic acid b) Catechin

c) Resveratrol d) All

74. Expand GABA
    a) Guanidine Alpha Butryic Acid
    b) Glycine amma Amino Butryic Acid
    c) Gamma Amino Butryic Acid
    d) Gamma Alcohol Butryic Acid
75. GABA is an amino acid that is found in
    a) Muscles b) Nervous system
    c) Lymph d) All
76. The long-chain Omega-3 Fatty Acid from fish
    a) Eicosapentaenoic acid (EPA)
    b) Docosahexaenoic acid (DHA)
    c) Docosapentaenoic acid (DPA)
    d) a&b
77. β-glucan is a
    a) Homopolysaccharide b) Heteropolysaccharide
    c) Disaccharide d) Nonosaccharide
78. Bond present in β-glucan's monomers attached in linear
    a) (1-4)-b-D-glucosyl b) (1-4)-b-L-glucosyl
    c) (1-6)-b-D-glucosyl d) None
79. Bond present in β-glucan's monomers branching point
    a) (1-2)-linkage b) (1-3)-linkages
    c) (1-4)-linkages d) (1-5)-linkages
80. β-glucan is present in
    a) Oats b) Barley
    c) Wheat d) All
81. β-carotene amount present in the golden rice
    a) 1.6 mg/g b) 2.6 mg/g
    c) 3.6 mg/g d) 4.6 mg/g
82. Which of the following is an antioxidant?
    a) Carotenoids b) Tocols
    c) γ-oryzanols d) All

83. β-glucan content present in barley

a) 5–11% b) 3–7%

c) 1-3% d) 1%

84. β-glucan content present in oats

a) 1% b) 1-3%

c) 3–7% d) 5–11%

85. β-glucan content present in wheat grains

a) 5–11% b) 3–7%

c) 1-3% d) 1%

86. Most common anthocyanidin found in plants is

a) Cyanidin b) Delphinidin

c) Peonidin d) Pelargonidin

87. Secoisolariciresinol diglucoside (SDG) is a

a) Chelating agent b) Antioxidant

c) Flavour d) Humectants

88. Richest source of Secoisolariciresinol diglucoside (SDG)

a) Soy bean b) Flaxseed

c) Safflower d) Sunflower

89. Amount of Secoisolariciresinol diglucoside (SDG) in flaxseed

a) 7–14 mg/g b) 10–18 mg/g

c) 7–24 mg/g d) 17–34 mg/g

90. Precursor of serotonin is

a) Glycine b) Tryptophan

c) Alanine d) Lysine

91. Whole-grain maize and whole-grain rye rich sources of

a) Vitamin C b) Vitamin E

c) Vitamin D d) Vitamin K

92. Whole-grain wheat, barley, oat, millet, sorghum and buckwheat contain

a) Less vitamin E b) More vitamin E

c) No vitamin E d) More vitamin C

93. Folates (vitamin B9) reduce the concentration of plasma homocysteine.

a) Glucose b) Cysteine

c) Homocysteine d) Sterols

94. Folates content in cereals are a moderate source of as they contain from ___ μg folates/100 g edible portion, depending on the type of cereal.

a) 16 - 143 μg/100 g b) 61 - 140 μg/100 g

c) 76 - 143 μg/100 g d) 96 - 143 μg/100 g

95. Function of glutathione peroxidase and thioredoxine reductase are is depends on

a) Mn b) Zn

c) Se d) Cu

96. Function of superoxide dismutase is depends on

a) Mn b) Zn

c) Cu d) All

97. Function of catalase is depends on

a) Zn b) Fe

c) Cr d) Ni

98. γ-Oryzanol is a mixture of

a) Phytosteryl ferulates b) Amino acids

c) Fatty acids d) All

99. γ-Oryzanol is extracted from

a) Rice bran b) Rice endosperm

c) Rice germ d) Broken rice

100. γ-Oryzanol is a powerful inhibitor of iron driven ________ radical formation.

a) Hydroxyl b) Oxide

c) Keto d) Aldehyde

## Answers

| | | | | | | | | | |
|---|---|---|---|---|---|---|---|---|---|
| 1. a | 2. b | 3. a | 4. d | 5. d | 6. d | 7. d | 8. d | 9. d | 10. d |
| 11. d | 12. d | 13. d | 14. d | 15. d | 16. d | 17. d | 18. b | 19. d | 20. d |
| 21. a | 22. d | 23. d | 24. d | 25. d | 26. c | 27. d | 28. d | 29. d | 30. d |
| 31. d | 32. d | 33. c | 34. a | 35. b | 36. d | 37. a | 38. c | 39. d | 40. d |
| 41. a | 42. d | 43. c | 44. d | 45. d | 46. b | 47. b | 48. b | 49. d | 50. d |
| 51. c | 52. d | 53. d | 54. d | 55. d | 56. a | 57. a | 58. b | 59. a | 60. d |
| 61. a | 62. a | 63. b | 64. d | 65. c | 66. c | 67. d | 68. c | 69. d | 70. a |
| 71. d | 72. d | 73. d | 74. c | 75. b | 76. d | 77. a | 78. a | 79. b | 80. d |
| 81. a | 82. d | 83. a | 84. c | 85. d | 86. a | 87. b | 88. b | 89. c | 90. b |
| 91. b | 92. a | 93. c | 94. a | 95. c | 96. d | 97. b | 98. a | 99. a | 100. a |

1. Free radicals are formed
   a) When oxygen is reduced in mitochondria
   b) When ATP produced
   c) By a variety of enzymatic reactions
   d) All
2. Whole-grain cereals have potentially antioxidant properties due to presence of
   a) Fisher ring b) Haworth ring
   c) Glycoside bond d) Aromatic phenolic ring
3. Alkylresorcinols are phenolic
   a) Lipids b) Carbohydrates
   c) Proteins d) All
4. At what carbon number long hydrocarbon chain is positioned in phenolic ring of alkylresorcinols?
   a) First b) Third
   c) Fifth d) Sixth
5. γ-Oryzanol is a powerful inhibitor of _____driven hydroxyl radical formation.
   a) Copper b) Iron
   c) Zinc d) Aluminum
6. Choline content in 100 g of toasted wheat germ
   a) 152 mg b) 252 mg
   c) 352 mg d) 452 mg
7. Betaine content in 100 g of toasted wheat germ
   a) 1420 mg b) 1024 mg
   c) 1240 mg d) 240 mg
8. Betaine percentage in wheat bran
   a) 1.0 % b) 1.5 %
   c) 2.5 % d) 3.0%
9. Which of the following is an essential amino acid?
   a) Methionine b) Valine
   c) Tryptophan d) All

10. Which of the following is a non-essential amino acid
    a) Cysteine b) Alanine
    c) Serine d) All
11. Cysteine is synthesized from
    a) Glycine b) Methionine
    c) Alanine d) Valine
12. Sulphur containing amino acid
    a) Methionine b) Cysteine
    c) Taurine d) All
13. Methionine content in wheat is
    a) 0.1-0.5 g /100 g protein b) 0.5-0.9 g /100 g protein
    c) 1.2-1.3 g /100 g protein d) 2.2-2.3 g /100 g protein
14. Cysteine content in wheat is
    a) 0.1-0.5 g /100 g protein b) 0.5-0.9 g /100 g protein
    c) 1.3-1.5 g /100 g protein d) 2.3-2.5 g /100 g protein
15. Major phosphorus storage compound in most seeds and cereal grains is
    a) Cellulose b) Phytic acid
    c) Lignin d) Amylose
16. Phytic acid is inositol hexaphosphate
    a) Inositol monophosphate b) Inositol diphosphate
    c) Inositol pentaphosphate d) Inositol hexaphosphate
17. Phytic acid content in most seeds and cereal grains is about ___% of their dry weight.
    a) 0.5-2 % b) 1-5%
    c) 1-7% d) 2-8%
18. In wholegrain cereals the phytic acid is mainly located in
    a) Testa b) Aleurone layer
    c) Endosperm d) Germ
19. Phytic acid is considered as an antinutrient because
    a) Forms colour compounds
    b) Strongly chelates minerals
    c) Strongly chelates trace elements
    d) b&c

20. Which of the following is nutritionally inert within the digestive tract and has potential antioxidant capacities?
    a) Cellulose b) Hemicellulose
    c) Lignin d) Carotene
21. Potential antioxidant capacities of lignin is due to presence of
    a) Trace elements b) Hydroxyl group
    c) Polyphenolic structure d) None
22. Lignins are a major component of whole-grain cereals, and may account for _____% of the bran fraction.
    a) 2-4 % b) 2-6 %
    c) 3-7% d) 4-8%
23. Wheat germ contains about _____mg carotenoids/kg dry weight.
    a) 0.5 b) 2.5
    c) 4.5 d) 5.5
24. Lutein and zeaxanthin are most abundant in
    a) Carrot b) Maize
    c) Sorghum d) Raddish
25. Trolox Equivalent is an unit to measure
    a) Quantity of resins b) Activity of flavonoids
    c) Quality of flavonoids d) Activity of antioxidants
26. Fraction was 8500 μmol TE/100 g and that of the fraction was 5000 μmol TE/100 g while that of was estimated to reach The antioxidant capacity of rice bran
    a) 22,300 μmol TE/100 g b) 23,300 μmol TE/100 g
    c) 24,300 μmol TE/100 g d) 25,300 μmol TE/100 g
27. The antioxidant capacities of the wheat bran and germ are ____ & ____ respectively.
    a) 8500 μmol TE/100 g & 6500 μmol TE/100 g
    b) 8500 μmol TE/100 g & 5000 μmol TE/100 g
    c) 5000 μmol TE/100 g & 7500 μmol TE/100 g
    d) 8500 μmol TE/100 g & 2000 μmol TE/100 g
28. Most abundant free phenolic acid present in wheat
    a) Ferulic acids b) Vanillic acids
    c) P-coumaric acids d) All

29. Major compound involved in the antioxidant activity of red rice is
    a) Ferulic acid b) P-coumaric acid
    c) Vanillic acid d) Procyanidin
30. Major antioxidant compounds in black rice
    a) Cyanidin-3-glucoside b) Peonidin 3- glucoside
    c) Ferulic acid d) a&b
31. Avenanthramides are typically found in oats and have antioxidant activity in
    a) *In vitro* b) *In vivo*
    c) Both *in vitro* and *in vivo* d) None
32. Avenanthramides are typically found in oats and have antioxidant activity in
    a) *In vivo* only b) *In vitro* only
    c) Both *in vivo & in vitro* d) None of the above
33. Which of the following has more powerful antioxidant property
    a) Avenanthramid b) As ferulic acid
    c) P-hydroxybenzoic acid d) Phytic acid
34. free phenolic acids content in Oats
    a) 20.6 mg/kg b) 8.7 mg/kg
    c) 57.0 mg/kg d) 27.0 mg/kg
35. soluble phenolic acids content in Oats
    a) 20.6 mg/kg b) 8.7 mg/kg
    c) 57.0 mg/kg d) 27.0 mg/kg
36. Insoluble phenolic acids content in Oats
    a) 20.6 mg/kg b) 8.7 mg/kg
    c) 57.0 mg/kg d) 27.0 mg/kg
37. Ferulic acid is the most abundant phenolic acid in barley, with a concentration of ____μg/gdry weight.
    a) 359-624 b) 359-724
    c) 359-824 d) 359-924
38. Which of the following statement is correct?
    a) The anthocyanin content of sorghum bran is 3 to 4-fold higher than that of the whole grain

b) Sorghum grains and bran were also shown to have a greater antioxidant capacity

c) Sorghum based foods can be suggested for celiac disease

d) All

39. The phenolic compounds in cereals are in

a) 6% free form b) 17-30% soluble form

c) 66-80% bound form d) All

40. Antioxidant capacity of cereals (<0.1-34 μmol TE/g product).

a) <1-24 μmol TE/g b) <0.1-34 μmol TE/g

c) <0.1-44 μmol TE/g d) <0.1-54 μmol TE/g

41. Antioxidant capacity of sorghum grains

a) 25-300 μmol TE/g b) 52-400 μmol TE/g

c) 62-500 μmol TE/g d) 82-600 μmol TE/g

42. Type of vitamin E found in oats and barley

a) α-tocotrienol b) β-tocotrienol

c) δ-tocopherol d) None

43. Type of vitamin E found in corn

a) α-tocotrienol b) β-tocotrienol

c) δ-tocopherol d) None

44. Type of vitamin E found in wheat

a) α-tocotrienol b) β-tocotrienol

c) δ-tocopherol d) None

45. Plant sterols and plant stanols are collectively termed as

a) Phytosterols b) Phytostanols

c) Photostanols d) Phytophenyls

46. The phytosterol that has double bond in sterol ring is

a) Plant sterols b) Plant stanols

c) Both a&b d) None

47. Which of the following is a plant sterol?

a) Sitosterol b) Campesterol

c) Stigmasterol d) All

48. Which of the following is a plant stanol ?
    a) Sitostanol b) Campestanol
    c) Stigmastanol d) All
49. For vegetarians a normal diet could have an even higher dietary intake of plant sterols andstanols reaching about
    a) 200 mg/ g b) 300 mg/ g
    c) 400 mg/ g d) 500 mg/ g
50. Phytosterols inhibit cholesterol absorption in the intestinal lumen by
    a) Compete with amino acids for protien formation
    b) Compete with fatty acids for foam formation
    c) Compete with cholesterol for micelle formation
    d) Compete with choline for foam formation
51. Natural storage carbohydrate in chicory and artichoke is
    a) Glycogen b) Starch
    c) Inulin d) Amylose
52. Natural storage form of carbohydrate in leek, onion and asparagus
    a) Glycogen b) Inulin
    c) Starch d) Amylose
53. Inulin is a natural storage form of carbohydrate in found in
    a) Rye b) Garlic
    c) Bananas d) All
54. The bond between fructose units in inulin is a β-(2→1) glycosidic linkage.
    a) α-(2→1)glycosidic b) β-(2→1)glycosidic
    c) β-(1→4) glycosidic d) None
55. Inulin, when ingested, acts as a
    a) Probiotic b) Prebiotic
    c) Synbiotic d) Symbiotic
56. For which of the following microorganisms, inulin is preferred as food?
    a) *Lactobacilli* b) *Bifidobacteria*
    c) *Pseudomonas* d) a&b
57. The starch that can resist upper intestinal digestion and pass into the lower intestine to be fermented by microflora in the colon is called
    a) Iso starch b) Aniso starch
    c) Resistant starch d) All

58. Which of the following is one types of resistant starch?
    a) Resistant starch granules b) Trapped starch
    c) Retrograded starch d) All
59. Which of the following is correct statement regarding starch?
    a) Glucose units are joined by glycosidic bond
    b) High-amylose starch retrogrades faster than high-amylopectin starch
    c) It is important source of our diet
    d) All
60. Resistant starch is fermented by microflora in the colon to produce
    a) Short-chain fatty acids b) Methyl compound
    c) Ethyl compound d) All
61. Resistant starch is fermented by microflora in the colon to produce
    a) Propionate b) Butyrate
    c) Acetate d) All
62. Edible parts of plants or analogous carbohydrates that are resistant to digestion and absorption in the human small intestine with complete or partial fermentation in the large intestine are called as
    a) Dietary starch b) Resistant starch
    c) Dietary fibre d) None
63. Which of the following is a dietary fibre?
    a) Mucilage b) Guar gum
    c) Hemicellulose d) All
64. Which of the following is soluble dietary fibre?
    a) Pentosans b) Pectins
    c) Mucilage d) All
65. Which of the following is insoluble dietary fibre?
    a) Cellulose b) Hemicellulose
    c) Lignin d) All
66. The first genetically engineered food is
    a) BT cotton seed b) Tomato
    c) Maize d) Soybean

67. Name of tomato that was approved as genetically engineered food for human consumption at first

a) FlavrSavr™ b) Roundup ready™

c) Pusa d) None

68. Ethylene resistant tomato developed by DRDO (Defense Research Development Organization) of India?

a) Pusa joint b) Pusa rubi

c) Pusa lal d) Pusa red

69. Maillard reaction is occur in between amino acid and

a) Reducing sugar b) Non reducing sugar

c) Esters d) Fatty acids

70. Which of the following is a nonreducing sugar?

a) Sucrose b) Lactose

c) Maltose d) All

71. Which of the following is a nonreducing sugar?

a) Glucose b) Maltose

c) Lactose d) All

72. Maillard reaction is a browning reaction that occurs in between ______groups of lysine and reducing sugar.

a) α-amino b) β-amino

c) δ-amino d) ε-amino

73. The amino acid formed during acid hydrolysis of the Amadori compounds

a) Fuchsine b) Furosine

c) Lysine d) Tryptophan

74. Indicator of the degree of damage during the initial steps of the Maillard reaction (MR) in cereal products such as pasta, commercial baby cereals, bread and breakfast cereals is

a) Fuchsine b) Melanin

c) Furosine d) Tryptophan

75. Bioactive compound present in peanuts

a) Isoflavones b) Trans-resveratrol

c) Cis-resveratrol d) a&b

76. Which of the following has been claimed to be able to prevent obesity?
    a) Acylglycerols b) Diacylglycerols
    c) Triacylglycerols d) Polyacylglycerols
77. Menthol that present in in peppermint essential oil is a
    a) Monoterpene b) Diterpene
    c) Sesquiterpene d) Triterpene
78. Nutraceutical diterpene in coffee beans
    a) Cafestol b) Kahweol
    c) Menthol d) a&b
79. Carotenoids percentage present in crude palm oil
    a) 0.2% b) 0.5%
    c) 0.8 % d) 2.0 %
80. Glycosaminoglycans (GAG) are often referred to as
    a) Mucopolyamines b) Mucopolysaccharides
    c) Mucopolysacyls d) All
81. Example/s of glycosaminoglycan
    a) Hyaluronic acid b) Chondroitin sulfate
    c) Kahweol d) a&b
82. Hyaluronic acid is a component of the ground substance found in
    a) Synovial fluid of joints b) Connective tissue
    c) Lymph d) a&b
83. Hyaluronic acid is composed of repeating disaccharides of
    a) α–glucuronic acid and N-acetyl-Dglucosamine
    b) β–glucuronic acid and N-acetyl- L glucosamine
    c) β–glucuronic acid and N-acetyl-D glucosamine
    d) α–glucuronic acid and N-acetyl- L glucosamine
84. Chondroitin sulfate is composed of
    a) β–gluconic acid and N-acetylgalactosamine sulfate
    b) β–glucuronic acid and N-acetylglucosamine sulfate
    c) β–glucuronic acid and N-acetylgalactosamine sulfate
    d) α–glucuronic acid and N-acetylgalactosamine sulfate

85. Chondroitin sulfate is the major organic component of the ground substance of

    a) Cartilage b) Connective tissue

    c) Bone d) a&c

86. Pectins is present in

    a) Middle laminae b) Primary wall

    c) Secondary wall d) a&b

87. Food Sources of pectin

    a) Legumes b) Cabbage

    c) Root vegetables d) All

88. Pectins are rich in

    a) Galacturonic acid b) Rhamnose

    c) Arabinose d) All

89. Monomer units present in gums

    a) Pentose b) Hexose

    c) Pectic acid d) a&b

90. Gum type fibers are present in

    a) Oatmeal b) Dried beans

    c) Some legumes d) All

91. Monomer unit/s present in cellulose

    a) Glucose b) Fructose

    c) Galactose d) a&b

92. Gel-formation properties of soluble fibers (e.g pectin) developed during ripening of fruit are depend upon

    a) Degree of mcthylation

    b) Linkage/ bond present in units of pectin

    c) Number of monomer units

    d) All

93. Which of the following compound can hold water?

    a) Pectins b) Cellulose

    c) Hemicellulose d) All

94. The fiber that can increases colonic fermentation is

a) Lignin
b) Cellulose
c) Hemicellulose
d) Pectins

95. Mode of action of Cellulose

a) Holds water, reduces colonic pressure, reduced transit time of digestion
b) Holds water, increases stool bulk, bind bile acids, reduces transit time
c) Holds water, may bind trace minerals and may increase fecal steroid levels
d) Slow gastric emptying, bind bile acids, increase colonic fermentation

96. Mode of action of Hemicellulose

a) Holds water, reduces colonic pressure, reduced transit time of digestion
b) Holds water, increases stool bulk, bind bile acids, reduces transit time
c) Holds water, may bind trace minerals and may increase fecal steroid levels
d) Slow gastric emptying, bind bile acids, increase colonic fermentation

97. Mode of action of Lignin

a) Holds water, reduces colonic pressure, reduced transit time of digestion
b) Holds water, increases stool bulk, bind bile acids, reduces transit time
c) Holds water, may bind trace minerals and may increase fecal steroid levels
d) Slow gastric emptying, bind bile acids, increase colonic fermentation

98. Chylomicron is example for

a) Lipid
b) Protein
c) Lipoprotein
d) Glycoprotein

99. Which of the following lipoprotein is a cholesterol transporter

a) Chylomicrons and VLDL
b) VLDL and LDL
c) LDL and HDL
d) All

100. Example for cruciferous vegetable

a) Bok choy
b) Kale
c) Turnip
d) All

## Answers

| | | | | | | | | | |
|---|---|---|---|---|---|---|---|---|---|
| 1. d | 2. d | 3. a | 4. c | 5. b | 6. a | 7. c | 8. a | 9. d | 10. d |
| 11. b | 12. d | 13. c | 14. d | 15. b | 16. d | 17. c | 18. b | 19. d | 20. c |
| 21. c | 22. c | 23. d | 24. b | 25. d | 26. c | 27. b | 28. d | 29. d | 30. d |
| 31. c | 32. c | 33. a | 34. a | 35. b | 36. c | 37. a | 38. d | 39. d | 40. b |
| 41. b | 42. a | 43. c | 44. b | 45. a | 46. a | 47. d | 48. d | 49. d | 50. c |
| 51. c | 52. b | 53. d | 54. b | 55. b | 56. d | 57. c | 58. d | 59. d | 60. a |
| 61. d | 62. c | 63. d | 64. d | 65. d | 66. b | 67. d | 68. b | 69. a | 70. a |
| 71. d | 72. d | 73. b | 74. c | 75. d | 76. b | 77. a | 78. d | 79. a | 80. b |
| 81. d | 82. d | 83. c | 84. c | 85. d | 86. d | 87. d | 88. d | 89. d | 90. d |
| 91. d | 92. a | 93. d | 94. d | 95. a | 96. b | 97. c | 98. c | 99. c | 100. d |

1. Dietary fibre is highest in
   a) Cereals b) Millets
   c) Legumes d) Oil seeds
2. High dietary fibre containing cereal grain is
   a) Barley b) Rye
   c) Rice d) Wheat
3. Arrange the cereals fibre content in decreasing order
   a) Rye > Triticale > Corn > Oats > Wheat
   b) Rye > Corn > Oats > Triticale > Wheat
   c) Rye > Oats > Corn > Triticale > Wheat
   d) Rye > Corn > Triticale > Oats > Wheat
4. Total dietary fibre content in legumes
   a) 10.6-20.9 % b) 11.6-22.9 %
   c) 13.6-28.9 % d) 18.6-29.9 %
5. Total dietary fibre content in rye
   a) 15.5 % b) 16.5%
   c) 17.5% d) 18.5%
6. Total dietary fibre content in rye
   a) 12% b) 13%
   c) 14% d) 15%
7. The RS1 type of resistant starch is
   a) Physically inaccessible to α–amylase
   b) Chemically inaccessible to β–amylase
   c) Physically inaccessible to α–amylase
   d) Chemically inaccessible to β–amylase
8. Which of the following is a RS2 type of resistant starch
   a) Amylopectin crystals of ungelatinized starch granules
   b) Retrograded amylase crystals
   c) Chemically modified starch
   d) Starch that is inaccessible to α–amylase

9. Which of the following is a RS3 type of resistant starch
   a) Chemically modified starch
   b) Retrograded amylase crystals
   c) Starch that is inaccessible to α–amylase
   d) Amylopectin crystals of ungelatinized starch granules
10. Which of the following is a RS4 type of resistant starch
    a) Amylopectin crystals of ungelatinized starch granules
    b) Retrograded amylase crystals
    c) Chemically modified starch
    d) Starch that is inaccessible to α–amylase
11. RS3 type of resistant starch is analyzed as an
    a) Insoluble dietary starch
    b) Insoluble dietary fiber
    c) Soluble dietary starch
    d) Soluble dietary fiber
12. RS3 type of resistant starch is acts as soluble dietary fiber
    a) Insoluble dietary starch b) Insoluble dietary fiber
    c) Soluble dietary starch d) Soluble dietary fiber
13. Fermentation of RS3 type of resistant starch in the colon results in production of high concentration of short chain
    a) Oligosaccharides b) Amino acids
    c) Fatty acids d) All
14. A requisite for a product to be considered probiotic is that the number of bacteria at the time of consumption is approximately _____ viable cells per gram.
    a) $10^6$ b) $10^7$
    c) $10^8$ d) $10^9$
15. A requisite for a product to be considered probiotic is that millilitre of product being theminimum recommended daily dose _____ cells.
    a) $10^4$ to $10^6$ b) $10^6$ to $10^7$
    c) $10^7$ to $10^8$ d) $10^8$ to $10^9$

16. Microorganisms involved in Yogurt preparation
    a) *S. thermophilus* b) *L. bulgaricus*
    c) *L. lactis* d) a&b
17. The predominant microorganism found in African mahewu is
    a) *S. thermophilus* b) *L. bulgaricus*
    c) *S. lactis* d) *Lactococcus lactis*
18. Characteristic of ideal probiotic microorganism
    a) Pathogenicity b) Virulence
    c) Flatulence d) Acid tolerance
19. Which of the following is a characteristic of ideal probiotic microorganism
    a) Capability for easy proliferation *in vivo.*
    b) Capability for easy proliferation *in vitro.*
    c) High survival rate
    d) All
20. Cancer preventing agent present in broccoli
    a) Sulfoxide b) Sulfamide
    c) Sulforaphane d) Bromide
21. The agent present in flax seed that lowers the risk of cardiovascular disease is : from
    a) Alpha-linolenic acid b) Beta-linolenic acid
    c) Gama-linolenic acid d) All
22. Antioxidant property containing agent found in berries is
    a) Alpha carotene b) Beta carotene
    c) Anthocyanin d) Chlorophyll
23. Lentils are rich in folates which are helpful in prevention of
    a) Birth defects b) Heart diseases
    c) Bone decay d) a & b
24. Vitamin E is an antioxidant and helps in the production of red blood cells. Vitamin E rich food is
    a) Avocado b) Tomato
    c) Brussels sprouts d) All

25. Vitamin K plays a vital role in the blood clotting process in the body and helps to maintain strong bones. The best food sources of Vitamin K is

a) Meat b) Liver

c) Egg yolk d) All

26. Vitamin P, also known as

a) Biotin b) Bioflavonoid

c) Phylloquinone d) Phytonadione

27. The vitamin that helps in blood circulation by strengthening the capillaries is

a) Vitamin K b) Vitamin P

c) Vitamin D d) Vitamin A

28. Best food sources of vitamin P is

a) Oranges b) Garlic

c) Buckwheat d) All

29. The mineral essential for healthy nerve and muscle function is

a) Ca b) Mn

c) Mg d) Fe

30. Manganese containing food sources

a) Whole grains b) Nuts

c) Leafy vegetables d) All

31. The mineral essential for cell reproduction, normal growth and development in children, wound healing, production of sperm and testosterone is

a) Ca b) Mg

c) Al d) Zn

32. Great food for upping the Zinc levels in body is

a) Beef b) Spinach

c) Shrimp d) Watermelon

33. Lipotropic agent necessary for amino acid transport and movement of potassium and sodium is

a) Choline b) Inositol

c) Taurine d) Ergo sterol

34. Food sources of inositol

a) Lecithin granules b) Wheat germ
c) Liver d) All

35. The minor components of food contribute about _____% of the dry weight of our diet.

a) 0.5 b) 1.0
c) 1.5 d) 2.0

36. Which of the following is a minor mineral for human body

a) Sodium b) Iron
c) Potassium d) Calcium

37. The major phytochemical used as neutraceutical is

a) Choline b) Lecithin
c) Spingolipid d) Glucosinolate

38. Which of the following phytochemical can inhibits tumor growth?

a) Lycopene b) B-carotene
c) A-carotene d) Lutein

39. Lycopene is present in

a) Tomato b) Red pepper
c) Water melon d) All the above

40. Which of the following compound acts as antioxidant and protecting lung tissues from free oxygen?

a) Lycopene b) Carotenoids
c) Limonoids d) Xanthophylls

41. Neutraceutical compound present in citrus peel

a) Flavanoids b) Limonoids
c) Carotenoids d) All

42. Pink skin of groud nut and red wine are rich in

a) Ferulic acid b) Ellagic acid
c) Resveratrol d) Lecithin

43. Isoflavones from Pueraria labata plant is usesd for treatment of

a) Cancer b) Alcoholism
c) CVD d) Blood presure

44. Brinjal and dark chocolates are rich in

    a) Lutein b) Zeaxanthin

    c) Flavonoids d) All

45. Fermentation of dietary fibre results in production of ______lowering the intestinal pH.

    a) Alkaloid b) Polysaccharide

    c) Flavanoids d) Short chain F.A

46. Brain and nervous tissue membrane lipids are rich sourse of

    a) DHA b) EPA

    c) Isoflavone d) All

47. Egg, red meat and oysters are good source of

    a) Iron b) Zinc

    c) Calcium d) Magnesium

48. Good source of chromium which can control the blood sugar?

    a) Whole grains b) Wheat germ

    c) Brewer's yeast d) All

49. Which of the following dish is good source of protein for vegetarians

    a) Fresh vegetables b) Fruits

    c) Chapathi & Dhal d) a & b

50. Banana is rich sourse of

    a) Vitamin $B_2$ b) Vitamin $B_4$

    c) Vitamin $B_5$ d) Vitamin $B_6$

## Answers

| | | | | | | | | | |
|---|---|---|---|---|---|---|---|---|---|
| 1. c | 2. b | 3. d | 4. c | 5. a | 6. c | 7. a | 8. a | 9. b | 10. c |
| 11. b | 12. d | 13. c | 14. a | 15. d | 16. d | 17. d | 18. d | 19. d | 20. c |
| 21. a | 22. c | 23. d | 24. d | 25. d | 26. b | 27. b | 28. d | 29. b | 30. d |
| 31. d | 32. a | 33. b | 34. d | 35. d | 36. b | 37. d | 38. c | 39. d | 40. c |
| 41. b | 42. c | 43. b | 44. c | 45. d | 46. a | 47. b | 48. d | 49. c | 50. d |

# 9

# Food Process Engineering

1. Which of the following is not fluid

   a) Powders  b) Gases

   c) Liquids  d) None

2. The ratio between volume of air ($m^3$) and volume of bulk sample ($m^3$) termed as

   a) Porosity  b) Density

   c) Angle of repose  d) Mass density

3. The density of liquids can be expressed as (SG)

   a) Specific gravity  b) Porosity

   c) Angle of repose  d) None

4. Units for gas constant

   a) N (m)/(K)  b) N (m)/(kmole)

   c) N /(kmole K)  d) N (m)/(kmole K)

5. Which of the following is Non- Newtonian fluid

   a) Water  b) Sugars solution

   c) Salts solution  d) Emulsions

6. Tomato ketchup is

   a) Bingham plastic fluid  b) Thixotropic fluid

   c) Rheopectic fluid  d) Dilatant fluid

7. The following one is not a Viscoelastic material

   a) Dough  b) Cheese

   c) Gelled foods  d) Cream

8. Kinematic viscosity is

   a) Dynamic viscosity/$\rho$  b) $\rho$/dynamic viscosity

   c) Viscosity/$\rho$  d) $\rho$/viscosity

9. Water percentage present in wheat flour
   a) 11-13% b) 8-11%
   c) 5-7% d) 3-1%
10. Colloids which having great affinity for water are cAlled
   a) Lyophobic colloids b) Lyophilic colloids
   c) a & b d) None
11. Freeze-drying also known as
   a) Lyophilisation b) Cryodesiccation
   c) Lyophilization d) All the above
12. Colloids can be separated from crystalloids by the process known as
   a) Distillation b) Dialysis
   c) Coagulation d) None
13. Process by which water is drawn into a solution through a semi permeable membrane is
   a) Diffusion b) Osmosis
   c) Distillation d) All the above
14. In reverse osmosis, solvent can be removed from a solution using a pressure _____ than the osmotic pressure.
   a) Less b) Greater
   c) Equal d) None
15. A sol can be transformed into a gel under following condition
   a) Chemical alteration of gelling agent
   b) Lowering of temp
   c) Adjustment of pH or addition of salt
   d) All the above
16. Spontaneous exudation of liquid from a gel is called
   a) Syneresis b) Regormortis
   c) Imbibation d) All the above
17. The opposite process of syneresis is
   a) Syneresis b) Regormortis
   c) Imbibation d) All the above

18. Syneresis in curd is due to
    a) Imbalance in intermolecular forces
    b) Sudden removal of the hydrophilic macropeptides
    c) Starting of bonds between hydrophobic sites
    d) All the above
19. characteristic of emulsifier with HLB values ranging from 11-20 is
    a) Intermediate b) Lipophilic
    c) Hydrophilic d) All the above
20. Which of the following is foaming agent
    a) Phospholipids b) Saponins
    c) Proteins d) All the above
21. Which of the following is antifoaming agent
    a) Poly dimethylsiloxane b) Ethylene bis stearamide
    c) Mineral oil d) All the above
22. Stability changes in food emulsions can occur through the process of
    a) Flocculation b) Creaming
    c) Coalescence. d) All the above
23. In food foams, the structure of the foam may be stabilised by
    a) Heating b) Freezing
    c) Gelation d) All the above
24. Reynolds number can be used to determine
    a) Power requirements for blenders
    b) Power requirements for mixers
    c) Power requirements for pumps
    d) All the above
25. Boiler operation is improved by:
    a) Returning condensate as feed water
    b) Pre-heating air for combustion of fuel
    c) Insulating the boiler
    d) All the above

26. Resistance of bulk grain to air flow is a function of
    a) Kernel size
    b) Porosity
    c) a & b
    d) None
27. Angle of repose is not influenced by
    a) Size & shape
    b) Orientation of the particles
    c) Moisture content
    d) None
28. The rate of microbial destruction is follows
    a) Zero-order reaction
    b) First-order reaction
    c) Second-order reaction
    d) Third-order reaction
29. Method employed for the determination of the coefficient internal friction is by
    a) Tri-axial compression test apparatus
    b) Shear test apparatus
    c) a & b
    d) None
30. The time needed to destroy 90% of the micro-organisms (to reduce their numbers by a factor of 10) is referred to as
    a) D value
    b) F value
    c) Z value
    d) T value
31. The number of degrees Celsius required to bring about a ten-fold change in decimal reduction time is
    a) D value
    b) F value
    c) Z value
    d) T value
32. The process which is lethality or the time in minutes, at a specific temperature required to destroy a certain number of viable cells is
    a) D value
    b) F value
    c) Z value
    d) T value
33. The heat which causes an increase or decrease in the temperature of a body without changing its state is known as
    a) Latent heat
    b) Sensible heat
    c) Specific heat
    d) None

34. In older method of lye peeling, concentration of lye is

a) 0-1% b) 1-2%

c) 2-3% d) 3-4%

35. In dry caustic peeling method of lye peeling, concentration of lye is

a) 5% b) 10 %

c) 20% d) 30 %

36. Lower angle of repose represents__________ flowability.

a) Easier b) Difficult

c) No d) None

37. In comminution is a unit operation in which the average size of solid pieces of food is reduced by

a) Compression b) Grinding

c) Impact forces d) All the above

38. In Hammer mills size of solid pieces of food is reduced by mainly

a) Compression b) Grinding

c) Impact forces d) All the above

39. In Attrition mills size of solid pieces of food is reduced by mainly

a) Impact forces b) Crushing

c) Shear d) b & c

40. Frequency of sound waves used in Ultrasonic homogenisers

a) 1-15 kHz b) 18–30 kHz

c) 50-100 kHz d) 100-200 kHz

41. Margarine is ____ emulsion

a) Oil in water b) Water in Oil

c) Air in oil d) Air in water

42. Which of the following seed property is used for designing of the seed grader

a) Size b) Shape

c) a & b d) None

43. In milk, homogenisation reduces the average size of fat globules from __ to less than__ μm

a) 4; 1 b) 10; 5

c) 20; 10 d) 30; 10

44. Chocolate is a example for
    a) Viscoelastic food b) Dilatant food
    c) Pseudoplastic food d) Bingham food
45. The temperature of most extraction operations is limited to _______ 100°C by economic considerations
    a) Equal to b) Less than
    c) Greater than d) None
46. Reverse osmosis is also called
    a) Hypofiltration b) Hyperfiltration
    c) Loose reverse osmosis d) None
47. Nano filtration also called
    a) Hypofiltration b) Loose reverse osmosis'
    c) Hyperfiltration d) None
48. Processing method for the separation of mixtures of liquids by partial vaporization through a non-porous or porous membrane is called
    a) Evaporation b) Pervaporation
    c) Dialysis d) Osmosis
49. Which of the following is fermented sausage
    a) Salami b) Bologna
    c) Medwurst d) All of the above
50. Fermented sausage are pasteurised at
    a) 30-50°C/ 1 h b) 85-98°C/8 h
    c) 65-68°C for 1-2 h d) 65-68°C for 4-8 h
51. Requirement for commercial enzyme production from micro-organisms is
    a) Inexpensive substrate
    b) High yield of enzyme
    c) Simple enzyme recovery method
    d) All of the above
52. Advantage of enzyme immobilization is
    a) Re-use b) Continuous processing
    c) Control of pH & temperature d) All of the above

53. HFCS means

a) High-fructose corn solution b) High-fructose corn syrup

c) High-fructose corn sugar d) None

54. In HFCS-42 the number 42 indicates

a) Fructose % b) Water %

c) Total sugar % d) TSS

55. Application of Glucose isomerase is

a) Dextrose production

b) Fruit juice clarification

c) Removal of glucose from egg white

d) All of the above

56. The degree of hydrolysis of starch is measured by

a) Glucose equivalent b) Maltose equivalent

c) Dextrose equivalent d) None

57. Acid proteases are produced mostly by

a) Bacteria b) Yeast

c) Fungi d) Virus

58. Neutral and alkaline proteases are produced by

a) Fungi b) Bacteria

c) Both a & b d) Yeast

59. First enzyme to become commercially available from genetically engineered is

a) Chymosin b) Papain

c) Ficin d) Bromelain

60. Heat-resistant enzyme which is found in most vegetable

a) Catalase b) Peroxidase

c) Lipase d) a & b

61. IQB means

a) Individual quick boiling b) Individual quick blanching

c) Individual quick breaking d) None

62. IQF means
    a) Individual quick filling
    b) Individual quick frying
    c) Individual quick freezing
    d) None
63. Loss of vitamins during blanching is due to
    a) Leaching
    b) Thermal destruction
    c) Oxidation
    d) All of the above
64. To retain chlorophyll of green vegetables________ added to blancher water
    a) Sodium carbonate
    b) Zinc oxides
    c) Zinc chloride
    d) All of the above
65. Naturally occurring enzyme in raw milk which has a similar D value to heat-resistant pathogens is
    a) Alkaline lipase
    b) Alkaline phosphatase
    c) Alkaline catalase
    d) None
66. Test used to know effectiveness of liquid-egg pasteurisation is based on
    a) Residual β–amylase activity
    b) Residual α–amylase activity
    c) Residual lactase activity
    d) Residual glucose isomerase activity
67. Most heat resistant spores have a z value of around
    a) 1°C
    b) 10°C
    c) 100°C
    d) 110°C
68. In canning air is removed from the retort by the incoming steam is called as
    a) Venting
    b) Exhausting
    c) a & b
    d) None
69. Typical minimum time–temperature conditions needed to destroy *Cl. botulinum* ($F_0$ = 3) are
    a) 141°C/ 2 sec
    b) 141°C/1.8 sec
    c) 121°C/1.8 sec
    d) 111°C/1.8 sec
70. In canning removal of the air from can before it is sealed is called
    a) Venting
    b) Exhausting
    c) a & b
    d) None

71. Retort temperatures for sterilisation processes are generally between

a) 90 and 100°C b) 100 and 110°C

c) 110 and 120°C d) 110 and 130°C

72. Generally in Plate heat exchangers, the plates are of polished stainless steel of in _______ thickness

a) 5–12.5mm b) 0.5–12.5mm

c) 0.05–1.25mm d) 0.5–1.25mm

73. Generally in Plate heat exchangers, the plates are of polished stainless steel are separated by

a) 0.1–1mm b) 1–3 mm

c) 3–6 mm d) 5–10 mm

74. The condition at which all three states of material may exist together is called the

a) Sharp point b) Freezing point

c) Flash point d) Triple point

75. Triple point of water is at

a) 0.0098°C and 0.64 kPa b) 0.0098°C and 4.8 mm of Hg

c) 0.98°C and 0.4 kPa d) a & b

76. heat exchanger is also called as

a) Pump b) Calandria

c) Ejector d) None

77. Most widespread evaporators for juice concentration are

a) Raising film evaporators b) Falling film evaporators

c) Forced circulation evaporators d) None

78. TVP means

a) Thermal volatile product b) Thermal vegetable protein

c) Texturised vegetable protein d) None

79. Expand BVO

a) Brominated volatile oil b) Brominated vegetable oil

c) Brominated vegetable odour d) None

80. Which of the following is a measure of the extent of protein denaturation

a) Sulphur solubility index b) Phosphorus solubility index

c) Carbon solubility index d) Nitrogen solubility index

81. Extrusion cooking is also called
    a) LTLT b) HTST
    c) UHT d) None
82. Time–temperature conditions for Extrusion cooking are
    a) 120–160°C/40–60sec b) 180–190°C/40–60 sec
    c) 180°C/4 sec d) 180–190°C/20–40 sec
83. Stability of butter is mostly due to
    a) Its semi-solid nature b) Action of an emulsifying agent
    c) a & b d) None
84. Edible oils are
    a) Newtonian fluids. b) Non-Newtonian fluids.
    c) a & b d) None
85. Pectins are
    a) Newtonian fluids. b) Non-Newtonian fluids.
    c) a & b d) None
86. Most creams are
    a) Pseudoplastic fluids b) Dilatant fluids
    c) Bingham fluids d) Thixotropic fluids
87. In smoked fish dispersed phase is
    a) Solid b) Gas
    c) Liquid d) None
88. In smoked meat continuous phase is
    a) Solid b) Gas
    c) Liquid d) None
89. In milk dispersed phase and continuous phase are____ & ______ respectively
    a) Gas & liquid b) Solid & liquid
    c) Liquid & liquid d) None
90. In true solution soluble particle size is
    a) < 1 mμ in diameter b) = 1 mμ in diameter
    c) > 1 mμ in diameter d) None

91. In colloidal solution particle size is

a) < 1 mμ in diameter  b) =1 mμ in diameter

c) > 1 mμ in diameter  d) None

92. Which of the following is a lyophilic colloid

a) Clay particles  b) Protein

c) Starch  d) b & c

93. If the particles size of emulsion is from 1 mμ to 0.1 mμ then the appearance of that emulsion in light is

a) Milky white.  b) Blue white

c) Grey semi transparent  d) Grey white

94. If the particles size of emulsion is from 0.1 mμ to 0.05 mμ then the appearance of that emulsion in light is

a) Milky white.  b) Blue white

c) Grey semi transparent  d) Grey white

95. Characteristics of foams are

a) Large amounts of entrapped gas

b) Extensive surface area

c) Rigid or semi rigid and elastic walls

d) All of the above

96. Which of the following is not an example for buffer system

a) Acetic acid and sodium acetate mixture.

b) Citric acid and sodium citrate mixture

c) Disodium phosphate and protactinium acid phosphate mixture.

d) Acetic acid and sodium chloride mixture.

97. Characteristics of bound water are

a) Cannot be frozen

b) Can not be removed under pressure from the tissue

c) Appreciable vapour pressure

d) All of the above

98. Reynolds number (Re) is

a) $\mu/Dv\rho$  b) $D\rho/\mu v$

c) $Dv/\mu\rho$  d) $Dv\rho/\mu$

99. Reynolds number of less than 2100 describes
    a) Laminar flow
    b) Streamline flow
    c) Turbulent flow
    d) Transitional flow

100. Reynolds number of more than 4000 describes.
    a) Laminar flow
    b) Streamline flow
    c) Turbulent flow
    d) Transitional flow

## Answers

| | | | | | | | | | |
|---|---|---|---|---|---|---|---|---|---|
| 1. d | 2. a | 3. a | 4. d | 5. d | 6. a | 7. d | 8. a | 9. c | 10. b |
| 11. d | 12. b | 13. b | 14. b | 15. d | 16. a | 17. c | 18. d | 19. c | 20. d |
| 21. d | 22. d | 23. d | 24. d | 25. d | 26. c | 27. d | 28. b | 29. c | 30. a |
| 31. c | 32. b | 33. b | 34. b | 35. b | 36. a | 37. d | 38. c | 39. d | 40. b |
| 41. b | 42. c | 43. a | 44. b | 45. b | 46. b | 47. b | 48. b | 49. d | 50. d |
| 51. d | 52. d | 53. b | 54. a | 55. d | 56. c | 57. c | 58. c | 59. a | 60. d |
| 61. b | 62. c | 63. d | 64. a | 65. b | 66. b | 67. b | 68. a | 69. b | 70. b |
| 71. d | 72. d | 73. c | 74. d | 75. d | 76. b | 77. b | 78. c | 79. b | 80. d |
| 81. b | 82. d | 83. a | 84. a | 85. b | 86. d | 87. b | 88. c | 89. a | 90. a |
| 91. c | 92. d | 93. b | 94. c | 95. d | 96. d | 97. d | 98. d | 99. b | 100. c |

1. The most widespread evaporator for juice concentration is
   a) Open -pan evaporators b) Closed-pan evaporators
   c) Raising film evaporators d) Falling film evaporators
2. Thermally accelerated short-time evaporator' (TASTE) is example for
   a) Open -pan evaporators b) Closed-pan evaporators
   c) Falling film evaporators d) Raising film evaporators
3. Attrition mill is also called
   a) Burr mill b) Plate mill
   c) Hammer mill d) a & b
4. Pressure homogenizers are operated at the pressures
   a) 100–700X$10^3$ Pa b) 1000–7000 X $10^3$ Pa
   c) 10000–70000 X $10^3$ Pa d) 10 –70 X $10^3$ Pa
5. Ultrasonic homogenizer uses sound waves and is operated at the frequency
   a) 1-18 kHz b) 18-30 kHz
   c) 30-60 kHz d) 60-100 kHz
6. Ultrasonic homogenizers are used for
   a) Dispersing powders in liquids. b) Baby foods
   c) Essential oil emulsions. d) All
7. In milk, homogenisation reduces the average size of fat globules
   a) From 40 μm to <10 μm b) From 4 μm to <1 μm
   c) From 0.4 μm to <0.1 μm d) From 0.04 μm to <0.01 μm
8. Butter is a
   a) w/o emulsion b) o/w emulsion
   c) g/o emulsion d) g/w emulsion
9. Margarine is a
   a) w/o emulsion b) o/w emulsion
   c) g/o emulsion d) g/w emulsion
10. Ice cream is a
    a) w/o emulsion b) o/w emulsion
    c) g/o emulsion d) g/w emulsion

11. The phenomenon refers to the amount of air that is whipped into the ice cream mixture is
    a) Cavitation b) Overrun
    c) Spreading d) Agitation
12. An overrun of 100 percent would mean for every gallon of ice cream mix, you get______ gallons of finished ice cream
    a) One b) Two
    C) Three d) Four
13. Commercial ice creams have overruns of ____________%.
    a) 1-6 b) 6-10
    c) 10-60 d) 60-100
14. The degree of mixing that is achieved depends on
    a) Size, shape and density
    b) Tendency of the materials to aggregate
    c) Moisture content
    d) All
15. Mixers used for coating applications are
    a) Ribbon mixers b) Tumbling mixers
    c) Vertical-screw mixers d) None
16. Solvent extraction depends on
    a) Temperature of extraction
    b) Surface area of solids exposed to the solvent
    c) Viscosity of the solvent
    d) All
17. The mixer used for the incorporation of small quantities of ingredients into a bulk of material is
    a) Ribbon mixers b) Tumbling mixers
    c) Vertical-screw mixers d) None
18. Which of the following is a membrane separation technique, in which a liquid feed mixture is separated by partial vaporisation through a non-porous, selectively permeable membrane
    a) Electrophoresis b) Dialysis
    c) Electrodialysis d) Pervaporation

19. In Electrodialysis
    a) Anions that migrate towards the anode pass through anionmembranes
    b) Anions that migrate towards the anode are rejected by cation membranes
    c) Cations that migrate towards the anode pass through anionmembranes
    d) a & b
20. Which of the following is a fermented sausage
    a) Salami b) Bologna
    c) Pepperoni d) All
21. Which of the following is a correct pattern of fermented sausage preparation
    a) Casings filling → Pasteurization → Fermentation → Drying
    b) Casings filling → Fermentation → Drying → Pasteurization
    c) Casings filling → Drying → Pasteurization → Fermentation
    d) Casings filling → Fermentation → Pasteurization → Drying
22. Micro-organism/s used for $\alpha$-amylase enzyme production is/are
    a) *B.licheniformis* b) *Bacillus subtilis*
    c) *Aspergillus oryzae* d) All
23. Micro-organism/s used for Amyloglucosidase enzyme production is/are
    a) *B.licheniformis* b) *Bacillus subtilis*
    c) *Aspergillus oryzae* d) *A. niger*
24. Micro-organism/s used for Cellulases enzyme production is/are
    a) *B.licheniformis* b) *Bacillus subtilis*
    c) *Aspergillus oryzae* d) *A.niger*
25. Micro-organism/s used for Glucoamylases enzyme production is/are
    a) *Rhizopus* spp. b) *Bacillus subtilis*
    c) *A.niger* d) a & c
26. Micro-organism/s used for Glucose isomerase enzyme production is/are
    a) *Streptococcus* spp. b) *Lactobacillus* spp.
    c) *Escherichia* spp. d) All
27. Micro-organism/s used for Glucose oxidase enzyme production is/are
    a) *B.licheniformis* b) *Bacillus subtilis*
    c) *Aspergillus oryzae* d) *A.niger*

28. Micro-organism/s used for Invertase enzyme production is/are
    a) *Aspergillus oryzae* b) *S.cerevisiae*
    c) *Kluyveromyces fragilis* d) b & c
29. Micro-organism/s used for Lactase enzyme production is/are
    a) *K.fragilis* b) *Aspergillus* spp.
    c) *Mucor pusillus* d) All
30. Micro-organism/s used for Pectinase enzyme production is/are
    a) *B.licheniformis* b) *Bacillus subtilis*
    c) *Aspergillus oryzae* d) *A.niger*
31. Rennet is produced by
    a) *Mucor pusillus* b) *Mucor miehei*
    c) Bovine (chymosin) d) All
32. main advantage/s of enzyme immobilisation is/are that, there isand.Immobilisation is at present used when an enzyme is difficult to isolate or expensive to prepare.
    a) Enzymes are re-used
    b) Continuous processing
    c) Closer control of pH and temperature to achieve optimum activity
    d) All
33. α-amylases liquefy the starch and produce
    a) Maltose b) Galactose
    c) Sucrose d) Fructose
34. α-amylases randomly hydrolyse ________ linkages to liquefy starch and produce maltose.
    a) $\alpha$–(1 → 6) b) $\alpha$–(1 → 4)
    c) $\alpha$–(1 → 3) d) $\alpha$–(1 → 2)
35. The enzyme that hydrolyses sucrose to glucose and fructose.
    a) Amylase b) Invertase
    c) Lactase d) All
36. The enzyme that hydrolyses lactose to D-glucose and D-galactose
    a) α-D-galactosidase b) β-D-galactosidase
    c) β-D-glucosidase d) None

37. The enzyme used for the production of high-fructose corn syrup (HFCS) is

a) Amylase b) Glucose isomerase
c) Lactase d) All

38. Glucose isomerase functioning as

a) Hydrolyses sucrose to glucose and fructose.
b) Makes fructose from glucose
c) Hydrolyses lactose to D-glucose and D-galactose
d) None

39. In HFCS-42, the number 42 means

a) Glucose % b) Fructose %
c) Syrup % d) None

40. The products widely used as an alternative to sucrose in the bakery and confectionery are

a) HFCS-42 &HFCS-55 b) HFCS-42 & HFCS-66
c) HFCS-42 & HFCS-77 d) None

41. Cyclodextrins are molecules containing _______ glucose units, linked together in a ring

a) 6-8 b) 5-6
c) 3-5 d) 1-3

42. CGT-ase means

a) Cyclodextrinase
b) Cycloglycotransferase
c) Cyclodextrin glycosyltransferase
d) None

43. Cyclodextrins are produced from starch using the enzyme

a) CDT-ase b) CGT-ase
c) CT-ase d) None

44. Which of the following is not degrade the pectin

a) Pectin esterase b) Polygalacturonase
c) Pectin lyase d) None

45. The enzyme that hydrolyses pectin molecules to form low-methoxyl pectin and polygalacturonic (pectic) acid is
 a) Pectin esterase b) Polygalacturonase
 c) Pectin lyase d) None
46. The enzyme that hydrolyses polygalacturonic acid to produce oligogalacturonans and galacturonic acid is
 a) Pectin esterase b) Polygalacturonase
 c) Pectin lyase d) None
47. Acid proteases are produced mostly by
 a) Bacteria b) Fungi
 c) Both fungi and bacteria d) Actinomycetes
48. Neutral proteases are produced mostly by
 a) Bacteria b) Fungi
 c) Both fungi and bacteria d) Actinomycetes
49. Alkaline proteases are produced mostly by
 a) Bacteria b) Fungi
 c) Both fungi and bacteria d) Actinomycetes
50. The first enzyme to become commercially available from genetically engineered is *Kluyveromyces lactis* and *Escherichia coli*
 a) Chymosin b) Lactase
 c) Cellulase d) Amylase
51. Chymosin enzyme is first time genetically engineered from
 a) *Kluyveromyces lactis* b) *Escherichia coli*
 c) a & b D) None
52. Which of the following is not a protease
 a) Chymosin b) Ficin
 c) Elastase d) None
53. The enzyme used to de-sugar, and hence to stabilise egg products is
 a) Amylase b) Glucose isomerase
 c) Lactase d) Glucose oxidase
54. Which of the following enzyme decomposes hydrogen peroxide to form water and oxygen
 a) Catalase b) Amylase
 c) Ficin d) Elastasc

55. The enzyme used to provide oxygenfor de-sugaring egg products by glucose oxidase is

a) Ficin b) Amylase

c) Catalase d) Elastase

56. Diacetyl reductase converts the flavour compound diacetyl to flavourless______

a) Ketones b) Diethyl

c) Acetoin d) None

57. Limonin is present in

a) Grapes b) Banana

C) Pineapple d) Orange

58. Naringin is present in

a) Grapes b) Orange

C) Pineapple d) Banana

59. Enzyme/s used for debittering of fruit products is/are

a) Lipase b) Limoninase

C) Nariniginase d) b & c

60. Enzyme used to reduce bread staling and to reduce dough viscosity is

a) Pentosanase b) Amylase

c) Lipase d) Protease

61. The enzyme pentosanase converts pentosans to

a) D-lyxose and L-arabinose b) D-xylose and L-arabinose

c) D-xylose and L-glucose d) None

62. The enzyme retards the development of oxidised flavours in milk is

a) Lypase b) Trypsin

c) Amylase d) None

63. The enzyme/s soften and tenderise connective tissues in meat is/are

a) Collagenase b) Elastase

c) Trypsin d) a & b

64. Heat-resistant enzymes found in most vegetables are

a) Peroxidase b) Catalase

c) a & b d) None

65. Test for effective pasteurization is

a) Phosphatas b) Peroxidase

c) Catalase d) Lipase

66. Test for effective blanching/sterilization

a) Phosphatas b) Peroxidase

c) Catalase d) Lipase

67. IQB means

a) Individual quick bleaching b) Individual quick breaking

c) Individual quick blanching d) None

68. Main purpose of pasteurization for fruit juice is

a) Enzyme inactivation

b) Destruction of spoilage microorganisms

c) Destruction of pathogens

d) None

69. Main purpose of pasteurization for Beer is

a) Enzyme inactivation

b) Destruction of spoilage microorganisms

c) Destruction of pathogens

d) None

70. In general, bacteria in the ____ phase of growth are more barosensitive (sensitive to high pressures)

a) Lag b) Log

c) Stationary d) Death

71. Which of the following does not require blanching

a) Peas b) Green peppers

c) Onions d) b & c

72. Main purpose of blanching

a) Removal of dust b) Removal of insects

c) Enzyme inactivation d) None

73. The following factor does not influence the blanching time is

a) Blanching temperature b) Method of heating.

c) Size of pieces d) None

74. Chemicals often added to blancher water to protect chlorophyll are
   a) Sodium carbonate b) Calcium oxides
   c) NaCl d) a & b
75. Sodium carbonate content added to blancher water to protect chlorophyll is
   a) 0.5% w/w b) 0.25% w/w
   c) 0.125% w/w d) 1.25% w/w
76. Enzymic browning of cut apples and potatoes is prevented by holding the food in dilute (2%w/w) _______ prior to blanching.
   a) Brine b) Lime
   c) sodium carbonate d) All
77. Extent of the heat treatment required to stabilize a food is determined by the D value
   a) D value b) F value
   c) Z value d) T value
78. Milk pasteurization is based on __________ reduction in the numbers of *C.burnetii*,
   a) D63 & a 11 logarithmic cycle b) D40 & a 11 logarithmic cycle
   c) D50 & a 12 logarithmic cycle d) D60 & a 12 logarithmic cycle
79. LTLT Pasteurization operates at
   a) 63°C / 30 min b) 71.7 /15 sec
   c) 137.5/1 sec d) None
80. HTST Pasteurization operates at
   a) 63°C / 30 min b) 71.7 /15 sec
   c) 137.5/1 sec d) None
81. UHT Pasteurization operates at
   a) 63°C / 30 min b) 71.7 /15 sec
   c) 137.5/1 sec d) None
82. Most heat resistant spores have a z value of around ________ °C
   a) 100°C b) 90°C
   c) 10°C d) 30°C
83. Thermophilic bacteria grow at
   a) 25-35 °C b) 35–55°C
   c) 55-65 °C d) 65-70 °C

84. Mesophilic bacteria grow at

a) 10-40 ºC b) 25-35 ºC

c) 35–55ºC d) 55-65 ºC

85. Which of the following is a heat-sterilisable container

a) Glass jars or bottle b) Metal can

c) Flexible pouch d) All

86. Scraped surface heat exchangers are used for

a) Paneer b) Ice cream

c) Yoghurt d) None

87. Boiling of water occurs more at ___________ pressures.

a) Lower b) Higher

c) Atmospheric d) All

88. Evaporation of water occurs more at ___________ pressures.

a) Lower b) Higher

c) Atmospheric d) All

89. The ratio of the partial pressure of the water vapour in the air to the partial pressure of saturated water vapour at the same temperature is called

a) Dew point b) Relative humidity

c) Boiling point d) None

90. Mode of Heat is transferred in Fluidized Bed Dryers is

a) Conduction b) Convection

c) Radiation d) All

91. During evaporation ________ is transferred from steam to the food, to raise the temperatureto its boiling point.

a) Sensible heat b) Latent heat

c) Specific heat d) a & b

92. Which of the following heat supplied by the steam to form bubbles of vapour, which leave the surface of the boiling liquid is

a) Sensible heat b) Latent heat

c) Specific heat d) a & b

93. In commercial vacuum evaporators the boiling point may be reduced to as low as _____ ºC.

a) 100°C b) 90°C

c) 70°C d) 40°C

94. For low-viscosity foods ______________ evaporator is used
   a) Falling-film
   b) Climbing-film
   c) Scraped-surface
   d) All
95. For high-viscosity foods______________ evaporator is used
   a) Falling-film
   b) Climbing-film
   c) Scraped-surface
   d) All
96. Cold extrusion is used to mix and shape foods such as and
   a) Pasta
   b) Meat products
   c) Sauce
   d) a & b
97. Low pressure extrusion is used to produce
   a) Liquorice
   b) Fish pastes
   c) Surimi
   d) All
98. The cooking temperature in extrusion is________°C
   a) 300–350°C
   b) 200–250°C
   c) 180–190°C
   d) 110–120°C
99. The cooking time in extrusion usually ______________
   a) 180-120 seconds
   b) 120-60 seconds
   c) 20–40 seconds.
   d) 20-10 seconds
100. Flavor reversion is very high in______________
   a) Safflower
   b) Sunflower
   c) Soybean
   d) Sesame

## Answers

| | | | | | | | | | |
|---|---|---|---|---|---|---|---|---|---|
| 1. d | 2. c | 3. d | 4. c | 5. b | 6. d | 7. b | 8. a | 9. a | 10. b |
| 11. b | 12. b | 13. d | 14. d | 15. b | 16. d | 17. c | 18. d | 19. d | 20. d |
| 21. d | 22. d | 23. d | 24. d | 25. d | 26. d | 27. d | 28. d | 29. d | 30. d |
| 31. d | 32. d | 33. a | 34. b | 35. b | 36. b | 37. b | 38. b | 39. b | 40. a |
| 41. a | 42. c | 43. b | 44. d | 45. a | 46. b | 47. b | 48. c | 49. c | 50. a |
| 51. c | 52. d | 53. d | 54. a | 55. c | 56. c | 57. d | 58. a | 59. d | 60. a |
| 61. b | 62. b | 63. d | 64. c | 65. a | 66. b | 67. c | 68. a | 69. b | 70. b |
| 71. d | 72. c | 73. d | 74. d | 75. c | 76. a | 77. a | 78. d | 79. a | 80. b |
| 81. c | 82. c | 83. b | 84. a | 85. d | 86. b | 87. b | 88. a | 89. b | 90. b |
| 91. a | 92. b | 93. d | 94. b | 95. a | 96. d | 97. d | 98. c | 99. c | 100. c |

1. The application of heat under controlled conditions to remove the majority of the water present in a food to preset level by evaporation is called
   a) Dehydration b) Drying
   c) Ashing d) Icing
2. The application of heat under controlled conditions to remove the majority of the water present in a food to bone dry level by evaporation is called
   a) Dehydration b) Drying
   c) Icing d) None
3. The study of inter-related properties of air–water vapour systems is called
   a) Rhealogy b) Pneamology
   c) Psychrometry d) None
4. The study of flow and deformation of systems is called
   a) Rhealogy b) Pneamology
   c) Psychrometry d) None
5. The dryer in which air is distributed at a uniform velocity all around the material
   a) Bin drier b) Cabinet drier
   c) Tunnel drier d) Fluidised-bed drier
6. The properties food to be feed into Spray drier are
   a) Fine dispersion b) Pre-concentrated
   c) Diluted d) a & b
7. The pre-concentration of food to be feed into Spray drier is to
   a) 20–30% moisture b) 30–40% moisture
   c) 40–60% moisture d) None
8. In spray dryer fine food droplets sprayed into heated air at__________ °C.
   a) 100–150 °C b) 150–200 °C
   c) 150–300 °C d) None
9. Heart of spray dryer is
   a) Pump b) Atomizer
   c) Feeder d) Drying chamber

10. Function of the atomizer is
    a) To form fine droplets sprayed into a co- or counter-current flow of heated air
    b) To spray fine droplets into heated air
    c) To create vacuum
    d) a & b
11. In which of the following atomizer, wider range of droplet sizes is produced?
    a) Two-fluid nozzle atomizer b) Pressure nozzle atomizer
    c) Ultrasonic nozzle atomizer d) Centrifugal atomiser
12. In which of the following is a two-stage atomizer
    a) Two-fluid nozzle atomizer b) Pressure nozzle atomizer
    c) Ultrasonic nozzle atomizer d) Centrifugal atomiser
13. The breakdown product of oil produced at high temperatures, which forms a blue haze above the oil
    a) Acetolin b) Acrelein
    c) Amides d) All
14. The frying suitable for foods of all shapes is
    a) Shallow frying b) Deep-fat frying
    c) Contact frying d) All
15. The fatty acids of naturally occurring lipids have even-number carbon atom chains with a typical length of ________
    a) 11 and 12 b) 14 and 16
    c) 16 and 18 d) 18 and 24
16. Important characteristics of industrial frying oils are
    a) High oxidative stability b) High smoke point,
    c) Low melting point d) All
17. Antioxidant precursor present in sesame oil is
    a) Sesamolin b) Sesamol
    c) Sesaminol d) All
18. Antioxidant activity having compound present in rice bran oil is
    a) Orygene b) Oryzanol
    c) Orygano d) All

19. Which of the following compound possess moderate antioxidant activities
    a) Squalene b) Phospholipids
    c) Triglycerides d) a & b

20. Squalene, possess moderate antioxidant activities is present in
    a) Shark liver oil b) Onions
    a) Palm oil a) Coconut oil

21. The rate of free fatty acid development depends on
    a) Moisture content of food b) Temperature of frying oil
    c) Rate of turnover of the oil d) All

22. The rate of rancidity depends on
    a) Amount of FFA in oil b) Amount of di-glycerides in oil
    c) Moisture content of food d) All

23. Key elements of good frying oil are
    a) Bland flavour b) Pale colour
    c) Good oxidative stability d) All

24. minimum smoke point for frying oils to provide adequate resistance to oxidation/degradation during continuous use in a fryer is
    a) 120°C b) 220°C
    c) 280°C d) 380°C

25. The safe level of lechithins in a frying medium is generally considered to be less than ___oil.
    a) 1 mg/100g b) 10 mg/100g
    c) 100 mg/100g d) 10 mg/1000g

26. Emulsifiers, e.g. mono- and di-glycerides (E 471) used in foods are escape into the oil during frying operations and they will cause
    a) Antioxidant property b) Darkening
    c) Foaming d) All of the above

27. The main difficulty with soya-bean and rapeseed oils is _____acid present, as this makes these two oils prone to oxidation and off-flavour development.
    a) Linolenic b) Palmetic
    c) Stearic d) Capric

28. In Ohmic heating
    a) Electrical resistance of foods to directly convert electricity to heat
    b) Molecular friction in water molecules to produce heat
    c) Dipole movement of the water to produce heat
    d) All
29. In Microwave heating
    a) Electrical resistance of foods to directly convert electricity to heat
    b) Molecular friction in water molecules to produce heat
    c) Dipole movement of the water to produce heat
    d) b & c
30. Microwaves produced in Microwave oven by
    a) Electro plate b) Megatron
    c) Gigatran d) None
31. Ohmic heating is also called
    a) Microwave heating b) Electroheating
    c) Resistance heating d) b & c
32. Advantages of microwave and D.E heating
    a) Uniform heating b) Rapid process control
    c) Efficient energy conservation d) All
33. The first generation membranes were made up of
    a) Polyanides b) Polysulphones
    c) Cellulose acetate d) PVC
34. Advantage of membrane system is
    a) Fouling
    b) Thermal degradation of nutrients
    c) Very high energy requirement
    d) Removal of water is accomplished without a change in it's state
35. Disadvantage of membrane system is
    a) Limitations of total solids (<50%)
    b) High viscosity of the concentrate units the process
    c) Reaction with cleaning agents
    d) All

36. Which of the following is not a use of membranes in food industry
    a) Separation of Milk fractions
    b) Removal of bacteria from milk
    c) Defatting of skim milk
    d) None
37. Which of the following is a dewatering technique
    a) Reverse osmosis b) Nano Filtration
    c) Ultra Filtration d) Micro Filtration
38. Microwave radiation is that part of the electromagnetic spectrum that falls between thefrequency of___________
    a) 3-3,00MHz b) 300-300,000 MHz
    c) 300-3,000 MHz d) 30-3,000 MHz
39. Microwave radiation is that part of the electromagnetic spectrum that falls between thewavelength of___________
    a) 1mm - 0.1 meter b) 00.1 mm - 00.1meter
    c) 0.1 mm - 0.1 meter d) 1 mm - 1meter
40. Electromagnetic spectrum frequency in Dielectric heating is
    a) 195 MHz b) 591 MHz
    c) 915 MHz d) 9150 MHz
41. Electromagnetic spectrum frequency in Microwave heating is
    a) 5420 MHz b) 4250 MHz
    c) 2450 MHz d) 245 MHz
42. Nano Filtration is also known as
    a) Loose Reverse osmosis b) Hyper filtration
    c) Hypo filtration d) All
43. Factors affecting efficiency of membranes
    a) Pressure b) Feed concentration
    c) Temperature d) All
44. Second generation membrane is
    a) Teflon b) Nylon
    c) PVC d) All

45. Third generation membrane is
    a) Zirconium oxide
    b) PVC
    c) Polyethylene
    d) PP

46. Typical pressure ranges for the Reverse osmosis membrane system for industry use are:
    a) 200-600 psi
    b) 20-60 psi
    c) 2-6 psi
    d) 0.2-0.6 psi

47. Typical pressure ranges for the Reverse osmosis membrane system for home use are
    a) 50 psi
    b) 5 psi
    c) 0.5 psi
    d) 0.05 psi

48. Typical pressure ranges for the Nano Filtration membrane system for industry use are
    a) 25-150 psi
    b) 100-200 psi
    c) 10-25 psi
    d) 1-2 psi

49. Typical pressure ranges for the Ultra Filtration membrane system for industry use are
    a) 25-150 psi
    b) 100-200 psi
    c) 10-25 psi
    d) 1-5 psi

50. Typical pressure ranges for the Micro Filtration membrane system for industry use are
    a) 25-150 psi
    b) 100-200 psi
    c) 10-25 psi
    d) 1-7 psi

51. Micro Filtration will remove ____________ from a water stream
    a) Salts
    b) Viruses
    c) Proteins
    d) None

52. For processing of protein gelatin in meat industry by products_____ system is used.
    a) Micro Filtration
    b) Ultra Filtration
    c) Nano Filtration
    d) Reverse osmosis

53. Dairy processors use ______ to concentrate skim milk for ice cream.
    a) Micro Filtration
    b) Ultra Filtration
    c) Nano Filtration
    d) Reverse osmosis

54. Membrane pore size in Micro-filtration is
 a) 0.1μ to 0.5μ  b) 1μ to 5μ
 c) 10μ to 50μ  d) 10μ to 100μ
55. Particles molecular weight (Dolton) for Ultra-filtration (UF) are
 a) 1,000 to 1,00,000  b) 1,00 to 1,00,00
 c) 1,0 to 1,00,0  d) 1 to 1,00
56. Particles molecular weight (Dolton) for Nano Filtration are
 a) 25 to 40  b) 250 to 400
 c) 2500 to 4000  d) 25000 to 40000
57. Particles molecular weight (Dolton) for Reverse osmosis are
 a) 125  b) 152
 c) 251  d) 215
58. The filtration technology providing an intermediate step between, UF and RO is
 a) Osmosis  b) Micro-filtration
 c) Nano-filtration  d) b & c
59. The filtration technology used to retain lactose in a whey stream is
 a) Osmosis  b) Micro-filtration
 c) Nano-filtration  d) b & c
60. Nano-filtration membrane can retain up to ______ % of lactose in a whey stream
 a) 66 %  b) 77 %
 c) 88 %  d) 99 %
61. The filtration technology used to purification of seawater is
 a) Osmosis  b) Micro-filtration
 c) Nano-filtration  d) Reverse osmosis
62. Typical application of Reverse osmosis (R.O) is purification of
 a) Brackish well water  b) City water
 c) Waste water  d) All
63. Typical rejection rate of R.O filters is ____% of NaCl as well as many other impurities.
 a) 9 to 9.9%  b) 90 to 99%
 c) 90 to 99.9%  d) None

64. Primary area where membranes can be applied on processing of fruit juices is

a) Deacidification b) Clarification

c) Concentration d) All

65. In the production of sparkling clear beverages ______ filtration technology used.

a) Osmosis b) Micro-filtration

c) Nano-filtration d) Reverse osmosis

66. Heating food from the frozen state is not recommended, because

a) Crystal formation is more

b) Unthawed areas are over cooked

c) Thawed areas are under cooked

d) Areas of solid mass that become thawed absorb more energy

67. Which of the following statement is not true about microwave heating

a) The microwaves penetrate deeper into ice than into water

b) Microwave absorption rate is much higher for water than for ice.

c) Microwave ovens are more efficient than standard electric ovens and gas oven.

d) Effect of microwave absorption is less than effect of penetration

68. The osmotic pressure of solution is proportional to

a) Absolute temperature b) Concentration

c) Flow rate d) a & b

69. In membrane system flux means: volumetric flow rate per unit area of membrane surface

a) Flow rate

b) Volumetric flow rate

c) Volumetric flow rate per unit area

d) None

70. Model for membrane transport is/are

a) Capillary flow model b) Solution diffusion model

c) Concentration model d) a & b

71. In capillary flow model of membrane transport
    a) Solute is rejected by a filtering
    b) Solute is rejected by sieving action
    c) Liquid flows through a micro porous structure which prevents the passage of larger molecules
    d) All
72. In Solution - diffusion model of membrane transport
    a) Material across the followed by molecular diffusion across the barrier.
    b) Solute is rejected by sieving action
    c) Liquid flows through a micro porous structure which prevents the passage of larger molecules
    d) Solute is rejected by a filtering
73. Material used in commercial reverse osmosis (RO) applications is
    a) Cellulose acetate b) Cellulose polyamide
    c) Modified form of nylon d) All
74. Expand IMF
    a) Intermediate Meat Foods b) Intermediate Milk Foods
    c) Integrated Moisture Foods d) Intermediate Moisture Foods
75. IMF foods contain moisture of the order of ______% by weight.
    a) 10-60% b) 20-50%
    c) 30-70% d) 20-60%
76. Which of the following is a intermediate moisture food example
    a) Jelly b) Jam
    c) Honey d) All
77. Which of the following food requires refrigeration to prevent microbial deterioration?
    a) Jelly b) Jam
    c) Honey d) Meat
78. Which of the following food requires refrigeration to prevent microbial deterioration?
    a) Fruit cake b) Sweetened condensed milk
    c) Dates d) Fresh milk

79. When a food is in moisture equilibrium with its environment then the $a_w$ of the food will bequantitatively equal to the RH/100

a) 100 b) RH

c) RH/100 d) 100/RH

80. Which of the following utensils are the best conductors of heat

a) Copper b) Steel

c) Aluminum d) Wood

81. Correct decreasing order of thermal conductivity is

a) Copper < steel < Aluminum < Glass

b) Copper < Aluminum < steel < Glass

c) Copper < Aluminum < Glass < steel

d) Copper < Aluminum < steel < Glass

82. Liquids in glass type thermometer can measure the temperature upto____

a) 110°C b) 310°C

c) 410°C d) 510°C

83. Operating temperatures range for mercury in glass type thermometer is

a) -10°C to 110°C b) 0°C to 150°C

c) 0°C to 100°C d) -37°C to 100°C

84. For food industrial purpose, ______in glass type thermometer is acceptable to prevent chances of chemical hazards.

a) Mercury b) Alcohol

c) Hexane d) All

85. Gas used in Gas- filled thermometers usually is

a) Hydrogen b) Nitrogen

c) Oxygen d) All

86. Operating temperatures range for nitrogen - filled thermometer is

a) -200° to 800° F b) -20° to 80° F

c) -20° to 80° C d) -200° to 800° C

87. Operating temperatures range for thermometer with vapor – pressure system is

a) -20° to 60° F b) -20° to 600° F

c) -20° to 600° C d) -2° to 60° C

88. Which of the following is used to measure temperature
   a) Anemometer b) Pyrometer
   c) Thermometer d) b & c
89. Which of the following is used to measure wind speed
   a) Anemometer b) Pyrometer
   c) Thermocouple d) Manometer
90. Indicator used in pyrometer is
   a) Thermocouple b) Nitrogen gas
   c) Mercury d) Alcohol
91. Operating temperatures range for resistance thermometers is
   a) -40 to 180° C b) -400 to 1800° C
   c) -400 to 1800° F d) -40 to 180° F
92. Platinum (Pt) - Pt (alloy) thermocouples are preferred for pyrometry,
   a) Platinum (Pt) - Pt (alloy) b) Ag- Ag alloy
   c) Pt - Ag alloy d) Pt - Ni alloy
93. Platinum (Pt) - Pt (alloy) thermocouples are preferred for pyrometry, because
   a) They are more resistance to oxidation at high temperature
   b) They are more resistance to oxidation at low temperature
   c) They are more resistance to conduction at high temperature
   d) They are more resistance to conduction at low temperature
94. The approximate operation teperature range of thermistors is
   a) -15 to 50° F b) -150 to 500° F
   c) -15 to 50° C d) -150 to 500° C
95. A hydroscopic salt used in electric hygrometer is
   a) Sodium chloride b) Potasium chloride
   c) Lithium chloride d) Zinc chloride
96. The cost of items that usually not directly related to the amount of use is called
   a) Fixed cost b) Operating costs
   c) Variable cost d) Final cost

97. The cost of items that usually directly related to the amount of use is called

a) Fixed cost b) Operating costs

c) Variable cost d) a & b

98. Operating cost is also called

a) Fixed cost b) Variable cost

c) Final cost d) Operational cost

99. Which of the following is a Fixed cost

a) Depreciation b) Interest on the investment

c) Housing d) All

100. Which of the following is a variable cost

a) Fuel, power and utilities b) Maintenance

c) Labour d) All

## Answers

| | | | | | | | | | | | | | | | | | | | |
|---|---|---|---|---|---|---|---|---|---|---|---|---|---|---|---|---|---|---|---|
| 1. | b | 2. | a | 3. | c | 4. | a | 5. | d | 6. | d | 7. | c | 8. | c | 9. | b | 10. | d |
| 11. | a | 12. | c | 13. | b | 14. | b | 15. | c | 16. | d | 17. | a | 18. | b | 19. | d | 20. | a |
| 21. | d | 22. | d | 23. | d | 24. | b | 25. | b | 26. | c | 27. | a | 28. | a | 29. | d | 30. | b |
| 31. | d | 32. | d | 33. | c | 34. | d | 35. | d | 36. | d | 37. | a | 38. | b | 39. | d | 40. | c |
| 41. | c | 42. | a | 43. | d | 44. | d | 45. | a | 46. | a | 47. | a | 48. | b | 49. | a | 50. | c |
| 51. | d | 52. | b | 53. | b | 54. | a | 55. | a | 56. | b | 57. | a | 58. | c | 59. | c | 60. | d |
| 61. | d | 62. | d | 63. | c | 64. | d | 65. | d | 66. | d | 67. | d | 68. | d | 69. | c | 70. | d |
| 71. | d | 72. | a | 73. | d | 74. | d | 75. | b | 76. | d | 77. | d | 78. | d | 79. | c | 80. | a |
| 81. | b | 82. | d | 83. | d | 84. | b | 85. | b | 86. | a | 87. | b | 88. | d | 89. | a | 90. | a |
| 91. | c | 92. | a | 93. | a | 94. | b | 95. | c | 96. | a | 97. | d | 98. | b | 99. | d | 100. | d |

1. Decrease in value of a piece of property during a period of time is called
   a) Esculation b) Depreciation
   c) Intensification d) Augmentation
2. The depreciated value also called as
   a) Actual value b) Book value
   c) Null value d) Fixed value
3. The frying method in which fried food is partially immersed in oil.
   a) Shallow fat frying b) Deep fat frying
   c) Pan frying d) a & b
4. In shallow fat frying the layer of frying oil in a pan is about ____ mm thick.
   a) 1-2 mm b) 1-5 mm
   c) 1-10 mm d) 10-15 mm
5. In shallow fat frying, frying time is about __________
   a) 10-15 min b) 5-10 min
   c) 1-5 min d) 25-30 min
6. In deep fat frying the layer of frying oil in a pan is about ____ mm thick.
   a) 25-30 min b) 20-200 mm
   c) 250-300 min d) 225-330 min
7. In deep fat frying, frying time is about __________
   a) 10-15 min b) 5-10 min
   c) 1-5 min d) 25-30 min
8. During frying fat or oil is preheated to temperature of ______°C
   a) 50-150°C b) 180-200°C
   c) 150-180°C d) 80-100°C
9. During frying the temperature in inner layer of fat or oil reaches to ______°C
   a) 50-80°C b) 150-180°C
   c) 80-100°C d) 180-200°C
10. During frying at approximately ___°C, as oils thermally and oxidative decomposes, volatile and non volatile products are formed.
    a) 19°C b) 109°C
    c) 190°C d) 290°C

11. During frying
    a) Water is continuously released from the food into the hot oil
    b) Steam distillation effect occurs
    c) Sweeping volatile oxidative products from the oil
    d) All
12. The oil is subjected to ________ reaction during deep frying.
    a) Hydrolytic b) Oxidative
    c) Pyrolysis d) All
13. In deep frying, _____________ are formed as primary reaction products.
    a) Protons b) Hydro peroxides
    c) Super oxides d) Halides
14. Hydro peroxides in deep frying are formed, but their content is not exceeds 1% because they are
    a) Unstable at high temperature b) Unstable at high oil content
    c) Unstable at high $p^H$ d) Unstable at low $p^H$
15. Antioxidant that is more efficient in frying oils is
    a) Gallates b) Citric acid
    c) Ascorbyte, d) Palmitate
16. Defenation for synergist:_________________________
    a) Action of antioxidant with chelation
    b) Synergist is like symbiosis
    c) Oposite actionon antioxidant
    d) None
17. Which of the following are acts as synergist for tocopherol.
    a) Palmitate b) Citric acid
    c) Ascorbyte, d) All
18. Which of the following are also found to be active in frying oils for preventing oxidation.
    a) Rosemary b) Sage resins
    c) a & b d) None
19. Oxidation reactions can be inhibited by polysiloxans and polysiloxans are considered safe for human consumption because
    a) They form a very thin layer on the surface of the frying oil
    b) Prevent the access to oxygen
    c) They are not absorbed in the intestines
    d) All

20. Chilling is the unit operation in which the temperature of a food is reduced to between

a) 0°C & 5°C  b) 0°C & 8°C

c) -1°C & 8°C  d) 1°C & 10°C

21. Chilling cause

a) Reduction in the rate of biochemical changes

b) Reduction in the rate of microbiological changes

c) Minimal changes to sensory characteristics and nutritional properties

d) All

22. Chilling temperature for fresh meat is

a) -1°C to +1°C  b) 0°C to +5°C

c) 0°C to +8°C  d) 1°C to +10°C

23. Chilling temperature for pasteurized canned meat is

a) -1°C to +1°C  b) 0°C to +5°C

c) 0°C to +8°C  d) 1°C to +10°C

24. Chilling temperature for fully cooked meat is

a) -1°C to +1°C  b) 0°C to +5°C

c) 0°C to +8°C  d) 1°C to +10°C

25. Chilling temperature for fresh smoked meat is

a) -1°C to +1°C  b) 0°C to +5°C

c) 0°C to +8°C  d) 1°C to +10°C

26. Chilling temperature for milk, cream, yoghurt is

a) -1°C to +1°C  b) 0°C to +5°C

c) 0°C to +8°C  d) 1°C to +10°C

27. Chilling temperature for butter, margarine is

a) -1°C to +1°C  b) 0°C to +5°C

c) 0°C to +8°C  d) 1°C to +10°C

28. Chilling temperature for cooked rice is

a) -1°C to +1°C  b) 0°C to +5°C

c) 0°C to +8°C  d) 1°C to +10°C

29. Chilling temperature for fruit juices and soft fruits is

a) -1°C to +1°C  b) 0°C to +5°C

c) 0°C to +8°C  d) 1°C to +10°C

30. Chilling temperature for unbaked dough is

a) -1°C to +1°C  b) 0°C to +5°C

c) 0°C to +8°C  d) 1°C to +10°C

31. The Fruit which undergo short ripening and abrupt increase in the rate of respiration is called

a) Climacteric fruit  b) Non-Climacteric fruit

c) Critical fruits  d) False fruit

32. The Fruit which is not undergoes ripening after harvesting & slow in the rate of respiration is

a) Climacteric fruit  b) Non-Climacteric fruit

c) Critical fruits  d) False fruit

33. Which of the following is a Climacteric fruit

a) Apple  b) Apricot

c) Avocado  d) All

34. Which of the following is a Non-Climacteric fruit

a) Cherry  b) Cucumber

c) Fig  d) All

35. Which of the following is a Climacteric fruit

a) Peach  b) Pear

c) Plum  d) All

36. Which of the following is a Non-Climacteric fruit

a) Grapefruit  b) Pineapple

c) Strawberry  d) All

37. Undesirable changes to some fruits and vegetables occur when the temperature is reduced below a specific optimum for the individual fruit is termed as

a) Cooling injury  b) Crushing injury

c) Chilling injury  d) Biological injury

38. Characteristic of chilling injury

a) Internal or external browning b) Failure to ripen

c) Skin blemishes d) All

39. Chilling injury occurs in apples at the temperature of ____°C.

a) Less than 2–3°C b) Less than 12°C

c) Less than 13°C d) Less than 14°C

40. Chilling injury occurs in avocados at the temperature of ____°C.

a) Less than 16°C b) Less than 13°C

c) Less than 14°C d) Less than 15°C

41. Chilling injury occurs in banana is at the temperature of ____°C.

a) Less than 15°C b) Less than 13°C

c) Less than 14°C d) Less than 12–13°C

42. Chilling injury occurs in lemons is at the temperature of ____°C.bananas

a) Less than 17°C b) Less than 16°C

c) Less than 15°C d) Less than 14°C

43. Chilling injury occurs in mangoes is at the temperature of ____°C.

a) Less than 10–13°C b) Less than 13–14°C

c) Less than 14°C d) Less than 15°C

44. Chilling injury occurs in pineapple is at the temperature of ____°C.

a) Less than 7–10°C b) Less than 13–14°C

c) Less than 14°C d) Less than 15°C

45. Chilling injury occurs in melon is at the temperature of ____°C.

a) Less than 7–10°C b) Less than 13–14°C

c) Less than 14°C d) Less than 15°C

46. Chilling injury occurs in tomato is at the temperature of ____°C.

a) Less than 7–10°C b) Less than 13–14°C

c) Less than 14°C d) Less than 15°C

47. Arrange the order of steps in rigor mortis formation

a) Aerobic respiration → animal slaughtering → Anaerobic respiration → decline in pH

b) Animal slaughtering → aerobic respiration → Anaerobic respiration → decline in pH

c) Aerobic respiration → animal slaughtering → decline in pH→ Anaerobic respiration

d) Anaerobic respiration → animal slaughtering → aerobic respiration → decline in pH

48. Cooling during anaerobic respiration of meat is necessary to

a) Produce the required texture of meat

b) Produce the required colour of meat

c) Reduce bacterial contamination

d) All

49. Undesirable changes, caused by cooling meat before rigor mortis has occurred, are termed

a) Chilling injury b) Cold shortening

c) Heat shock d) Cold shock

50. After harvesting of fruits and vegetables, they are access to shade places to remove

a) Field heat b) Latent heat

c) Specific heat d) None

51. Field heat is also known as

a) Sensible heat b) Latent heat

c) Specific heat d) None

52. To chill fresh foods it is necessary to remove

a) Heat generated by respiratory activity

b) Field heat

c) Latent heat

d) a & b

53. Basic element of mechanical refrigerator is

a) Evaporator b) Condenser & compressor

c) Expansion valve d) All

54. Heart of mechanical refrigerator is

a) Evaporator b) Condenser

c) Expansion valve d) Compressor

55. In the evaporator of mechanical refrigerator the liquid refrigerant evaporates under

a) Reduced pressure
b) High pressure
c) Atmospheric pressure
d) None

56. Arrange the elements of mechanical refrigerator in order of flow of refrigerant

a) Evaporator → Compressor → Condenser → Expansion valve
b) Evaporator → Condenser → Compressor → Expansion valve
c) Evaporator → Compressor → Expansion valve → Condenser
d) Condenser → Compressor → Expansion valve → Evaporator

57. The pressure in compressor is____ compared to condenser of mechanical refrigerator

a) Decreased
b) Increased
c) Not changed
d) Depends on refrigerant

58. In the expansion valve of mechanical refrigerator the pressure of refrigerant is

a) Decreased
b) Increased
c) Not changed
d) Depends on refrigerant

59. Refrigeration is the action of

a) Moving heat from low temperature to high temperature
b) Cooling by removal of heat
c) Discarding of heat at a higher temperature
d) All

60. Formula for conversion of °C to ° F is

a) (°F-32)/9=°C/5
b) (°F-32)/5=°C/9
c) (°F-32)/180=°C/100
d) a&c

61. Triple point of water is at

a) 0.0098°C and 0.64 kPa
b) 0.0098°C and 4.8 mm of Hg
c) 0.98°C and 0.4 kPa
d) a & b

62. Triple point of ice is

a) 0.1°C & 6.1 mbar
b) 0.01°C &61 mbar
c) 0.01°C & 6.1 mbar
d) None

63. The point where the solid, liquid, and gas phases of a substance are in thermodynamic equilibrium is called

    a) Single point
    b) Double point
    c) Triple point
    d) None

64. Efficiency of Refrigeration system is measured in terms of

    a) COP
    b) COB
    c) COD
    d) BOD

65. Expand COP

    a) Coefficient of performance
    b) Coefficient of percentage
    c) Coefficient of packaging
    d) Coefficient of pump

66. If a change of enthalpy can be sensed as a change of temperature, it is called

    a) Latent heat
    b) Specific heat
    c) Sensible heat
    d) Entropy

67. If there is no change of temperature but a change of state, it is called.

    a) Latent heat
    b) Specific heat
    c) Sensible heat.
    d) Entropy

68. The process in which the solid can be changed directly to a gas is called

    a) Imbibation
    b) Sublimation
    c) Gasification
    d) None

69. Sublimation occurs at a pressure _____ the triple point pressure

    a) Below
    b) Above
    c) Equal
    d) None

70. Boyle's Law states that

    a) pV = constant
    b) V/T = constant
    c) p/T = constant
    d) R/T = constant

71. Charles' Law states that

    a) pV = constant
    b) V/T = constant
    c) p/T = constant
    d) R/T = constant

72. Gay-Lussac law states that

    a) pV = constant
    b) V/T = constant
    c) p/T = constant
    d) R/T = constant

73. Ideal gas equation is

a) TV = mRp
b) pm = VRT
c) pV = mRT
d) RV = mpT

74. The law states, "total pressure of the mixture is equal to the sum of the individual pressures, if each gas separately occupied the spac" is

a) Cornot's Law
b) Pearson's Law
c) Boyle's Law
d) Dalton's Law

75. Mode of heat transfer in which direct from one body touching the other is

a) Conduction
b) Convection
c) Radiation
d) All

76. Mode of heat transfer by means of a heat-carrying fluid moving between one and the other is

a) Conduction
b) Convection
c) Radiation
d) All

77. Mode of heat transfer through radiation is

a) Conduction.
b) Convection.
c) Radiation
d) All

78. Which of the following is the Reynolds number

a) $\rho vx/\mu$
b) $hx/k$
c) $Cp\mu/K$
d) $\beta g\ \rho 2\ x3\ \theta/\mu 2$

79. Which of the following is the Nusselt number

a) $\rho vx/\mu$
b) $hx/k$
c) $Cp\mu/K$
d) $\beta g\ \rho 2\ x3\ \theta/\mu 2$

80. Which of the following is the Prandtl number

a) $\rho vx/\mu$
b) $hx/k$
c) $Cp\mu/K$
d) $\beta g\ \rho 2\ x3\ \theta/\mu 2$

81. Which of the following is the Grashof number

a) $\rho vx/\mu$
b) $hx/k$
c) $Cp\mu/K$
d) $\beta g\ \rho 2\ x3\ \theta/\mu 2$

82. Which of the following dimensionless number is deals with Forced flow in pipes

a) Reynolds
b) Nusselt
c) Prandtl
d) Grashof

83. Which of the following dimensionless number is deals with Convection heat transfer rate

a) Reynolds b) Nusselt
c) Prandtl d) Grashof

84. Which of the following dimensionless number is deals with Fluid properties

a) Reynolds b) Nusselt
c) Prandtl d) Grashof

85. Which of the following dimensionless number is deals with Natural convection

a) Reynolds b) Nusselt
c) Prandtl d) Grashof

86. Expand the LMTD

a) Long mean temperature difference
b) Low mean temperature difference
c) Logarithmic mean temperature difference
d) Logic mean temperature difference

87. A process in which the heat of compression raises the enthalpy of the gas is termed ____compression

a) Adiabatic b) Ideal
c) Critical d) Vapour

88. Ideal property for a refrigerant is

a) High latent heat of vaporization
b) High suction gas density
c) Critical temperature and triple point well outside the working range
d) All

89. Which of the following is not a ideal property for a refrigerant

a) Non-corrosive, non-toxic and non-flammable
b) High dielectric strength
c) Environmentally friendly
d) High cost

90. R (Refrigerant) number for ammonia is

a) R717 b) R177
c) R170 d) R117

91. As a refrigerant, ammonia cannot be used with _______ metal.

a) Copper b) Copper alloys

c) Aluminum d) a & b

92. As a refrigerant, ammonia can be used with _______ metal.

a) Copper b) Steel

c) Aluminum d) b & c

93. Zeotropic mixtures of refrigerant are assigned an identifying number in the _____ series.

a) 400 b) 500

c) 600 d) 700

94. Azeotropic mixtures of refrigerant are assigned an identifying number in the _____ series.

a) 400 b) 500

c) 600 d) 700

95. Miscellaneous refrigerants are assigned an identifying number in the _____ series.

a) 400 b) 500

c) 600 d) 700

96. During freezing

a) Bacterial degradation reactions are reduced

b) Structural changes occur in the product

c) Ice crystals are formed

d) All

97. Quick frozen produce is generally frozen in _____ minutes in an air blast freezer.

a) 1–5 b) 5–10

c) 10–15 d) 15–20

98. Arrange the freeze drying process steps in order

1. Placing the product in contact freezing plates
2. Freeze the product to about -25°C to -50°C depend on product
3. Controlling of vacuum and heat to provide the latent heat of sublimation
4. Water is driven off from product then product is collapses to a dry powder

a) $1 \rightarrow 4 \rightarrow 3 \rightarrow 2$ b) $1 \rightarrow 3 \rightarrow 2 \rightarrow 4$

c) $1 \rightarrow 4 \rightarrow 2 \rightarrow 3$ d) $1 \rightarrow 2 \rightarrow 3 \rightarrow 4$

99. Characteristic of freeze dried product is ________________ in nature.

a) Hygroscopic b) Lipophillic

c) Halophillic d) None

100. The refrigeration load is depends on

a) Heat of respiration of vegetables and fruits

b) Heat gain through the walls, ceiling and floor

c) Evaporator fans, pumps, lighting

d) All

## Answers

| | | | | | | | | | |
|---|---|---|---|---|---|---|---|---|---|
| 1. b | 2. b | 3. a | 4. c | 5. b | 6. b | 7. b | 8. c | 9. c | 10. c |
| 11. d | 12. d | 13. b | 14. a | 15. a | 16. a | 17. d | 18. c | 19. d | 20. c |
| 21. d | 22. a | 23. b | 24. c | 25. a | 26. b | 27. c | 28. c | 29. c | 30. b |
| 31. a | 32. b | 33. d | 34. d | 35. d | 36. d | 37. c | 38. d | 39. a | 40. b |
| 41. d | 42. d | 43. a | 44. a | 45. a | 46. a | 47. a | 48. d | 49. b | 50. a |
| 51. a | 52. d | 53. d | 54. d | 55. a | 56. a | 57. c | 58. a | 59. d | 60. d |
| 61. d | 62. c | 63. c | 64. a | 65. a | 66. c | 67. a | 68. b | 69. a | 70. a |
| 71. b | 72. c | 73. c | 74. d | 75. a | 76. b | 77. c | 78. a | 79. b | 80. c |
| 81. d | 82. a | 83. b | 84. c | 85. d | 86. c | 87. a | 88. d | 89. d | 90. a |
| 91. d | 92. d | 93. a | 94. b | 95. c | 96. d | 97. b | 98. d | 99. a | 100. d |

1. Which of the following is used as insulating material
   a) Polyisocyanurate (PIR)
   b) Polyurethane foam(PUF)
   c) Cyanide
   d) a & b
2. Most efficient insulating material is
   a) Polyisocyanurate (PIR)
   b) Polyurethane foam(PUF)
   c) Glass
   d) Wood
3. The ability of an insulant to reduce heat flow is expressed as _____
   a) Eesistivity
   b) Conductivity
   c) Flow rate
   d) Overall heat transfer
4. The units of the heat flow resistivity are
   a) mW/K
   b) KW/m
   c) W/mK
   d) K/mW
5. Milk comes from the cow at about 37°C and must be cooled to 4°C within
   a) 30 min
   b) 1 hour
   c) 2 hours
   d) 3 hours
6. In pasteurizer raw milk at 4°C is heated by the outgoing milk up to about _____°C
   a) 41°C
   b) 51°C
   c) 61°C
   d) 71°C
7. In pasteurizer the milk that is heated by the outgoing milk, finally heated by hot water up to the pasteurizing temperature of ___°C
   a) 61°C
   b) 65°C
   c) 71°C
   d) 75°C
8. Holding time for HTST process pasteurizer is
   a) 15 sec
   b) 15 min
   c) 30 sec
   d) 30 min
9. In pasteurizer the milk holded for holding time then it is cooled by the incoming milk, down to about ____°C.
   a) 40°C
   b) 30°C
   c) 20°C
   d) 10°C

10. Butter must be stored at 5°C to

    a) Crystallize the fat  b) Crystallize the water
    c) Crystallize the sugar  d) None

11. For long-term storage of butter it is stored at ____°C.

    a) 5°C  b) - 5°C
    c) - 15°C  d) - 25°C

12. The most usual refrigerant for ice-cream continuous freezers is

    a) CFC  b) Methane
    c) Ammonia  d) Ethylene

13. Evaporating temperature of ammonia is ____°C.

    a) - 5°C to - 3°C  b) - 35°C to - 30°C
    c) - 53°C to - 40°C  d) - 75°C to - 50°C.

14. Ice cream must then be hardened by cooling down to a storage temperature of ___°C

    a) 5°C  b) -5°C
    c) -25°C  d) 25°C

15. The optimum temperature of fermentation of red wine is ___°C.

    a) 16°C  b) 29°C
    c) 39°C  d) 49°C

16. The optimum temperature of fermentation of white wine is ___°C.

    a) 16°C  b) 29°C
    c) 39°C  d) 49°C

17. Fermented wine is traditionally matured and stored in caves or cellars at about ___°C.

    a) 10°C  b) 16°C
    c) 29°C  d) 39°C

18. In carbonation, each litre of water will have dissolved ____litres of. carbon dioxide gas.

    a) 0.5–3  b) 3–3.5
    c) 3.5–5  d) 5–5.3

19. For short-term cooling of airline passenger meals____ is used.

    a) Water ice cubes  b) Dry ice
    c) Wet ice  d) None

20. Dry ice is nothing but__________.
   a) Solid nitrogen
   b) Solid carbon dioxide
   c) Liquid carbon dioxide
   d) All
21. For vacuum cooling, the pressure is reduced to approximately ___ kPa
   a) 0.5 kPa
   b) 1 kPa
   c) 1.5 kPa
   d) 2 kPa
22. Direct immersion of fruit and Vegetables in chilled water to remove field heat is called
   a) Preheating
   b) Hydration
   c) Hydrocooling
   d) Freezing
23. cheese is often cooled by direct immersion in refrigerated _____.
   a) Acid
   b) Sugar
   c) Brine
   d) Alcohol
24. A cryogen is a refrigerant that changes phase by absorbing ____heat to cool the food.
   a) Specific
   b) Sensible
   c) Latent
   d) None
25. Cryogenic chillers use________ for chilling
   a) Solid carbon dioxide
   b) Liquid carbon dioxide
   c) Liquid nitrogen
   d) All
26. Latent heat of sublimation removed by Solid carbon dioxide is ___ at -78°C.
   a) 352 kJ /kg
   b) 358 kJ /kg
   c) 352 J /kg
   d) 352 Cal /kg
27. Liquid cryogens remove latent heat of vaporization is _______ at -196°C.
   a) 352 kJ /kg
   b) 358 kJ /kg
   c) 352 J /kg
   d) 352 k.cal /kg
28. Most widely used thermocouples is/ are
   a) Type K
   b) Type T
   c) Type S
   d) Both a & b
29. Nickel-chromium and nickel-aluminium is a_______type thermocouples.
   a) Type K
   b) Type T
   c) Typc S
   d) Type A

30. Copper and copper-nicke is a _______ type thermocouples.

    a) Type K
    b) Type T
    c) Type S
    d) Type A

31. Balancing the gas composition for long storage of foods is called

    a) CAS
    b) MAP
    c) MAS
    d) CAP

32. Modifying the gas composition for long storage of foods is

    a) CAS
    b) MAP
    c) MAT
    d) CAP

33. Controlled-atmosphere storage (CAS) is mostly used for apples

    a) Wood apples
    b) Pine apples
    c) Apples
    d) Guava

34. Oxygen and carbon dioxide percentages in air are ____ & ______ respectively.

    a) 27% and 0.035%
    b) 29% and 0.035%
    c) 21% and 0.35%
    d) 21% and 0.035%

35. Oxygen and carbon dioxide percentages in controlled-atmosphere storage (CAS) are ____& ______respectively.

    a) 0% and 0.5%
    b) 0% and 5%
    c) 0% and 2%
    d) 0% and 20%

36. Widely used gas in Modified-atmosphere packaging (MAP) is

    a) Helium
    b) Nitrogen
    c) Hydrogen
    d) All

37. The two most important technical parameters of packs for MAP is/are

    a) Gas permeability
    b) Moisture vapour permeability
    c) Strength
    d) a & b

38. Oxygen permeability in low barrier packaging material intended for MAP is

    a) $>300$ cc/m$^2$
    b) 50–300 cc/m$^2$
    c) 10–50 cc/m$^2$
    d) $\leq 10$ cc/m$^2$

39. MAP packaging material with high oxygen permeability is can be used for___

a) Fresh meat b) Dried meat

c) Smoked meat d) Cooked meat

40. Oxygen permeability in medium barrier packaging material intended for MAP is

a) >300 cc/m$^2$ b) 50–300 cc/m$^2$

c) 10–50 cc/m$^2$ d) $\leq$ 10 cc/m$^2$

41. Oxygen permeability in high barrier packaging material intended for MAP is

a) >300 cc/m$^2$ b) 50–300 cc/m$^2$

c) 10–50 cc/m$^2$ d) $\leq$ 10 cc/m$^2$

42. Oxygen permeability in ultra high barrier packaging material intended for MAP is

a) >300 cc/m$^2$ b) 50–300 cc/m$^2$

c) 10–50 cc/m$^2$ d) $\leq$ 10 cc/m$^2$

43. Ethylene scavenger used in active packaging systems to slow down the ripening of fruits is

a) Potassium permanganate b) Silica gel

c) Agar gel d) All

44. Carbon dioxide scavenger used in active packaging systems is

a) Ferrous carbonate b) Calcium hydroxide

c) Iron oxide d) All

45. Which of the following foods freezes at -0.5°C

a) Milk b) Egg

c) Meat d) a & b

46. For ice crystal formation, a ________ of water molecules must be present.

a) Nucleus b) Segregation

c) Isolation d) All

47. The nucleation in which the chance orientation and combination of water molecules occurs in ice crystal formation is

a) Homogeneous b) Heterogeneous

c) Mixed d) a & b

48. The nucleation in which formation of a nucleus around suspended particles or at a cell wall occurs in ice crystal formation is

a) Homogeneous b) Heterogeneous

c) Mixed d) a & b

49. Heterogeneous nucleation is more likely to occur in foods & takes place during _______.

a) Cooling b) Hydro cooling

c) Super cooling d) a & b

50. The temperature at which a crystal of an individual solute exists in equilibrium with theunfrozen liquor and ice is called eutectic temperature

a) Freezing point b) Boiling point

c) Normal temperature d) Eutectic temperature

51. In blast freezers, air is recirculated over food at temperature between____

a) 3°C and 4°C b) -3°C and -4°C

c) -30°C and -40°C d) -40°C and -60°C

52. In blast freezers, air is recirculated over food at between -30°C and -40°C at a velocity of _____ m/s.

a) 1–2 m/s b) 1.5–6.0 m/s

c) 15–30 m/s d) 25–60 m/s

53. Fluidised-bed freezers are modified blast freezers in which air at temperature between

a) -25°C and -35°C b) -2°C and -3°C

c) -30°C and -40°C d) -40°C and -60°C

54. Fluidised-bed freezers are modified blast freezers in which air at between -25°C and -35°C ispassed at a high velocity ______m/s.

a) 1–2 m/s b) 2–6 m/s

c) 6–8 m/s d) 8–10 m/s

55. In immersion freezers, packaged food is passed through a bath of refrigerated___solution.

a) Propylene glycol b) Glycerol

c) Calcium chloride d) All

56. Plate freezers consist of a vertical or horizontal stack of hollow plates, through whichrefrigerant is pumped at ____°C

a) -10°C b) -20°C

c) -30°C d) -40°C

57. freezer used for ice cream manufacturing is

a) Scraped-surface b) Plate

c) Blast d) All

58. In baked goods a ____ proportion of amylopectin is needed in the starch to preven tretrogradation and staling during slow freezing and frozen storage.

a) Low b) High

c) Medium d) None

59. The main change/s to frozen foods during storage is/are

a) Degradation of pigments b) Loss of vitamins

c) Oxidation of lipids d) All

60. Physical changes to ice crystals (for example changes in their shape, size or orientation) are collectively known as

a) Thawing b) Crystallisation

c) Recrystallisation d) All

61. Which one of the followings is a recrystallisation type.

a) Isomass b) Accretive

c) Migratory d) All

62. The recrystallisation type in which changes in surface shape or internal structure occur is

a) Isomass b) Accretive

c) Migratory d) All

63. The recrystallisation type in which two adjacent ice crystals join together to form a largercrystal and cause an overall reduction in the number of crystals is

a) Isomass b) Accretive

c) Migratory d) All

64. The recrystallisation type in which increase in the average size and a reductionin the average number of crystals are occur is

a) Isomass b) Accretive

c) Migratory d) All

65. The humidity in cold stores compared to its outside atmosphere is

a) Equal b) More

c) Low d) None

66. Freezer burn is a particular problem in foods that have a ____surface-area-to-volume ratio.

a) Small b) Large

c) Medium d) None

67. Cold stores have a low humidity because

a) Moisture is added from the air by the refrigeration coils

b) Moisture is removed from the air by the refrigeration coils

c) Air is removed from the air by the refrigeration coils

d) Air is added from the air by the refrigeration coils

68. IQF foods are example for _____ surface-area-to-volume ratio foods

a) Small b) Large

c) Medium d) None

69. In Thawing of food

a) food is thawed in air or water

b) Surface ice melts to form a layer of water.

c) Surface layer of water reduces the rate at which heat is conducted to the frozen interior

d) All

70. The term 'Tempura' means

a) Batter b) Rigid batter

c) Viscous batter d) All

71. The sugar content in sugar cane juice is

a) 15-20 % b) 80-90%

c) 70-80% d) 5-10%

72. In Hard coating
    a) Centres are coated with a sweetener solution
    b) Sweetener solution is added at a rate of 10–15% of the weight of the centres
    c) hard coating is built up around the centres
    d) All
73. Coatings are traditionally made from ________% sucrose or dextrose syrups.
    a) 30–45% b) 40–55%
    c) 60–65% d) 70–85%
74. The function/s of packaging is/are
    a) Containment b) Communication
    c) Convenience d) All
75. Which of the packaging function is deals with performance on production lines for high speed filling, closing and collating.
    a) Containment b) Communication
    c) Convenience d) Machinability
76. Corrugated fiber board cases are the most widely used shipping container for ___kg loads,
    a) 1–5 kg kg b) 5–20 kg
    c) 20–30 kg d) 20–100 kg
77. In packaging IBC means
    a) International bulk containers b) Intermediate bulk containers
    c) Internal bulk containers d) None
78. Packaging material which has/have a low thermal conductivity is/are
    a) Paperboard b) Polystyrene
    c) Polyurethane d) All
79. Example for Intermediate bulk container is
    a) Combi-bins b) Metal boxes
    c) Corrugated fibre board d) All
80. Packaging material which has ability to reflect radiant heat is
    a) Paperboard b) Polystyrene
    c) Aluminium foil d) Polyurethane

81. The strength of polymer and paper or board materials can be assessed by measuring

a) Yield strength b) Young's modulus

c) Impact strength d) All

82. The factors that cause deterioration of foods during storage are

a) Pilferage b) Adulteration

c) Tampering d) All

83. Metal Cans are permeable to

a) Light b) Moisture

c) Odours d) None

84. Three-piece can, means it consists ___ can body and __ end pieces respectively.

a) 1, 2 b) 0, 3

c) 2, 1 d) 3, 0

85. Three-piece cans are made from

a) Steel b) Mild steel

c) Cast iron d) Copper

86. Three-piece cans are made from mild steel that is rolled to a strip _____mm thick.

a) 1.0 mm b) 1.2 mm

c) 2.1 mm d) 1.8 mm

87. In manufacturing of three-piece cans, mild steel is rolled to a strip and then dipped into hot dilute

a) HCl b) KCl

c) NaCl d) $H\text{-}_2SO_4$

88. Thickness of the Tin applied by electroplating on each side of the steel in tin cans manufacturing is ___ mm.

a) 0.1–0.3 mm b) 0.3–0.5 mm

c) 0.7–0.8 mm d) 0.8–0.9 mm

89. Amount of the Tin applied by electroplating on each side of the steel in tin cans manufacturing is ___ $g/m^2$.

a) 2.8–1.2 $g/m^2$ b) 1.8–2.2 $g/m^2$

c) 2.9–2.2 $g/m^2$ d) 4.8–5.2 $g/m^2$

90. The tin is coated with _______ lacquer compound to resistant to acids

a) Zinc oxide  b) Epoxy-phenolic
c) Butadiene  d) a & b

91. The tin is coated with _______ lacquer compound to resistant to acids as well as alkalis

a) Vinyl  b) Epoxy-phenolic
c) Zinc oxide  d) Butadiene

92. Vinyl compounds used as tin can lacquers are used in cans intended for canning of ____.

a) Beers  b) Wines
c) Fruit juices  d) All

93. Phenolic compounds used as tin can lacquers are used in cans intended for canning of ____.

a) Fish products  b) Fruits
c) Soups  d) All

94. Epoxy-phenolic compounds used as tin can lacquers are widely used in cans intended for canning of ____.

a) Meat  b) Fish
c) Fruit & vegetables  d) All

95. The tin is coated with _______ lacquer compound to prevent discolouration

a) Vinyl  b) Epoxy-phenolic
c) Zinc oxide  d) Butadiene

96. Butadiene compounds used as tin can lacquers are used in cans intended for canning of ____.

a) Beer  b) Soft drinks.
c) Meat  d) a & b

97. Properties of Epoxy amine lacquers are

a) Good adhesion  b) Heat resistance
c) Expensive  d) All

98. Properties of Alkyd lacquers are

a) Varnish able over inks  b) Off-flavour
c) Low cost  d) All

99. Properties of Oleoresinous lacquers are

    a) Gold coloured coatings  b) May incorporate zinc oxide

    c) Low cost  d) All

100. Oleoresinous lacquers are widely used in cans intended for canning of ____.

    a) Beans  b) Soups

    c) Meats  d) All

## Answers

| | | | | | | | | | |
|---|---|---|---|---|---|---|---|---|---|
| 1. d | 2. a | 3. a | 4. c | 5. c | 6. d | 7. d | 8. d | 9. d | 10. a |
| 11. d | 12. a | 13. b | 14. c | 15. b | 16. a | 17. a | 18. c | 19. b | 20. b |
| 21. a | 22. c | 23. c | 24. c | 25. d | 26. a | 27. b | 28. d | 29. a | 30. b |
| 31. a | 32. b | 33. c | 34. d | 35. d | 36. b | 37. d | 38. a | 39. a | 40. b |
| 41. c | 42. d | 43. a | 44. d | 45. d | 46. a | 47. a | 48. b | 49. c | 50. d |
| 51. c | 52. b | 53. a | 54. b | 55. d | 56. d | 57. a | 58. b | 59. d | 60. d |
| 61. d | 62. a | 63. b | 64. c | 65. c | 66. b | 67. b | 68. b | 69. d | 70. c |
| 71. a | 72. d | 73. c | 74. d | 75. d | 76. b | 77. b | 78. d | 79. c | 80. c |
| 81. d | 82. d | 83. d | 84. a | 85. b | 86. d | 87. d | 88. a | 89. a | 90. d |
| 91. a | 92. d | 93. d | 94. d | 95. d | 96. d | 97. d | 98. d | 99. d | 100. d |

1. In can manufacturing process, DRD means
   a) Drive and- redrive
   b) Drive and- repeated
   c) Draw and- redraw
   d) None
2. In can manufacturing process, DWI means
   a) Drive-and-wall-in
   b) Draw-and-wall-iron
   c) Drive-and-wall-iron
   d) None
3. The process used to produce aluminium cans for carbonated beverages is
   a) DWI
   b) DRD
   c) DRI
   d) None
4. Advantage of two-piece cans is
   a) Greater integrity
   b) Uniform lacquer coverage
   c) Savings in metal
   d) All
5. Thickness of the Aluminium foil used for wraps is (0.009 mm), bottle caps (0.05 mm) and trays for frozen and ready meals (0.05–0.1 mm). If foil is to be used to contain acid or salty foods it is normally coated with nitrocellulose.
   a) 0.009 mm
   b) 0.05 mm
   c) 0.05–0.1 mm
   d) 0.5–1 mm
6. Thickness of the Aluminium foil used for bottle caps is
   a) 0.009 mm
   b) 0.05 mm
   c) 0.05–0.1 mm
   d) 0.5–1 mm
7. Thickness of the Aluminium foil used for trays for frozen is
   a) 0.009 mm
   b) 0.05 mm
   c) 0.05–0.1 mm
   d) 0.5–1 mm
8. Thickness of the Aluminium foil used for ready meals is
   a) 0.009 mm
   b) 0.05 mm
   c) 0.05–0.1 mm
   d) 0.5–1 mm
9. Percentage of sand used in Glass jars and bottles manufacturing is 73%
   a) 63%
   b) 73%
   c) 83%
   d) 93%

10. Melting point of glass is
    a) 140–150°C b) 150–160°C
    c) 550–660°C d) 1350–1600°C
11. The glass colorant that imparts green colour to glass is
    a) Chromic oxide b) Iron
    c) Sulphur d) Carbon
12. The glass colorant that imparts amber colour to glass is
    a) Carbon b) Iron
    c) Sulphur d) All
13. The glass colorant that imparts blue colour to glass is
    a) Carbon b) Iron
    c) Sulphur d) Cobalt oxide
14. The glass colorant that produce Flint (clear) glass is
    a) Nickel b) Cobalt
    c) Sulphur d) a & b
15. The molten glass is shaped in a mould by __________ process.
    a) Blow-and-blow b) Press-and blow
    c) a or b d) None
16. The 'independent section (IS)' machine is now used universally for _____ making.
    a) Bottle b) Metal
    c) Pot d) All
17. Narrow neck containers are made by the ______ process.
    a) Blow-and-blow b) Press-and blow
    c) Press-and blow d) None
18. Wide neck containers are made by __________ process.
    a) Blow-and-blow b) Press-and blow
    c) Press-and blow d) None
19. The temperature glass bottles, when they leave the moulds is about ____°C.
    a) 250°C b) 350°C
    c) 450°C d) 550°C

20. Glass is annealed at ______ °C.
    a) 40–70°C b) 440–470°C
    c) 540–570°C d) 640–770°C
21. Removal of internal stresses in the unstable glass is done by
    a) Distortion b) Annealing
    c) Fracturing d) Moulding
22. Properties of glass are
    a) Transparent to microwaves b) Re-useable
    c) Resealable d) All
23. Glass surfaces often treated with _______ to increase their strength.
    a) Titanium b) Aluminium
    c) Zirconium d) All
24. condensation reaction between ethylene glycol and terephthalic acid produces
    a) LDPE b) PET
    c) PP d) HDPE
25. Flexible film is usually reserved for _______ plastic polymers.
    a) Fibrous b) Non-fibrous
    c) Elastic d) All
26. Thickness of Flexible film is
    a) 0.25 mm b) 0.25 mm
    c) = 0.25 mm d) None
27. In Greek 'plastikos' means
    a) To break b) To form
    c) To paste d) All
28. Most polymer films are made by ___________ process.
    a) Blow-and-blow b) Press-and blow
    c) Extrusion d) All
29. Cellulose films are produced by mixing sulphite paper pulp with _______.
    a) Table salt b) Caustic soda
    c) Baking soda d) HCl

30. In the production of Cellulose films, the mixture of sulphite paper pulp & caustic soda is allowed to 'ripen' for _____ to form sodium cellulose.

a) 2–3 hr  b) 5–8 hr

c) 2–3 days  d) 5–8 days

31. In the production of Cellulose films, sodium cellulose is converted to cellulose xanthate by treatment with ______.

a) Hydrogen sulphide  b) Carbon sulphide

c) Carbon disulphide  d) None

32. In the production of Cellulose films, the mixture of carbon disulphide & sodium cellulose is allowed to 'ripen' for _____ to form cellulose xanthate.

a) 1–2days  b) 2–3 days

c) 3–4 days  d) 4–5 days

33. Properties of Plain cellulose are

a) Transparent  b) Odourless

c) Biodegradable  d) All

34. Plain cellulose can biodegradable within approximately _____ days.

a) 10  b) 50

c) 100  d) 500

35. Kraft paper is produced by a ______ treatment process.

a) Salt  b) Caustic

c) Sulfite  d) Sulfate

36. Which of thc following is a strongest paper

a) Natural kraft  b) Brown kraft

c) Unbleached kraft  d) Bleached kraft

37. The natural kraft paper is used for

a) Bags  b) Wrapping

c) Books  d) Both a & b

38. Sulfite paper is used to make small bags or wrappers for packaging of ____________.

a) Confectionary  b) Biscuits

c) Salt  d) a & b

39. Greaseproof paper is made through a process known as ___.

a) Proofing  b) Heating
c) Beating  d) All

40. Greaseproof paper is resistance to____.

a) Wet agents  b) Oils agents
c) Oils but not wet agents  d) Both a & b

41. The greaseproof paper taken to an extreme (further hydration) to produce a very dense sheet with a highly smooth and glossy finish is ________.

a) Glassine  b) Parchment
c) Paperboard  d) None

42. Greaseproof paper is used to wrap _______________.

a) Snack food  b) Cookies
c) Candy bars  d) All

43. Glassine is used as a liner for______.

a) Biscuits  b) Cooking fats
c) Baked goods  d) All

44. Parchment paper is used to package of __________.

a) Vegetable fats  b) Butter
c) Lard  d) All

45. Paperboard is commonly used to make containers for shipping of______.

a) Boxes  b) Cartons
c) Trays  d) All

46. Solid board is also used to package of _______________.

a) Fruit juices  b) Soft drinks
c) Both a & b  d) None

47. Parchment paper is made from_________treated pulp.

a) Acid  b) Base
c) Distilled water  d) None

48. Parchment paper is good barrier to_____________.

a) Oxygen  b) Moisture
c) Carbon dioxide  d) None

49. Which of the following has multiple layers of bleached sulfate board
    a) Solid board  b) Chipboard
    c) Fiberboard  d) All
50. Solid fiberboard laminated with plastics or aluminum is used for ___________.
    a) Coffee  b) Milk powder
    c) Both a & b  d) None
51. OPP means_______________.
    a) Overall polypropylene  b) Orange polypropylene
    c) Oriented polypropylene  d) All
52. Oriented polypropylene property is
    a) Good optical properties  b) High tensile strength
    c) Puncture resistance  d) All
53. Oriented polypropylene property is
    a) Moderate permeability to moisture
    b) Moderate permeability to gases and odours
    c) Thermoplastic
    d) All
54. BOPP means_______________.
    a) Bio oriented polypropylene
    b) Biaxially overall protein permeability
    c) Biaxially oriented polypropylene
    d) Biaxially oriented polypentene
55. LDPE means_________.
    a) Low-density polyethane  b) Low-density polyeurythane
    c) Low-density polyethylene  d) None
56. LLDPE means___________.
    a) Large low-density polyethane  b) Large low-density polyeurythane
    c) Large low-density polyethylene  d) Linear low-density polyethylene
57. HDPE means___________.
    a) High dose polyethane  b) High density polyeurythane
    c) High dose polyethylene  d) High density polyethylene

58. PVC means ______________.
   a) Polyvinyl carbide
   b) Polyvinyl chloride
   c) Polyvinylidene chloride
   d) Polyvinyl carbon
59. PVdC means ________________.
   a) Polyvinyl carbide
   b) Polyvinylidene chloride
   c) Polyvinyl chloride
   d) Polyvinyl carbon
60. PET means
   a) Polyethane terephthalate
   b) Polyethylene tetraphthalate
   c) Polyethanol terephthalate
   d) Polyethylene terephthalate
61. WVTR means
   a) Water vapor turn rate
   b) Water vapor transmission
   c) Water vapor transmission rate
   d) Water vapor transmission route
62. Stretch-wrapping type packaging uses_________.
   a) LDPE
   b) LLDPE
   c) HDPE
   d) Both a & b
63. Thickness of LDPE in stretch-wrapping type packaging is __________.
   a) 25–38 μm
   b) 52–83 μm
   c) 17–24 μm
   d) 71–92 μm
64. Thickness of LLDPE in stretch-wrapping type packaging is ___________.
   a) 25–38 μm
   b) 52–83 μm
   c) 17–24 μm
   d) 71–92 μm
65. Which of the following material can be able to access heat sterilize.
   a) LDPE
   b) LLDPE
   c) HDPE
   d) Both a & b
66. Softening temperature of High-density polyethylene is ___ °C.
   a) 50°C
   b) 100°C
   c) 121°C
   d) 300°C
67. Sacks made from _____ mm HDPE have a high tear strength, tensile strength, penetrationresistance and seal strength.
   a) 0.03–0.15 mm
   b) 0.3–1.5 mm
   c) 3–15 mm
   d) 30-150 mm

68. Foamed HDPE is perforated with up to ____ holes/cm for use with fresh foods or bakeryproducts.

a) 50 holes/cm  b) 60 holes/cm

c) 70 holes/cm  d) 80 holes/cm

69. Rubber hydrochloride is similar to ___________.

a) PVC  b) PVdC

c) LDPE  d) HDPE

70. Melting temperature of Polypropylene is___°C.

a) 50°C  b) 150°C

c) 160°C  d) 121°C

71. Packaging material suitable for applications where thermal resistance required is______.

a) PVC  b) PVdC

c) LDPE  d) PP

72. Packaging material suitable for good mechanical properties in a wide temperature range is

a) Nylon  b) PVdC

c) LDPE  d) PP

73. The temperature range in which nylon can perform good mechanical properties is ______.

a) 6°C to 20°C  b) -6°C to 20°C

c) -60°C to 200°C  d) -60°C to 600°C

74. Polyethylene terephthalate ca be flexible at temperatures range ______.

a) -7°C to 35°C  b) -70°C to 135°C

c) -60°C to 200°C  d) -60°C to 600°C

75. Which of the following is a type of PET ?

a) APET  b) BPET

c) CPET  d) Both a&c

76. APET means ________.

a) Acid PET (APET)  b) Alkaline PET

c) Amorphous (APET)  d) None

77. CPET means

a) Carbide PET b) Critical PET

c) Crystalline PET d) None

78. APET is used for

a) Bottles for carbonated drinks b) Films for boil-in-bag

c) a & b d) None

79. CPET is used for

a) Microwave trays b) Semi-rigid containers

c) Tubs d) All

80. The most commonly used polyester in food packaging is

a) PET b) PVC

c) LDPE d) PP

81. PET is formed when terephthalic acid reacts with_______________.

a) Glycerol b) Ethylene glycol

c) Methylene glycol d) Ethylene glycerol

82. PET has good resistance to

a) Acids b) Bases

c) Solvents d) Both a&c

83. The main reasons for PET popularity is

a) Retention of carbonation b) Light weight

c) Transparency d) All

84. The major packaging applications of PET are

a) Containers

b) Semirigid sheets for thermoforming

c) Thin-oriented films

d) All

85. Polycarbonate used as a replacement for glass in items such as large / and.

a) Returnable water bottles b) Refillable water bottles

c) Sterilizable baby bottles d) All

86. Polycarbonate is a polymer of a sodium salt of bisphenol acid and ___________.

a) Carbon dioxide

b) Carbonyl dichloride

c) NaCl

d) None

87. PEN means_____________.

a) Polyethane nitrate

b) Polyethyl naphthol

c) Polyethylene naphthalate

d) None

88. PEN is a polymer of dimethyl naphthalene dicarboxylate and _______________.

a) Glycerol

b) Methylene glycol

c) Ethylene glycol

d) Ethylene glycerol

89. PVC is resistant to

a) Acids

b) Bases

c) Oils

d) All

90. PVC when co-polymerised with polyvinylidene chloride is known as ______________.

a) Saran

b) Rasan

c) Naras

d) Arnas

91. Polyvinylidene chloride (PVdC) is used for ____________.

a) Poultry,

b) Cured meats

c) Cheese

d) All

92. EA means ______________.

a) Ethanol vinyl acetate

b) Ethylene vinyl acetate

c) Ethylvinyl acetate

d) None

93. Ethylene vinyl acetate with less than 5% vinyl acetate, is used for _____applications.

a) Deep-freeze

b) Bag-inbox

c) Milkpouches

d) a & b

94. Ethylene vinyl acetate with 6-10% vinyl acetate, is used for _____ applications.

a) Deep-freeze

b) Bag-inbox

c) Milkpouches

d) a & b

95. Ethylene vinyl acetate with above 10% vinyl acetate, is used for _____applications.
    a) Deep-freeze    b) Bag-inbox
    c) Milk pouches    d) Hot-melt adhesive.
96. High-nitrile resins are copolymers of _________.
    a) Acrylonitrile-methyl acrylate    b) Acrylonitrilestyrene
    c) a & b    d) None
97. High-nitrile resins are used for_________.
    a) Meat,    b) Sheese
    c) Margarine    d) All
98. Depositing of vaporised aluminium particles onto the surface of a packaging film under vacuum is called _________.
    a) Metallisation    b) Forging
    c) Casting    d) Plating
99. Metallised films are suitable for __________ equipment.
    a) High-speed filling    b) Form–fill–seal
    c) a & b    d) None
100. Which of the following material can be metallised by depositing vaporised aluminium ?
    a) Cellulose    b) Polypropylene
    c) Polyester    d) All

## Answers

| | | | | | | | | | |
|---|---|---|---|---|---|---|---|---|---|
| 1. c | 2. b | 3. a | 4. d | 5. a | 6. b | 7. c | 8. c | 9. b | 10. d |
| 11. a | 12. d | 13. d | 14. d | 15. c | 16. a | 17. b | 18. b | 19. c | 20. c |
| 21. b | 22. d | 23. d | 24. b | 25. b | 26. b | 27. b | 28. c | 29. b | 30. c |
| 31. c | 32. d | 33. d | 34. c | 35. d | 36. a | 37. d | 38. d | 39. c | 40. b |
| 41. a | 42. d | 43. d | 44. d | 45. d | 46. c | 47. a | 48. d | 49)a | 50. c |
| 51. c | 52. d | 53. d | 54. c | 55. c | 56. d | 57. d | 58. b | 59. b | 60. d |
| 61. c | 62. d | 63. a | 64. c | 65. c | 66. c | 67. b | 68. d | 69. a | 70. c |
| 71. d | 72. a | 73. c | 74. b | 75. d | 76. c | 77. c | 78. c | 79. d | 80. a |
| 81. b | 82. d | 83. d | 84. d | 85. d | 86. c | 87. c | 88. c | 89. d | 90. a |
| 91. d | 92. b | 93. a | 94. d | 95. d | 96. c | 97. d | 98. a | 99. c | 100. d |

1. The degree of metallisation is expressed in__________ units.

   a) Density  b) Mass density

   c) Optical density  d) True density

2. The degree of metallisation is expressed in optical density units, up to a maximum of ___________units.

   a) 4  b) 5

   c) 6  d) 7

3. Hard-boiled sweets are generally twist wrapped individually in __________ plastics.

   a) Spin retaining  b) Twist retaining

   c) Wrap retaining  d) Coat retaining

4. Hard-boiled sweets are generally twist wrapped ____________ plastics.

   a) PVC  b) PP

   c) HDPE  d) All

5. Horizontal form-fill-seal (HFFS) is used for_____________.

   a) Chocolate bar  b) Meat

   c) Fish nuggets  d) Eggs

6. Wrapping material for Chocolate is _________.

   a) Polyester  b) OPP

   c) a & b  d) None

7. Nylons or polyamides are similar to

   a) Polyester  b) OPP

   c) PET  d) HDPE

8. Which of the following material has highest oxygen transmission rate?

   a) PVC  b) PP

   c) PET  d) HDPE

9. Which of the following material has highest water vapor transmission rate?

   a) HDPE  b) PET

   c) PP  d) PVC

10. The PET bottles cap is made out of ___________ polypropylene (for Pepsi and Coca Cola) or low densitypolyethylene for water bottles,
    a) LDPE b) PP
    c) EVA d) a & b
11. The liner in the PET bottles cap is made out of___________.
    a) PP b) LDPE
    c) EVA d) a & b
12. Label for PET bottles is made out of ___________ film.
    a) PP b) LDPE
    c) EVA d) HDPE
13. The density of EVA liner in the cap of PET bottles is ___________g/cc.
    a) 0.093 g/cc b) 0.93 g/cc
    c) 9.3 g/cc d) 93 g/cc
14. Example for thermoplastic film is ________.
    a) Polyethylene b) Polypropylene
    c) Polyvinyl alcohol d) PVC
15. Material used for milk jug manufacturing is ___________.
    a) PP b) LDPE
    c) EVA d) HDPE
16. Packaging material used for dry noodles is __________.
    a) PP b) LDPE
    c) EVA d) HDPE
17. Packaging material used for vitamin pills is __________.
    a) PVC b) LDPE
    c) EVA d) HDPE
18. Packaging material used for ketchup is __________.
    a) PET b) LDPE
    c) EVA d) HDPE
19. Packaging material used for honey is __________.
    a) PP b) LDPE
    c) EVA d) HDPE

20. Packaging material used for mustard is __________.
    a) HDPE b) LDPE
    c) EVA d) PP
21. Packaging material used for peanut butter is __________.
    a) HDPE b) PET
    c) EVA d) PP
22. Packaging material used for salad dressing is __________.
    a) HDPE b) LDPE
    c) EVA d) PP
23. Packaging material used for butter tubs is __________.
    a) PET b) LDPE
    c) EVA d) HDPE
24. Plastics are made by______________process.
    a) Polycondensation b) Polyaddition
    c) Polarization d) a & b
25. HDPE is suitable for milk jugs because it has ____________.
    a) Good water vapor barrier b) Milky white
    c) Excellent chemical resistance d) All
26. LDPE is suitable bags for dry noodles because it has ____________.
    a) Flexible b) Inexpensive
    c) Clear d) All
27. PVC is suitable for vitamin pills because it has ____________.
    a) Hard b) Durable
    c) Safety from tampering d) All
28. PET is suitable for ketchup because it has ____________.
    a) Good barrier to gases b) Light weight
    c) Cheaper to transport than glass d) All
29. PP is suitable for honey because it has ____________.
    a) Chemical resistance b) Stiffness
    c) Low odor/taste transfer d) All

30. LDPE is suitable for mustard because it has ____________.
   a) Flexible b) Ease of processing
   c) Chemical inertness d) All
31. PET is suitable for peanut butter because it has ____________.
   a) Good barrier to gases b) Light weight
   c) Cheaper to transport than glass d) All
32. HDPE is suitable for salad dressing because it has ____________.
   a) Good water vapor barrier b) Milky white
   c) Excellent chemical resistance d) All
33. HDPE is suitable for butter tubs because it has ____________.
   a) Colored for appeal b) Milky white
   c) Excellent chemical resistance d) All
34. The polymers that solidify or set irreversibly when heated and cannot be remolded are
   a) Thermosets b) Thermoplastics
   c) a & b d) None
35. The polymers that soften upon exposure to heat and return to their originalcondition at room temperature are
   a) Thermosets b) Thermoplastics
   c) a & b d) None
36. Which of the following polymers are ideal for food packaging ?
   a) Thermosets b) Thermoplastics
   c) a & b d) None
37. Substance used for ethylene absorption in active packaging system is
   a) Activated sodium b) Activated carbon
   c) Potassium permanganate d) b & c
38. Substance used for Ethanol emitting in active packaging system is
   a) Micro-encapsulated propanol b) Micro-encapsulated ethanol
   c) Methanol d) All
39. Substance used for Moisture absorption in active packaging system is
   a) Silica gel b) Polyvinyl alcohol encapsulation
   c) Clay-based d) All

40. Substance used for Antimicrobial agents releasing in active packaging system is
    a) Zeolites b) Mercurial compounds
    c) Chlorine dioxide d) All
41. Substance used for Antioxidant agents releasing in active packaging system is
    a) TBHQ b) Vitamin C
    c) Vitamin E d) All
42. Substance used for Flavor/odor absorbing in active packaging system is
    a) Activated carbon b) Sodium bicarbonate
    c) Zeolites d) a & b
43. Substance used for Chemical stabilization in active packaging system is
    a) Tocopherol b) Sodium bicarbonate
    c) Zeolites d) Activated carbon
44. The metal Aluminum is derived from______ ore.
    a) Bauxite b) Hematite
    c) Magnite d) All
45. Pure aluminum is used for light packaging of primarily
    a) Soft-drink cans b) Seafood
    c) Prethreaded closures d) All
46. The main disadvantage of aluminum is
    a) High cost b) Inability to weld
    c) a & b d) None
47. Tinplate produced from ____________steel.
    a) High-carbon b) Low-carbon
    c) Medium-carbon d) None
48. Tinplate containers are often lacquered with
    a) Epoxy phenolic groups b) Oleoresinous groups
    c) Vinyl resins. d) All
49. Tin-free steel is a steel electrolytically coated with
    a) Chromium b) Boron
    c) Iron d) Zinc

50. The property of anticorrosion of stain less steel is due to its
    a) Carbon  b) Iron
    c) Chromium  d) Oxygen
51. Tin-free steel is used for
    a) Food cans  b) Trays
    c) Bottle caps  d) All
52. Polyolefin is a collective term for polyethylene and polypropylene
    a) PE & PP  b) PE & PVC
    c) PVC & PP  d) EVA & PP
53. Laminated paper can be used to package of dried products such as, herbs, and spices.
    a) Dried herbs  b) Dried soups
    c) Dried spices  d) All
54. The antioxidant/solvent mixture between the packaging layers is at a concentration of about ________mg/ft$^2$.
    a) 0.1 to 3mg/ft$^2$  b) 1 to 30mg/ft$^2$
    c) 10 to 300mg/ft$^2$  d) 0.3 to 10mg/ft$^2$
55. An advanced form of data carrier (tag) for automatic product identification & traceability is
    a) RFID  b) RIDF
    c) RIDF  d) FIRD
56. RFID means
    a) Radio frequency identification detection
    b) Radar frequency identification
    c) Radiation frequency intimation
    d) Radio frequency identification
57. The RFID tags that have no battery and are powered by the energy supplied by the reader are called
    a) Passive tags  b) Active tags
    c) Dynamic tags  d) None

58. The RFID tags that have their own battery for powering the microchip's circuitry and broadcasting signals to the reader are called
    a) Passive tags b) Active tags
    c) Inert tags d) None
59. The more expensive active tags have a reading range of _______ feet or more.
    a) 10 feet b) 50 feet
    c) 80 feet d) 100 feet
60. Less expensive passive tags have a reading range of up to _______feet.
    a) 1 feet b) 5 fee
    c) 15 feet d) 150 feet
61. Advantage of glass is
    a) Impermeable to moisture b) Impermeable gases
    c) Non reactive (inert) d) All
62. Disadvantage of glass is
    a) Brittle b) Breakable
    c) Needs a separate closure d) All
63. Advantage of Aluminum is
    a) Impermeable to moisture b) Impermeable to gases
    c) Withstands heat processing d) All
64. Disadvantage of Aluminum is
    a) Cannot be welded b) Limited structural strength
    c) Impermeable to light d) a & b
65. Advantage of Tinplate is
    a) Strong and formable b) Resistant to corrosion
    c) Withstands heat proccssing d) All
66. Disadvantage of Tinplate is
    a) Can react with foods b) Coating required
    c) Coating is not required d) a & b
67. Advantage of Tin-free steel is
    a) Strong b) Good resistance to corrosion
    c) Withstands heat processing d) All

68. Disadvantage of Tin-free is
    a) Difficult to weld
    b) Requires removal of coating
    c) Less resistant to corrosion
    d) All
69. Advantage of Polyolefins is
    a) Strong
    b) Good moisture barrier
    c) Resistant to chemicals
    d) All
70. Disadvantage of Polyolefins is
    a) Not strong
    b) Good moisture barrier
    c) a & b
    d) Poor gas barrier
71. Low moisture barrier properties are exerted by _________ packaging material.
    a) PVC
    b) Al
    c) Glass
    d) Ethylene vinyl alcohol
72. Which of the following packaging material has poor barrier properties
    a) PVC
    b) Al
    c) Polyamide
    d) Polystyrene
73. The combination of Polyethylene terephthalate- aluminium– polypropylene is used for
    a) MAP
    b) Bag-in-box packaging
    c) Vacuum packs
    d) Retort pouches
74. Packaging material used for coffee is____________.
    a) Cellulose–polyethylene–cellulose
    b) BOPP/LDPE
    c) PET/LDPE
    d) All
75. In Boil-in-bag packaging the medium-density ethylene-butene is polymerized with
    a) PVC
    b) Al
    c) Nylon
    d) All
76. The main advantage coextruded films over other types of film is
    a) High barrier properties
    b) Layers cannot separate
    c) Production at a lower cost
    d) All

77. The copolymers used in coextruded films for adhesion is/are

a) Olefin b) Styrene

c) PVC d) All

78. Which of the following is a Olefin ?

a) Low-density polyethylene b) High-density polyethylene

c) Polypropylene d) All

79. For bag-in-box applications metallised polyester is replaced by_____ layer co-extrusion.

a) Two b) Three

c) Four d) Five

80. Polyhydrooxyalkonates (PHA) is good substitute for

a) Plastic b) Elastomer

c) Glass d) a & b

81. Polyhydrooxyalkonates (PHA) is synthesized by __________ microorganism.

a) *Cynobacteria* b) *Lactobacillus*

c) *Streptococci* d) *Aeromonas*

82. PHA is made of monomeric units of ___________.

a) Hydroxyalknoic acid b) Hydrochloric acid

c) Hydroxyketo acid d) None

83. PHA is made of monomeric units of Hydroxyalknoic acid linked together by ______bond.

a) Ester b) Peptide

c) Hydrogen d) Glycosidic

84. In PHA is made of monomeric units are presented in ________ configuration.

a) L (+) b) D (+)

c) L (-) d) D (-)

85. Packaging material used for pork & poultry pieces is

a) Methyl cellulose b) Hydroxy propyl cellulose

c) Dextrins d) CMC

86. Packaging material used for fruits like Bananas, Apples, Oranges & Fresh vegetables is

a) Methyl cellulose b) Hydroxy propyl cellulose

c) Dextrins d) CMC

87. Packaging material used for nuts is

a) Methyl cellulose b) Hydroxy propyl cellulose

c) Dextrins d) CMC

88. Packaging material used for

a) Methyl cellulose b) Hydroxy propyl cellulose

c) Dextrins d) CMC

89. Packaging material used for potato products

a) amylose b) Hydroxy propyl cellulose

c) Dextrins d) CMC

90. Packaging material used for freshly sliced apples

a) Amylose b) Hydroxy propyl cellulose

c) Dextrins d) CMC

91. Packaging material used for frozen fish and poultry pieces

a) Carrageenan b) Gelatin

c) Dextrins d) CMC

92. Packaging material used for tomatoes packing is

a) Carrageenan b) Corn-Zein

c) Dextrins d) CMC

93. Packaging material used for peanuts packing is

a) Whey Protein based b) Zein based

c) Collagen based d) CMC

94. CMC means

a) Carboxy methyl cellulose b) Carbonyl methyl cellulase

c) Carbon methyl cellulose d) Carboxy methyl cellulase

95. Characteristic of coextruded polymer is

a) Lower weight b) Easy to seal

c) Unbreakable d) All

96. Advantage of coextruded polymers over glass and metal is
    a) Can be produced at a lower temperatures
    b) Saves up to 40% in transport cost due to lower weight
    c) Can be moulded into a wider range of shapes than glass
    d) All
97. Trays for chocolates, eggs or soft fruit are made by _________method.
    a) Thermoforming b) Blow moulding
    c) Injection moulding d) Injection blow
98. Trays or cups or tubs for dairy products chocolates, eggs or soft fruit are made by __method.
    a) Thermoforming b) Blow moulding
    c) Injection moulding d) Injection blow
99. Wide mouthed containers (for example tubs and jars) and lids are made by ________method.
    a) Thermoforming b) Blow moulding
    c) Injection moulding d) Injection blow
100. Containers used for vegetable oils, vinegar beverages and sauces are made by ____method.
    a) Thermoforming b) Blow moulding
    c) Injection moulding d) Injection blow

## Answers

| | | | | | | | | | |
|---|---|---|---|---|---|---|---|---|---|
| 1. c | 2. a | 3. b | 4. d | 5. a | 6. c | 7. c | 8. c | 9. b | 10. d |
| 11. c | 12. a | 13. b | 14. a | 15. d | 16. b | 17. a | 18. a | 19. a | 20. b |
| 21. b | 22. a | 23. d | 24. d | 25. d | 26. d | 27. d | 28. d | 29. d | 30. d |
| 31. d | 32. d | 33. d | 34. a | 35. b | 36. b | 37. d | 38. b | 39. d | 40. d |
| 41. d | 42. d | 43. a | 44. a | 45. d | 46. c | 47. b | 48. d | 49. a | 50. c |
| 51. d | 52. a | 53. d | 54. b | 55. a | 56. d | 57. a | 58. b | 59. d | 60. c |
| 61. d | 62. d | 63. d | 64. d | 65. d | 66. d | 67. d | 68. d | 69. d | 70. d |
| 71. d | 72. d | 73. d | 74. d | 75. c | 76. d | 77. d | 78. d | 79. d | 80. d |
| 81. a | 82. a | 83. a | 84. d | 85. a | 86. d | 87. b | 88. c | 89. a | 90. c |
| 91. a | 92. b | 93. a | 94. a | 95. d | 96. d | 97. a | 98. a | 99. c | 100. b |

1. In packaging science the abbreviation 'HIPS' means
   a) High impact packaging b) High infused polystyrene
   c) High impact polysaccharide d) High impact polystyrene
2. In packaging science the abbreviation 'ABS' means
   a) Acryline buta styrene b) Acrylamide butadiene styrene
   c) Acrylonitrile butadiene styrene d) Acrylonitrile butadiene system
3. trays or cups or tubs for dairy products are thermoformed by using
   a) HIPS b) ABS
   c) PP d) a & b
4. Paper pulp is produced from wood chips by____________ the process.
   a) Kraft b) Sulphite
   c) Solvent d) a & b
5. Advantage of kraft process over sulfite process is
   a) More strengthened pulp b) Higher yields
   c) Recycling of chemicals d) All
6. In the sulphite process sulphur dioxide and calcium bisulphite are heated with wood chips at ___________ °C.
   a) 100°C b) 120°C
   c) 130°C d) 140°C
7. Which of the paper board is suitable for contact with food?
   a) White board b) Duplex board
   c) Chipboard d) Fibre board
8. Which of the paper board is made from recycled paper?
   a) White board b) Duplex board
   c) Chipboard d) Fibre board
9. Fibre board is used for packaging of foods in the form of____.
   a) Corrugated b) Solid
   c) Both a & b d) Liquid
10. Thickness of corrugated fibre board is __________.
    a) 11 mm b) 0.11 mm
    c) 0.11 cm d) 1 cm

11. Cartons used for ice cream, chocolate and frozen foods are
    a) White board b) Duplex board
    c) Chipboard d) Fibre board
12. Cartons used for packaging aseptically sterilised (UHT) foods are
    a) Laminated paper board b) Duplex board
    c) Chipboard d) Fibre board
13. Order of different packaging materials present in laminated paperboard cartons is
    a) LDPE–paper- surlyn - LDPE–aluminium - LDPE
    b) LDPE– surlyn - paper-LDPE–aluminium - LDPE
    c) LDPE–paper-HDPE–aluminium– surlyn- LDPE
    d) LDPE–paper-LDPE–aluminium– surlyn- LDPE
14. For $CO_2$ production in active packaging systems __________ is used.
    a) Calcium hydroxide b) Sodium hydroxide
    c) Potassium hydroxide d) All
15. In the process of printing on films and papers UPC means
    a) Unique Printing Cryptogram b) Unique Printing Code
    c) Universal Press Code d) Universal Print Code
16. The Universal Print Code (UPC) is printed on consumer packs for ____ reading at retail checkouts.
    a) Light b) Beam
    c) Laser d) None
17. Which one of following printing methods, electrically charged droplets of ink are deflected by charged deflector plates to create the image.
    a) Ink-jet b) Screen
    c) Offset d) Flexographic
18. The Universal Print Code (UPC) is also called as
    a) Bar code b) Beam code
    c) Check code d) All
19. Which one of following packages is the one which is in direct contact with the contained product.
    a) Primary b) Secondary
    c) Tertiary d) Quaternary

20. Corrugated case and corrugated boxes are example for _______ package.

    a) Primary b) Secondary

    c) Tertiary d) Quaternary

21. Stretch-wrapped pallet of corrugated cases are example for _______ package.

    a) Primary b) Secondary

    c) Tertiary d) Quaternary

22. The package generally a metal container up to 40 m in length which can be transferred to or from ships, trains is _____________.

    a) Primary b) Secondary

    c) Tertiary d) Quaternary

23. The degree of protection by package is measured a ________ for oxygen sensitive foods.

    a) OTR b) WVTR

    c) BTR d) ATR

24. The degree of protection by package is measured a ________ for moisture sensitive foods.

    a) WVTR b) OTR

    c) PTR d) QTR

25. Hot fill hold type packaging system is used for ___________ foods.

    a) Acid foods b) Beverages

    c) Alkali foods d) a & b

26. Packaging material used in Hot fill hold type Packaging system is

    a) Polyster laminates of PET

    b) Polyster laminates of Aluminium foil

    c) Polyethylene laminates of PET/Aluminium foil/PE

    d) All

27. Tetrapak type packaging system is used for ___________foods.

    a) Milk b) Non –carbonated beverages

    c) Edible oils d) All

28. Packaging material used in tetrapak type packaging system is
    a) PE coating b) PE adhesive
    c) Aluminium foil d) All
29. Brick-pack type packaging system is used for __________ foods.
    a) Liquid milk b) Fruit juices
    c) Carbonated beverages d) a & b
30. Packaging material used in brick-pack type packaging system is
    a) 18 micron PE b) Kraft paper
    c) 25 Micron PE d) Combinations of all
31. Which of the following is used for sterilization?
    a) $H_2O_2$ b) Co-extrusion heat process
    c) Ethylene oxide d) All
32. Which of the followings is not used for sterilization?
    a) Gamma-radiation b) UV light
    c) a & b d) Vis light
33. Which of the following sterilization method has residual problem?
    a) $H_2O_2$ b) Co-extrusion heat process
    c) Ethylene oxide d) Gamma-radiation
34. Which of the following sterilization method is chemically safe?
    a) $H_2O_2$ b) Co-extrusion heat process
    c) Ethylene oxide d) Gamma-radiation
35. Packaging material used in ethylene oxide sterilization method is?
    a) Glass b) Plastic containers
    c) Pre-formed plastic d) a & b
36. Which of the following combination of laminates for retortable pouch is correct ?
    a) Nylon 15μm/ Cast polypropylene 60 μm
    b) Nylon 15μm/ Cast polypropylene 70 μm
    c) Nylon 25μm/ Cast polypropylene 70 μm
    d) All

37. Requirements of sterilization temperatures for retort pouch is
    a) 100 – 111°C  b) 111 – 1116°C
    c) 116 – 145°C  d) 151 – 165°C
38. Oxygen permeability in retort pouch is ______cc/m$^2$/24 h/at atmosphere
    a) 0  b) 10
    c) 100  d) 1000
39. Moisture vapour transmission rate in retort pouch is ______ g/ M2/24 h
    a) 0  b) 0.1
    c) 1  d) 10
40. Seal strength in retort pouch is ______ Kg/100 mm.
    a) 0.02-0.035  b) 0.2-0.35
    c) 2-3.5  d) 20-35
41. Bond strength in retort pouch is ________g/10 ml.
    a) 0.15-0.50  b) 1.5-5.0
    c) 15-50  d) 150-500
42. Which of the following is a can coating?
    a) Fruit enamel  b) C-enamel
    c) Citrus enamel  d) All
43. Fruit enamel type of can coating is used for _____.
    a) Berries
    b) Cherries
    c) Fruits require protection from metallic salts
    d) All
44. C-enamel type of can coating is used for _____.
    a) Corn  b) Peas
    c) Sulphur bearing food products  d) All
45. Citrus enamel type of can coating is used for _____.
    a) Citrus scices  b) Citrus juice
    c) Citrus concentrate  d) All
46. Beverage can enamel type of can coating is used for _____.
    a) Vegetable juices  b) Red fruit juices
    c) Non carbonated beverages  d) All

47. In fruit enamel the coating material is

a) Oleoresinous b) Oleoresinous with zinc oxide

c) Modified Oleoresinous d) None

48. In Beverage can enamel the coating material is

a) Oleoresinous

b) Zinc oxide

c) Two coated with resinous base coat and Vinyl top coat

d) None

49. In C-enamel the coating material is

a) Oleoresinous b) Oleoresinous with zinc oxide

c) Modified Oleoresinous d) None

50. In citrus enamel the coating material is

a) Oleoresinous b) Oleoresinous with zinc oxide

c) Modified oleoresinous d) None

51. The type of steel base used for packaging food products is __________.

a) Type - MR b) Type - MS

c) Type - L d) All

52. The type of steel base used for strongly corrosive food products is ___________.

a) Type - MR b) Type - MS

c) Type - L d) Type - MC

53. Example for strongly corrosive food product/s is are__.

a) Berries b) Sauerkraut

c) Pickles d) All

54. The type of steel base used for moderately corrosive food products is ___________.

a) Type - MR b) Type - MS

c) Type - L d) Typc - MC

55. Example for moderately corrosive food product/s is are__.

a) Apricots b) Grape fruit

c) Peaches d) All

56. The type of steel base used for non-corrosive food products is __________.

    a) Type - MR
    b) Type - MS
    c) Type - L
    d) Type - MC or MR

57. Example for non-corrosive food product/s is are__.

    a) Dehydrated frozen foods
    b) Nuts
    c) Both a & b
    d) None

58. National Standards on Packaging code for food stuffs and perishables is according to the _____ code.

    a) FPO
    b) PFA
    c) BIS
    d) AGMARK

59. Pasteurized flavoured milk is packed in

    a) Glass bottles
    b) LDPE lined cartons
    c) Aseptic cartons
    d) All

60. Sterilized flavoured milk is packed in

    a) Poly papers
    b) Poly-laminated paper packs
    c) Glass bottles
    d) All

61. Sterilized milk is packed in

    a) Poly papers
    b) Poly-laminated paper packs
    c) Glass bottles
    d) All

62. The fermented milk products should be cooled before dispatch and maintained at a temperature below ______ °C, to develop good flavour and texture.

    a) 1°C
    b) 10°C
    c) 100°C
    d) 121°C

63. Milk powder is packed in

    a) Hermetically sealed containers
    b) Hermetically sealed tinplate containers
    c) Hermetically sealed tinplate containers in N2 or a mixture of N2 and $CO_2$ gas
    d) All

64. Cereal weaning foods are packed in
    a) Hermetically sealed containers
    b) Hermetically sealed tinplate containers
    c) Hermetically sealed tinplate containers in N2 or a mixture of N2 and $CO_2$ gas
    d) All
65. Raw vegetables and fruits are packed in
    a) Gunny bags b) Plastic crates
    c) Corrugated fiberboard boxes. d) All
66. Onion and garlic are packed in
    a) Gunny bags b) Wooden crates
    c) Corrugated fiberboard boxes d) All
67. Tomatoes are packed in
    a) Baskets b) Wooden boxes
    c) Corrugated fiberboard boxes. d) All
68. Chillies are packed in
    a) Gunny bags b) Bamboo crates
    c) Corrugated fiberboard boxes d) All
69. Juices, jams, jellies and marmalades are packed in
    a) Glass bottles b) Open top cans
    c) Tinplate containers d) All
70. Meat is wrapped in ____ sheets or bags and delivered in clean, rust free and closed containers.
    a) Polypropylene b) Polyethylene
    c) Polymethylene d) None
71. whole carcasses and cooked meat products may are vacuum packed using suitable wrapping materials such as ____________.
    a) PET b) LDPE
    c) HDPE d) a & b
72. Egg powder is gas packed in nitrogen in suitable tinplate containers or flexible packaging materials. such as ______.
    a) PET b) LDPE
    c) HDPE d) a & b

73. The fresh fish is packed in ___ lined insulated plywood.
    a) Polypropylene b) Polyethylene
    c) Polymethylene d) None
74. The thickness of insulation in plywood intended for fish packaging is ______ mm.
    a) 15-30 mm b) 30-55 mm
    c) 55-75 mm d) 75-100 mm
75. The packaging material used for bread is ____.
    a) LDPE coated poster paper b) Waxed paper
    c) Grease –proof paper d) All
76. The packaging material used for biscuits is ____.
    a) cello/LDPE b) BOPP/LDPE
    c) PET/LDPE d) All
77. The packaging material used for cakes is ____.
    a) Waxed paper b) Grease-proof polyethylene
    c) Glassine/LDPE d) All
78. The packaging material used for protein-rich extruded foods is ____.
    a) HDPE bags b) LDPE liner bag
    c) BOPP/LDPE pouch d) All
79. Peanut butter is packed in
    a) Glass jars b) Polystyrene tubs
    c) a & b d) None
80. Flours and starches are packed in
    a) LDPE coated jute bags b) LDPE coated raffia bags
    c) a & b d) None
81. Edible spray dried potato flour is packed in
    a) HDPE b) Metallized polyester bags
    c) BOPP/LDPE d) All
82. Oils and fats are packed in
    a) HDPE b) Food grade PVC
    c) PET d) All

83. Cereal grains are packed in
    a) Jute bags
    b) LDPE coated jute bags
    c) Raffia bags
    d) All
84. Sugar is packed in
    a) Polyethylene coated Hessian bags
    b) Polyethylenecoated raffia bags
    c) A-twill jute bags
    d) All
85. Cereal flours are packed in
    a) LDPE coated jute bag
    b) LDPE coated raffia bags
    c) A-twill or B-twill jute bags
    d) All
86. Cube sugar of net weight of _____ kg is wrapped in butter paper or kraft paper and packed in cartons.
    a) 0.5 kg
    b) 5 kg
    c) 50 kg
    d) 500 kg
87. Honey is packed in
    a) Acid resistant lacquered tin container
    b) Polyethylene
    c) Glass container
    d) All
88. Tea is packed in
    a) LDPE
    b) Paper coated LDPE
    c) BOPP/LDPE
    d) All
89. Roasted and ground coffee are packed in
    a) Paper/LDPE
    b) PET/LDPE
    c) BOPP/LDPE
    d) All
90. Chocolates is packed in
    a) Grease-proof paper
    b) Aluminium foil
    c) Laminates made of paper/LDPE, BOPP/LDPE
    d) All

91. Carbonated beverages should be filled in glass containers conforming to IS no. is

a) IS 1107:1896  b) IS 1107:1985

c) IS 1107:1986  d) IS 1107:1968

92. After filling of carbonated beverages, the containers should be hermetically sealed with clean, new crown corks conforming to IS no. is

a) IS 1994:1978  b) IS 1994:1986

c) IS 1994:1898  d) IS 1994:1987

93. Beer is filled in ______ ml glass bottles or in PET bottles

a) 650  b) 550

c) 450  d) 250

94. Beer is packed in bottle or cans. These bottles or cans should be packed in

a) Wooden cases  b) Corrugated fiber board boxes

c) a & b  d) None

95. The process in which surface fibers are reorients in the base sheet of paper through the use of pressure & which serves to smooth the surface, control surface texture and develop a glossy finish is called___.

a) Calendering  b) Candling

c) Canning  d) Cementing

96. Papers produced by calendering process are known as

a) Machine finished  b) Rough finished

c) Smooth finished  d) None

97. The containers made from sheets of paperboard which have been cut and scored for bending into desired shapes are called ________ cartons.

a) Open  b) Folding

c) Closed  d) Rolled

98. Thickness of paper boards used for preparation of folding cartons is _______ μm.

a) 3 - 10 μm  b) 30-110 μm

c) 300 - 1100 μm  d) 3000-11000 μm

99. Folding Carton Boxes are grouped into__________types.

a) 3  b) 4

c) 5  d) 6

100. Characteristics of Group-A type folding carton boxes are

a) Long seam glued
b) Rectangular surfaces
c) a & b
d) None

## Answers

| | | | | | | | | | |
|---|---|---|---|---|---|---|---|---|---|
| 1. d | 2. c | 3. d | 4. d | 5. d | 6. d | 7. a | 8. c | 9. c | 10. b |
| 11. a | 12. a | 13. d | 14. a | 15. d | 16. c | 17. a | 18. a | 19. a | 20. b |
| 21. c | 22. d | 23. a | 24. a | 25. d | 26. d | 27. d | 28. d | 29. d | 30. d |
| 31. d | 32. d | 33. c | 34. b | 35. d | 36. d | 37. c | 38. a | 39. a | 40. c |
| 41. d | 42. d | 43. d | 44. d | 45. d | 46. d | 47. a | 48. c | 49. b | 50. c |
| 51. d | 52. c | 53. d | 54. a | 55. d | 56. d | 57. c | 58. c | 59. d | 60. d |
| 61. d | 62. b | 63. d | 64. d | 65. d | 66. d | 67. d | 68. d | 69. d | 70. b |
| 71. d | 72. d | 73. b | 74. a | 75. d | 76. d | 77. d | 78. d | 79. c | 80. c |
| 81. d | 82. d | 83. d | 84. d | 85. d | 86. a | 87. d | 88. d | 89. d | 90. d |
| 91. c | 92. d | 93. a | 94. c | 95. a | 96. a | 97. b | 98. c | 99. d | 100. c |

1. Characteristics of Group-A type folding carton boxes are
   a) Long seam glued b) Rectangular surfaces
   c) a & b d) None
2. Characteristics of Group-B type folding carton boxes are
   a) Non-long seam glued b) Rectangular surfaces
   c) a & b d) None
3. Characteristics of Group-C type folding carton boxes are
   a) Long seam glued b) Non-rectangular surfaces
   c) a & b d) None
4. Characteristics of Group-D type folding carton boxes are
   a) Non-long seam glued b) Non-rectangular surfaces
   c) a & b d) None
5. Characteristics of Group-E type folding carton boxes are
   a) Long seam glued b) Rectangular surfaces
   c) Integrated folding d) None
6. Characteristics of Group-F type folding carton boxes are
   a) Long seam glued b) Rectangular surfaces
   c) Integrated folding d) None
7. Glass is a____________.
   a) Solid b) Liquid
   c) Amorphous d) None
8. Glass is a____________.
   a) Organic b) Inorganic
   c) a & b d) None
9. The principal ingredient of glass is silica derived from
   a) Sand b) Flint
   c) Quartz d) All
10. In glass manufacturing, silica is melted to temperatures of ___ °C to form fused silica glass.
    a) 1273°C b) 1372°C
    c) 7123°C d) 1723°C

11. Which of the following is a corrosion form of glass surface ?
    a) Etching b) Leaching
    c) Weathering d) All
12. The process in which glass surface is subject to alkaline attack which slowly destroys the silica network, releasing other glass components is called
    a) Etching b) Leaching
    c) Weathering d) None
13. The process in which glass surface is subject to acid attack in which hydrogen ions exchange for alkali or other positively charged mobile ions is called
    a) Etching b) Leaching
    c) Weathering d) None
14. The process in which glass surface is subject to surface bloom is called
    a) Etching b) Leaching
    c) Weathering d) None
15. The basic parts of a glass container are
    a) Finish b) Body
    c) Bottom d) All
16. The part of the bottle that holds the cap or closure, the glass surrounding the opening in the container is
    a) Finish b) Body
    c) Bottom d) All
17. Thc largest part of the container and lies between the finish and the bottom is
    a) Finish b) Body
    c) Bottom d) All
18. The part of container that is made in the bottom plate part of the glass-container mold is
    a) Finish b) Body
    c) Bottom d) All
19. The closures of glass bottles or jars consist of
    a) Cap b) Cork
    c) a & b d) None

20. The closures of glass bottles or jars, function of cork or plug is to _______ the jar or bottle.

a) Break b) Seal

c) Open d) None

21. Main function of closure of glass bottle or jar is

a) Provides an effective hermetic (air-tight) seal

b) Provides easy opening and resealing of the container

c) Provides evidence of inviolability

d) All

22. In packaging science, the abbreviation ROPP means

a) Round on pilfer proof b) Roll on pinch proof

c) Roll on pilfer proof d) All

23. In Roll on pilfer proof (ROPP) closures, a tale ring around the bottom of the neck of bottles _____opening.

a) After b) Before

c) While d) None

24. The flat center area in the top of the closure of bottle or jar is called

a) Panel b) Radius

c) Skirt d) Curl

25. The rounded area at the outer edge of the panel connecting he panel and the skirt of the closure of bottle or jar is called

a) Panel b) Radius

c) Skirt d) Curl

26. The flat, almost vertical portion on the side of the closure, which serves as the gripping surface and may be smooth, knurled or fluted of the closure of bottle or jar is called

a) Panel b) Radius

c) Skirt d) Curl

27. The round or rolled portion at the bottom of the skirt that adds rigidity to the cap and serves to protect the cut edge of the metal of the closure of bottle or jar is called

a) Curl b) Lug

c) Thread d) Face

28. A horizontal inward protrusion from the curl that is seated under the thread on the glass finish and holds the cap in position of the closure of bottle or jar is called
    a) Curl b) Lug
    c) Thread d) Face
29. The spiral groove on the skirt of a continuous thread closure that meshes with the tread on the glass finish of the closure of bottle or jar is called
    a) Curl b) Lug
    c) Thread d) Face
30. The part, outside of the cap of the closure of bottle or jar is called
    a) Curl b) Lug
    c) Thread d) Face
31. The part, inside of the cap of the closure of bottle or jar is called
    a) Reverse b) Lug
    c) Thread d) Face
32. The actual sealing component of the closure, which makes an intimate contact with the glass finish to form an effective seal is
    a) Gasket b) Liner
    c) a & b d) Thread
33. Gaskets are made from
    a) Rubber b) Plastisols
    c) Iron d) a & b
34. Resistance to _____of glass is determined by transferring glass containers from hot water bath to cold water bath.
    a) Observance b) Breakage
    c) Adherence d) All
35. Time and temperature combinations for hot water bath, for dipping of glass containers to determine the resistance to breaking of glass are
    a) 60°C for 5 min b) 63°C for 5 min
    c) 65°C for 5 min d) 68°C for 5 min
36. Glass is optically
    a) Isotropic b) Anisotropic
    c) Polytropic d) All

37. The optical properties of glass relate to the degree of penetration of light and the subsequent effect of that transmission which is a function of

    a) Wavelength  b) Frequency
    c) Energy  d) None

38. In packaging science, the abbreviation ECCS means

    a) Electrolytically chloride-coated strengthened
    b) Electrolytically chloride -coated steel
    c) Electrolytically chromium-coated strengthened
    d) Electrolytically chromium-coated steel

39. In packaging science, the abbreviation TFS means

    a) Titaniumn -full steel  b) Tin -full steel
    c) Tin-free steel  d) None

40. Tinplate refers to _____ carbon mild steel sheet.

    a) Low  b) High
    c) Free  d) None

41. Tinplate refers to low carbon mild steel sheet, varying in thickness from around ________ mm.

    a) 5 to 15  b) 1 to 1.5
    c) 0.15 to 0.5  d) 0.015 to 0.05

42. Tinplate refers to low carbon mild steel sheet, with a coating of tin between 2.8 to 17 gsm.

    a) 17 to 28  b) 2.8 to 17
    c) 0.28 to 0.17  d) 0.028 to 0.17

43. The principal method of electroplating is

    a) Acid stannous sulfate process  b) Halogen process
    c) Both a & b  d) None

44. The electroplating method, acid stannous sulfate process is also called as

    a) Ferro process  b) Halogen process
    c) Ferrostan process  d) None

45. The temperature range maintained for cathodic deposition in a dilute chromium plating electrolyte in the production of ECCS is

    a) 5 to 7°C  b) 50 to 70°C
    c) 500 to 700°C  d) 70 to 150°C

46. The concentration of $CrO_3$ for cathodic deposition in a dilute chromium plating electrolyte in the production of ECCS is

a) 0.05 g $L^{-1}$ b) 0.5 g $L^{-1}$

c) 5 g $L^{-1}$ d) 50 g $L^{-1}$

47. The concentration of $H_2SO_4$ for cathodic deposition in a dilute chromium plating electrolyte in the production of ECCS is

a) 0.5 g L-1 b) 0.5 g L-1

c) 0.5 g L-1 d) 0.5 g L-1

48. Aluminum foil is a thin-rolled sheet of alloyed aluminum varying in thickness from about __ μm.

a) 0.4 to 15 μm b) 4 to 150 μm

c) 40 to 1500 μm d) 400 to 15000 μm

49. Aluminum foil is essentially impermeable to gases and water vapor when it is thicker than__μm.

a) 0.254 μm b) 2.54 μm

c) 25.4 μm d) 254.0 μm

50. The aluminium tube is formed by the _____impact extrusion of an aluminum slug using a plunger.

a) Cold b) Hot

c) Lukewarm d) None

51. Compared to the metal can, the retort pouch enabling retorting times to be reduced by up to ____%

a) 30% b) 40%

c) 50% d) 60%

52. An outer layer of three-layer pouch structure is

a) PET b) PVC

c) HDPE d) LDPE

53. Thickness of PET layer in three-layer pouch structure is ____¼m.

a) 12 μm b) 11 μm

c) 10 μm d) 1 μm

54. The function of PET layer in three-layer pouch structure is

a) Strength b) Toughness

c) a & b d) None

55. An middle layer of three-layer pouch structure is
    a) Copper foil
    b) Aluminium foil
    c) Iron foil
    d) Carbon foil
56. Thickness of aluminium foil in three-layer pouch structure is ____μm.
    a) 70 to 100 μm
    b)700 to 1000 μm
    c) 60 to 100 μm
    d) 600 to 1000 μm
57. The function of aluminium foil in three-layer pouch structure is
    a) Heat sealability
    b) Strength
    c) Compatibility
    d) All
58. The term aseptic implies
    a) Absence or exclusion of any unwanted organisms from the product
    b) Absence or exclusion of any unwanted organisms from package
    c) Absence or exclusion of any unwanted organisms from specific areas
    d) All
59. The term hermetic implies term (strictly air tight) is used to indicate
    a) Mechanical properties to exclude the entrance of micro-organisms into a package
    b) Mechanical properties to exclude the entrance of gas into (or from) the package.
    c) Mechanical properties to exclude the entrance of water vapor into (or from) the package.
    d) All
60. The first aseptic packaging of food (specifically milk in metal cans) was carried out in Denmarkby ________before 1913.
    a) Nicolas
    b) Nielsen
    c) Naaren
    d) Nimpher
61. The first food that aseptically packed in metal cans
    a) Egg
    b) Milk
    c) Wine
    d) Water
62. Which of the following is a sterilization processes is
    a) HTST
    b) UHT
    c) LTLT
    d) Both a & b

63. After sterilization the sterile low viscous food product is cooled to an appropriate temperature, typically ______°C.

a) 20°C b) 30°C

c) 40°C d) 50°C

64. After sterilization the sterile high viscous food product is cooled to an appropriate temperature, typically ______°C.

a) 20°C b) 30°C

c) 40°C d) 50°C

65. The sterilization processes for packaging material is are in common use, eitherindividually or in combination:, and treatments.

a) Irradiation

b) Heat

c) Chemical

d) Any one of these options or in combination

66. Ionizing radiation technique uses gamma rays from

a) Cobalt-60 b) Cesium-137

c) a & b d) Cesium-173

67. A radiation dose of____ kGy or more is generally accepted to be sufficient to ensure sterility.

a) 0.25 b) 2.50

c) 25.0 d) 250

68. In Pulsed electric field the duration of the pulses ranges from ________sec. , and the flashes are typically applied at a rate of 1 to 20 flashesper second.

a) 1μsec to 0.1 sec b) 10μsec to 1 sec

c) 100μsec to 10 sec d) 1sec to 100 sec

69. In Pulsed electric field the flashes are typically applied at a rate of_____ flashes per second.

a) 100 to 200 b) 10 to 20

c) 1 to 2 d) 1 to 20

70. UV radiation has a wavelength of ______ nm.

a) 150 to 300 b) 200 to 315

c) 315 to 415 d) 415 to 515

71. Optimum wavelength of UV radiation for effectiveness destruction of microbes is _____nm.

a) 253.7 nm b) 235.7 nm

c) 225.7 nm d) 275.3 nm

72. Time temperature combinations for heat sterilization processes are

a) 121°C for 20 min b) 170°C for 60 min

c) Both a & b d) None

73. Time temperature combinations for steam heat sterilization processes are

a) 121°C for 20 min b) 170°C for 60 min

c) Both a & b d) None

74. Time temperature combinations for dry heat sterilization processes are

a) 121°C for 20 min b) 170°C for 60 min

c) Both a & b d) None

75. Example for chemical sterilizing agent is

a) Hydrogen peroxide b) Peracetic acid (PAA)

c) Both a & b d) None

76. A liquid sterilant, which is particularly effective against spores is

a) Hydrogen peroxide b) Peracetic acid (PAA)

c) Both a & b d) None

77. The criteria for an aseptic packaging system is

a) Cleaned properly after use

b) Effectively sterilized before use

c) Able to carry out the filling, sealing and critical transfer operations in a sterile environment.

d) All

78. The most widespread consumer package for aseptic products is

a) Paperboard laminate carton b) Parchment paper

c) Metal cans d) Butter paper

79. In aseptic packaging equipment the concentration $H_2O_2$ for sterilization purpose is _____%

a) 35% b) 45%

c) 55% d) 65%

80. In aseptic packaging equipment the temperature of the $H_2O_2$ bath for sterilization purpose is _____ °C.

a) 40°C
b) 50°C
c) 60°C
d) 70°C

81. In aseptic packaging equipment the temperatures applied to evaporate $H_2O_2$ on material is _____°C.

a) 80°C
b) 100°C
c) 125°C
d) 150°C

82. The test measures the ability of a paper or paper board to withstand pneumatic or hydraulic pressure built up is

a) Burst Strength
b) Tear Resistance
c) Tensile strength
d) Grease resistance

83. The test measures the ability of a paper or paper board to withstand tearing is

a) Burst Strength
b) Tear Resistance
c) Tensile strength
d) Grease resistance

84. The force applied parallel to the plane of the specimen of specified width and length under specified condition of loading is

a) Burst Strength
b) Tear Resistance
c) Tensile strength
d) Grease resistance

85. The resistance of packaging material measured by exposing one of the test specimen creased to grease containing red dye is

a) Burst Strength
b) Tear Resistance
c) Tensile strength
d) Grease resistance

86. The transmission rate that is normally determined by measuring the change in prcssurc at constant volume is

a) GTR
b) WVTR
c) OTR
d) None

87. In WVTR experiment the quantity of water vapour in grams that will permeate from one side to other side of the film of an area of one square meter in ___ hours.

a) 1 hour
b) 12 hours
c) 24 hours.
d) 48 hours

88. In WVTR experiment the quantity of water vapour in grams that will permeate from one side to other side of the film of an area of one square meter when the relative humidity difference between the two sides is maintained at ___ % gradient.

a) 75 % b) 80 %

c) 85 % d) 90 %

89. In WVTR experiment the quantity of water vapour in grams that will permeate from one side to other side of the film of an area of one square meter at the surrounding temperatures of 37.8°C.

a) 3.8°C b) 38°C

c) 37.8°C d) 38.7°C

90. The test designed to measure the ability of packaging material of the films to withstand fracture by shock the test is

a) Impact Strength b) Tear Resistance

c) Tensile strength d) Grease resistance

91. The test designed to measure the ability of packaging material to withstand surface wear and rubbing is

a) Abrasion Resistance b) Tear Resistance

c) Tensile strength d) Grease resistance

92. The requirement for the fast degradation process of bioplastics is

a) Temperature b) Humidity

c) Type of microbes d) All

93. Acceptable bioplastic is

a) Cellulose b) Starch

c) Poly-beta-hydroxyalkanoates d) All

94. Cellulose is fusible and soluble in hydrogen bond-breaking solvents such as

a) N-methylmorpholine-N-oxide b) Water

c) Alcohol d) All

95. Starch normally consists of

a) Amylose b) Amylopectin

c) Hemicellulose d) a & b

96. Starch normally consists of __________% amylopectin.

    a) 20-30% b) 70-80%

    c) 80-90% d) 90-95%

97. Starch normally consists of__________% amylase.

    a) 20-30% b) 70-80%

    c) 80-90% d) 90-95%

98. The property of Poly-beta-hydroxyalkanoate (PHB) is

    a) 100% resistance to water b) 100% biodegradability

    c) Thermoplastic process ability d) All

99. The property of Polylactide Acid (PLA) plastic is

    a) Excellent biodegradability b) Process ability

    c) Biocompatibility d) All

100. Self-cooling beer using _____technology makes it possible to drink cool beer anywhere.

    a) Zeolite b) CaO

    c) Li d) MgO

## Answers

| | | | | | | | | | |
|---|---|---|---|---|---|---|---|---|---|
| 1. c | 2. c | 3. c | 4. c | 5. c | 6. d | 7. c | 8. c | 9. b | 10. d |
| 11. d | 12. a | 13. b | 14. c | 15. d | 16. a | 17. b | 18. c | 19. c | 20. b |
| 21. d | 22. c | 23. a | 24. a | 25. b | 26. c | 27. a | 28. b | 29. c | 30. d |
| 31. a | 32. c | 33. d | 34. b | 35. b | 36. a | 37. a | 38. d | 39. c | 40. a |
| 41. c | 42. b | 43. c | 44. c | 45. b | 46. d | 47. a | 48. b | 49. c | 50. a |
| 51. d | 52. a | 53. A | 54. C | 55. b | 56. a | 57. d | 58. d | 59. d | 60. b |
| 61. b | 62. d | 63. a | 64. c | 65. d | 66. c | 67. c | 68. a | 69. d | 70. b |
| 71. a | 72. c | 73. a | 74. b | 75. c | 76. b | 77. d | 78. a | 79. a | 80. d |
| 81. c | 82. a | 83. b | 84. c | 85. d | 86. a | 87. c | 88. d | 89. c | 90. a |
| 91. a | 92. d | 93. d | 94. a | 95. d | 96. b | 97. a | 98. d | 99. d | 100. a |

1. Self-heating coffee container based on the ___exothermic reaction is becoming available.
   a) Zeolite
   b) CaO
   c) Li
   d) MgO
2. Packaging using electronics technology by using________ battery power sources that enhances the branding citation in a crowded consumer product category is an influencing feature.
   a) Be
   b) B
   c) Li
   d) Na
3. The common nanocomposites used in the food packaging industry are
   a) Polymer clay nanoclay
   b) Silica nanocomposites of nanosilver
   c) a & b
   d) None
4. The effect of nanoclay in polymers is
   a) Nucleating agent in foams
   b) Higher cell density
   c) Flame retardant
   d) All
5. Characteristic of nanoclay in polymers is
   a) Stiffness
   b) Strength
   c) Smaller cell size
   d) All
6. Nanosilver used in smart packaging is composed of __________water.
   a) Ionized
   b) De-ionized
   c) Mineral
   d) Demineral
7. Nanosilver used in smart packaging (with silver in suspension) has excellent___properties.
   a) Antifungal
   b) Antibacterial
   c) Antiviral
   d) None
8. Durethanda transparent plastic film containing nanoparticles of clay that block__from entering into the food particles.
   a) Oxygen
   b) Moisture
   c) Carbon dioxide
   d) All

9. Embedded nano crystals which create a molecular barrier that helps to prevent the escape of___________.

   a) Oxygen  b) Moisture

   c) Carbon dioxide  d) All

10. Property of antimicrobial-kodak packaging system is __________.

    a) Anti-microbial  b) Absorbs oxygen

    c) a & b  d) Absorbs moisture

11. nanoparticle films with embedded sensors in packaging in order to detect pathogens is called Electronic ___________.

    a) Tongue  b) Nose

    c) Eye  d) None

12. Absolute pressure devices measure the ______ pressure acting on the sensor.

    a) Atmosphere  b) Vacuum

    c) Gauge  d) Total

13. Gauge pressure devices measure the pressure relative to __________ pressure.

    a) Atmospheric  b) Vacuum

    c) Total  d) None

14. Capacity range of instant infusion of medium to high total solid content products is

    a) 50–20,00 l/h  b) 1000–20,00 l/h

    c) 50–20,000 l/h  d) 100–20,00 l/h

15. An instant infusion plant typically consists of

    a) Preheating section

    b) Instant infusion section

    c) Final cooling section

    d) All

16. In ohmic heating, ______ electrical current is passed through liquid-particulate food.

    a) Alternating  b) Direct

    c) a & b  d) None

17. In ohmic heating, the rate of heating is directly proportional to the______ of the electric field strength.

a) Root b) Square

c) Cube d) None

18. In ohmic heating, the electric field strength can be varied by adjusting the________.

a) Electrode gap b) Applied voltage.

c) a & b d) None

19. Factor effecting the rate of heating produced in ohmic heating is

a) Electrical conductivity of food b) Temperature dependence

c) Electrode gap d) All

20. In ohmic heating electroporation across cell walls occurs at frequency of _____ Hz.

a) 30–40 Hz b) 40–50 Hz

c) 50–60 Hz d) 60–70 Hz

21. Vegetative forms of yeasts and moulds are inactivated by pressures between______ MPa.

a) 100 and 200 b) 200 and 300

c) 300 and 400 d) 400 and 500

22. Gram negative bacteria can be inactivated by pressures of about ______ MPa.

a) 300 b) 400

c) 500 d) 600

23. Gram positive bacteria can be inactivated by pressures of about 400 MPa

a) 300 b) 400

c) 500 d) 600

24. Most pressure sensitive groups of bacteria is the Gram negative *Vibrio* species is______.

a) *V. paraheamolyticus* b) *V. gastroentiritis*

c) *V. cholera* d) None

25. *Vibrio paraheamolyticus* in buffer is completely inactivated by pressure of ________.

a) 170 MPa / 30 min
b) 170 MPa / 10 min
c) 70 MPa / 30min
d) 70 MPa / 10min

26. *Vibrio paraheamolyticus* in clam juice is inactivated by pressure of 170 MPa for 10 min.

a) 170 MPa / 30 min
b) 170 MPa / 10 min
c) 70 MPa / 30min
d) 70 MPa / 10min

27. Which of the following combination of enzyme & its role is correct

a) Polyphenoloxidase (PPO)-enzymatic browning
b) Lipoxygenase (LOX) - change in flavour, colour and nutritional value
c) Pectinmethylesterase (PME) -cloud estabilization & consistency changes
d) All

28. Which of the following combination of enzyme & its role is not correct

a) Polyphenoloxidase (PPO)-nonenzymatic browning
b) Lipoxygenase (LOX) - change in flavour, colour and nutritional value
c) Peroxidase (POD) - unfavourable flavours
d) Papain – tenderization

29. Super chilling is a process by which the temperature of a food product is lowered to ___°C below the initial freezing point of the product.

a) 1–2°C
b) 2–3°C
c) 3–4°C
d) 4–5°C

30. In cryogenic freezing by solid carbon dioxide the temperatures obtained are _____°C.

a) 78°C
b) -87°C
c) -78°C
d) -98°C

31. Ozone is a powerful antimicrobial substance due to the its potential ___________capacity.

a) Reducing
b) Neutralizing
c) Oxidizing
d) None

32. Advantage of ozone in the food industry is
   - a) Surface hygiene
   - b) Sanitation of equipment
   - c) Plant hygiene
   - d) All

33. Application of ozone in the food industry is
   - a) Reuse of waste water
   - b) Lowering biological oxygen demand (BOD) food plant waste
   - c) Lowering chemical oxygen demand (COD) food plant waste
   - d) All

34. Ozone solubility in water is ____ times that of oxygen at 0–30°C.
   - a) 12
   - b) 13
   - c) 14
   - d) 16

35. Ozone solubility in ______water is progressively more.
   - a) Cold
   - b) Hot
   - c) Lukewarm
   - d) None

36. In extraction knowledge the abbreviation SFE means
   - a) Super food extraction
   - b) Supercritical food extraction
   - c) Supercritical fluid extraction
   - d) None

37. CO2 acts as supercritical fluid at is above ___°C and ______ MPa.
   - a) 31.1°C & 7.38 MPa
   - b) 3.1°C & 73.8 MPa
   - c) 3°C & 7 MPa
   - d) 3.1°C & 7.38 MPa

38. Ice cold water (0°C) will dissolve ___ g of $CO_2$ per liter of water at atmospheric pressure.
   - a) 3.4 g
   - b) 4.5 g
   - c) 13.4 g
   - d) 18.4 g

39. The metal commonly used for the packaging of foods along with steel is
   - a) Aluminium
   - b) Tin
   - c) Chromium
   - d) All

40. Durethanda transparent plastic film containing nanoparticles of clay is
   - a) Durable
   - b) Heat resistant
   - c) Gas barrier
   - d) All

41. Absolute pressure devices measure approximately ____ bar.

a) 1 b) 10

c) 100 d) 1000

42. Gauge pressure devices measure a reading of 0 millibar – lower pressures are registered as ________values.

a) Zero b) Positive

c) Negative d) None

43. In instant infusion the maximum viscosity can use at sterilisation temperature is __ centipoises.

a) 200 cp b) 300 cp

c) 400 cp d) 500 cp

44. In instant infusion the viscosity range prior to sterilization is ______cp.

a) 1 to 1000 b) 10 to 100

c) 100 to 1000 d) 1000 to 10000

45. In instant infusion the maximum dry matter can use at sterilisation temperature is ___% TS.

a) 22% TS b) 32% TS

c) 42% TS d) 52% TS

46. Correct statement about material handling is

a) Convey materials by manual b) Convey materials by mechanical

c) Reduces the human drudgery d) All

47. In belt conveyor, belt's load supported on

a) Idlers b) Pulleys

c) Idelers d) a & b

48. Efficiency of belt conveyor is largely dependent on

a) Idlers b) Pulleys

c) Idelers d) a & b

49. The pitch of idlers in belt conveyor is determined by

a) Idler load rating b) Belt tension

c) Belt speed d) All

50. Which of the following conveying system is present in vertical position

a) Belt conveyor b) Bucket elevator

c) Screw conveyor d) Chain Conveyor

51. In belt conveyor, belt's cross section shape is

a) V-shaped b) Flat

c) a & b d) U-shaped

52. For transportation of grains by belt conveyor, the belt speed should not increase _____ m/s.

a) 35 m/s b) 3.5 m/s

c) 5.3 m/s d) 53 m/s

53. Recommended belt speed for grain conveying by belt conveyor is ________ m/s .

a) 2.5 to 2.8 m/s b) 5.2 to 5.8 m/s

c) 25 to 28 m/s d) 52 to 82 m/s

54. The flat belt idlers are used for granular materials having an angle of repose of not less than _____°.

a) 75° b) 65°

c) 55° d) 35°

55. In belt conveyor, troughing idlers with _____° trough is used for conveying all kinds of bulk materials.

a) 10° b) 20°

c) 30° d) 40°

56. In belt conveyor, troughing idlers with ________° trough angle is mainly used for transportation ofsmall particle light weight materials like grain, cotton seed etc.

a) 15° and 25° b) 25° and 35°

c) 35° and 45° d) 4 5° and 55°

57. The spacing between the idlers in belt conveyor should not exceed _______ metres.

a) 1-2 m b) 2-3 m

c) 4-5 m d) 5-6 m

58. In belt conveyor, the tension developed at the drive pulley in transmitting the required power to move the loaded belt is known as _________ tension.

a) Valuable b) Effective

c) Defective d) None

59. Bucket elevators are classified into _____.

a) Spaced bucket elevators
b) Continuous bucket elevators
c) a & b
d) None

60. Which of the followings is not a spaced bucket elevator

a) Centrifugal discharge elevator
b) Positive-discharge elevator
c) Marine leg elevator
d) Super capacity bucket elevator

61. Which of the followings is a continuous bucket elevator

a) Super capacity bucket elevator
b) Internal-discharge bucket elevator
c) a & b
d) None

62. Which of the following is not a material handling system

a) Bucket elevator
b) Screw conveyor
c) Chain Conveyor
d) None

63. The pitch of idlers in belt conveyor is determined by

a) Idler load rating
b) Belt speed
c) Belt tension
d) All

64. Which of the following material handling system is present in vertical position

a) Belt conveyor
b) Screw conveyor
c) Bucket elevator
d) Chain Conveyor

65. The capacity of bucket elevator's is determined by

a) Spacing
b) Way of loading and unloading the bucket
c) Characteristic of bulk material
d) All

66. In cryogenic freezing by liquid nitrogen the temperatures obtained are ________ °C.

a) 0
b) -169°C
c) -156°C
d) -196°C

67. What is the critical factor to be considered for capacity of bucket elevator
    a) Bucket size b) Conveying speed
    c) Bucket design d) Spacing
68. Which of the following conveyor and is more susceptible to wear and requires high power
    a) Belt conveyor b) Bucket elevator
    c) Screw conveyor d) Chain Conveyor
69. Bucket elevators with a belt carrier can be used at fairly high speeds of _______ m/s.
    a) 2.5 to 4 m/ s b) 25 to 40 m/ s
    c) 5.2 to 40 m/ s d) 25 to 45 m/ s
70. In bucket elevators, when the product is not thrown well clear of the buckets into thedischarge chute, it will fall in the down leg. This is called as
    a) Advance logging b) Forward logging
    c) Back logging d) None
71. In bucket elevators the total stretch of the belt under maximum load should not exceed ____ % of the belt length.
    a) 1-2 % b) 2-3 %
    c) 3-4 % d) 4-5 %
72. In bucket elevators the bucket top angle is generally taken as _____°.
    a) 20° b) 30°
    c) 20-30° d) 80°
73. In bucket elevators the bucket bottom angle is generally taken as _____°.
    a) 10-15° b) 15-20°
    c) 20-30° d) 80°
74. The pitch of a standard screw in screw conveyor is_________ to its diameter.
    a) More than b) Equal
    c) Less than d) None
75. If the pitch of a standard screw in screw conveyor is 10 cm then, its diameter is____cm.
    a) 10 cm b) 20 cm
    c) 30 cm d) 40 cm

76. Most common shape of trough of screw conveyor is

a) U shaped b) O shaped

c) V shaped d) L shaped

77. Necessary air velocities in the range of ______ m/ s for dispersion of bulk material by pneumatic conveyor.

a) 5-15 m/s b) 15-30 m/s

c) 35-40 m/s d) 45-50 m/s

78. In blowing or positive pressure systems of pneumatic conveying, the product is conveyed by using air pressures ______ than the atmospheric pressure.

a) Greater than b) Equal

c) Less than d) None

79. Cyclone is a device which removes the bulk of the product particle from the conveying airstream by ________ force.

a) Gravitational b) Centrifugal

c) Centripetal d) Viscous

80. By using permanent magnets or electromagnets ____________ metals are removed.

a) Aluminum b) Copper

c) Ferrous d) a & b

81. The main physical property used to sort foods is

a) Size b) shape

c) Colour d) All

82. Screens with apertures are used for ____ sorting.

a) Size b) Shape

c) Colour d) Weight

83. In tumbling mixers are run at speed of about ______ rpm.

a) 100 rpm b) 200 rpm

c) 300 rpm d) 400 rpm

84. The product from an intermediate grinder might pass a _____ mesh screen.

a) 30 mesh b) 40 mesh

c) 50 mesh d) 60 mesh

85. most of the product from a fine grinder would pass a 200 mesh screen with a _____ µm opening

a) 47 µm  b) 57 µm

c) 74 µm  d) 84 µm

86. An ultra fine grinder accepts feed particles no larger than _____ mm.

a) 5 mm  b) 6 mm

c) 7 mm  d) 8 mm

87. In a jaw crusher the feed is admitted between two jaws, set to form a ____open at the top.

a) U shaped  b) O shaped

c) V shaped  d) L shaped

88. In hammer mills for fine reduction, the peripheral speed of the hammer tips may reach _____ m/s.

a) 100 m/s  b) 110 m/s

c) 200 m/s  d) 220 m/s

89. In hammer mills for fine reduction feed rate is ________tons/h to sizes finer than 200-mesh.

a) 0.01 to 1.5 tons/h  b) 0.1 to 1.5 tons/h

c) 0.1 to 15 tons/h  d) 1 to 150 tons/h

90. Commercial Hammer mills consume one kilo watt energy to reduce _______ kg of solid per hour.

a) 30 to 90 kg  b) 40 to 160 kg

c) 60 to 240 kg  d) 80 to 360 kg

91. In a ball mill or pebble mill, most of the size reduction is done by ___________ forces.

a) Impact  b) Shear

c) Frictional  d) Compression

92. In a roller mill, most of the size reduction is done by___________ forces.

a) Impact  b) Shear

c) Frictional  d) Compression

93. In a hammer mill most of the size reduction is done by ___________forces.

a) Impact  b) Shear

c) Frictional  d) Compression

94. In a pin mill, most of the size reduction is done by ___________forces.

a) Impact b) Shear

c) a & b d) Compression

95. In a attrition mill, most of the size reduction is done by ___________ forces.

a) Pressure b) Frictional

c) Shear d) a & b

96. In a tumbling mill, most of the size reduction is done by ___________ forces.

a) Impact b) Rolling

c) a & b d) Compression

97. In a large ball mill the shell might be 3 m in diameter and ______ m long.

a) 1.25 m b) 2.25 m

c) 3.25 m d) 4.25 m

98. In a large ball mill the balls are ________ mm in diameter.

a) 5 to 25 mm b) 25 to 125 mm

c) 125 to 225 mm d) 225 to 325 mm

99. In a large pebble mill the pebbles are ________ mm in diameter.

a) 30 to 115 mm b) 40 to 145 mm

c) 50 to 175 mm d) 150 to 195 mm

100. Particles size obtained by ultra fine grinders is _______ μm.

a) 10 to 200 μm b) 10 to 40 μm

c) 10 to 2 μm d) 1 to 20 μm

## Answers

| | | | | | | | | | |
|---|---|---|---|---|---|---|---|---|---|
| 1. b | 2. c | 3. c | 4. d | 5. d | 6. b | 7. b | 8. d | 9. a | 10. c |
| 11. a | 12. d | 13. a | 14. c | 15. d | 16. a | 17. b | 18. c | 19. d | 20. c |
| 21. b | 22. a | 23. b | 24. a | 25. a | 26. b | 27. d | 28. a | 29. a | 30. c |
| 31. c | 32. d | 33. d | 34. b | 35. a | 36. c | 37. a | 38. a | 39. d | 40. d |
| 41. a | 42. c | 43. d | 44. a | 45. d | 46. d | 47. d | 48. a | 49. d | 50. b |
| 51. c | 52. b | 53. a | 54. d | 55. b | 56. c | 57. a | 58. b | 59. c | 60. d |
| 61. c | 62. d | 63. d | 64. c | 65. d | 66. d | 67. b | 68. c | 69. a | 70. c |
| 71. a | 72. d | 73. c | 74. b | 75. a | 76. a | 77. b | 78. a | 79. b | 80. c |
| 81. d | 82. a | 83. a | 84. b | 85. c | 86. b | 87. c | 88. b | 89. c | 90. c |
| 91. a | 92. d | 93. a | 94. c | 95. d | 96. c | 97. d | 98. b | 99. c | 100. d |

1. In filtration the resulting liquor after removal of insoluble solids from a suspension is termed as

a) Filtrate b) Filter cake

c) Feed slurry d) Press

2. In filtration the separated solids are termed as

a) Filtrate b) Filter cake

c) Feed slurry d) Press

3. Filtration is used to clarify _______ by the removal of small amounts of solid particles.

a) Liquid b) Gases

c) Solids d) a & b

4. Filtration can be used to

a) Wine b) Syrups

c) Oils d) All

5. In the plate-and-frame filter press design, cloth or paper filters are supported on _____ plates.

a) Vertical b) Horizontal

c) Inclined d) None

6. The filter press is widely used for the production of ___________.

a) Apple juice b) Cider

c) a & b d) None

7. The filter press has

a) Low capital costs b) High flexibility

c) Easily maintenance d) All

8. The screw expellers are widely used for extraction of_________.

a) Apple juice b) Cider

c) Oil d) None

9. In screw expeller, horizontal barrel containing a _________helical screw.

a) Stainless steel b) Iron

c) Copper d) Zinc

10. In screw expeller, the pitch of the screw flights gradually _____towards the discharge end, to increase the pressure on the material as it is carried through the barrel.

a) Increases b) Decreases

c) Constant d) None

11. In screw expeller, the ____section of the barrel is perforated to allow expressed liquid to escape.

a) First b) Middle

c) Last but one d) Final

12. In screw expeller, pressure in the barrel is regulated by adjusting the ________ of the discharge port.

a) Length b) Diameter

c) Height d) Depth

13. In screw expeller, frictional heat _________ the viscosity of the oil.

a) Increases b) Reduces

c) Constant d) None

14. In some types of expellers have supplementary ______ fitted to the barrel to improve yields.

a) Condensers b) Coolers

c) Heaters d) Desiccators

15. The oil cake has _______ % (w/w) residual oil, depending on the type of oil-seed and the operating conditions.

a) 1–8% b) 5–18%

c) 15–28% d) 25–38%

16. In solid-liquid extraction, the desired component to be removed is a

a) Solution b) Solvent

c) Solute d) All

17. Arrange the stages in solid-liquid extraction method.

i) The solution becomes dispersed in the bulk of the solvent.

ii) The solution moves through the particle of food to its surface

iii) The solute dissolves in the solvent

a) I, II, III, b) III, II, I

c) I, III, II d) II, I, III

18. Loose reverse osmosis is also called
    a) Nanofiltration  b) Ultrafiltration
    c) Microfiltration  d) None
19. In nanofiltration membrane removes materials having molecular weights in the order of _____ Da.
    a) 3–10 Da  b) 30–100 Da
    c) 300–1000 Da  d) 3000–10000 Da
20. The unit change, due to force, in the size or shape of a body is referred to as
    a) Strain  b) Stress
    c) Yield point  d) Bearing load
21. Strain is a __________quantity.
    a) Dimensional  b) Non dimensional
    c) a & b  d) None
22. Strain is frequently expressed in
    a) Cm/cm  b) M/m
    c) Mm/mm  d) All
23. The change per unit length due to force in an original linear dimension is referred to as
    a) Linear strain  b) Tensile strain
    c) Compressive strain  d) All
24. Linear strain in a plane parallel to the longitudinal axis of the specimen is referred to as
    a) Axial strain  b) Transverse strain
    c) Shear strain  d) True strain
25. Linear strain in a plane perpendicular to the axis of the specimen is referred to as
    a) Axial strain  b) Transverse strain
    c) Shear strain  d) True strain
26. The tangent of the angular change, due to force between two lines originally perpendicular to each other through a point in a body is referred to as
    a) Axial strain  b) Transverse strain
    c) Shear strain  d) True strain

27. Shear strain also called as

a) Axial strain
b) Transverse strain
c) Angular strain
d) True strain

28. In a body subjected to axial force, the natural logarithm of the ratio of the gauge length at the moment of observation to original gauge length is referred to as

a) True strain
b) Transverse strain
c) Angular strain
d) Axial strain

29. The mean strain over any finite gauge length of measurement large incomparison with inter atomic distances is referred to as

a) Macro strain
b) Transverse strain
c) Angular strain
d) Axial strain

30. The intensity of a point in a body of the internal forces or components of force that act on a given plane through the point is referred to as

a) Strain
b) Stress
c) Yield point
d) Bearing load

31. Stress is expressed in ____________.

a) Force per unit of area
b) Kg-force per square millimeter
c) a & b
d) None

32. The stress component perpendicular to a plane on which the forces act is referred to as

a) Torsional stress
b) Shear stress
c) Normal stress
d) True stress

33. Normal stress may be

a) Tensile stress
b) Compressive stress
c) Shear stress
d) a & b

34. Normal stress due to forces directed away from the plane on which they act is referred to as

a) Tensile stress
b) Compressive stress
c) Shear stress
d) a & b

35. Normal stress due to forces directed towards the plane on which they act is referred to as

a) Tensile stress
b) Compressive stress
c) Shear stress
d) a & b

36. The stress component tangential to the plane on which the forces act is referred to as

a) Torsional stress b) Shear stress

c) Normal stress d) True stress

37. The shear stress on a transverse cross section resulting from a twisting action is referred to as

a) Torsional stress b) Shear stress

c) Normal stress d) True stress

38. The axial stress in a tension or compression test, calculated on the basis of the instantaneous cross sectional area instead of the original area is referred to as

a) Torsional stress b) Shear stress

c) Normal stress d) True stress

39. The maximum or minimum value of the normal stress at a point in a plane considered with respect to all possible orientations of the considered plane is referred to as

a) Torsional stress b) Principal stress

c) Normal stress d) True stress

40. Shear stress on all possible orientations of principal planes the is ___.

a) 0 b) 1

c) 1/2 d) 100

41. The true normal stress on the minimum cross sectional area at the beginning of fracture is referred to as

a) Torsional stress b) Principal stress

c) Normal stress d) Fracture stress

42. A compressive load on an interface is called as

a) Bearing load b) Compressive strength

c) Elastic Limit d) Poisson's ratio

43. The maximum load during a compression test on original cross sectional area of the specimen is

a) Bearing load b) Compressive strength

c) Elastic Limit d) Poisson's ratio

44. The greatest stress which a material is capable of sustaining without any permanent strain remaining upon complete release of the stress is called

a) Bearing load  b) Compressive strength

c) Elastic Limit  d) Poisson's ratio

45. The absolute value of the ratio of transverse strain to the corresponding axial strain resulting from uniformly distributed axial stress below the proportional limit of the material is referred as

a) Bearing load  b) Compressive strength

c) Elastic Limit  d) Poisson's ratio

46. The maximum stress that a material is capable of sustaining is calculated from the maximum load during a tension test carried "to rupture and the original cross sectional area of the specimen is

a) Tensile strength  b) Yield point

c) Bio-yield point  d) Rupture point

47. The first stress in a material, less than the maximum attainable stress, at which an increase in strain occurs without an increase in stress is

a) Tensile strength  b) Yield point

c) Bio-yield point  d) Rupture point

48. In biological materials, rupture may cause puncture of shell or skin, cracking, or fracture planes. It may be stated as

a) Tensile strength  b) Yield point

c) Bio-yield point  d) Rupture point

49. In biological materials, the point corresponds to a failure in the microstructure is called as

a) Tensile point  b) Yield point

c) Bio-yield point  d) Rupture point

50. The resistance to applied force is a

a) Stress  b) Strain

c) Elasticity  d) Strength

51. Units for strength are_______________

a) Cm/sec  b) Kg/cm

c) M/h  d) None

52. The stress corresponding to the rupture point is
    a) Ultimate strength b) Bio-yield strength
    c) Toughness d) Rigidity or shear modulus
53. The stress corresponding to the bio-yield point is
    a) Ultimate strength b) Bio-yield strength
    c) Toughness d) Rigidity or shear modulus
54. The work required to cause rupture in the material is
    a) Ultimate strength b) Bio-yield strength
    c) Toughness d) Rigidity or shear modulus
55. The ratio between shearing stress and the elastic shearing strain is called
    a) Ultimate strength b) Bio-yield strength
    c) Toughness d) Rigidity or shear modulus
56. The units for ultimate strength are
    a) Cm/sec b) Kg/cm
    c) M/h d) $kg/cm^2$
57. The units for rigidity or shear modulus are
    a) Cm/sec b) Kg/cm
    c) M/h d) $kg/cm^2$
58. The stress which is responsible for change in shape is
    a) Deviatoric stress component b) Isotropic stress component
    c) a & b d) None
59. The stress which is responsible for change in volume is
    a) Deviatoric stress component b) Isotropic stress component
    c) a & b d) None
60. A combined solid-like and liquid-like behavior in which the stress-strain relationship is time dependent is called
    a) Visco-elasticity b) Creep
    c) Retardation time d) None
61. Deformation with time when the material is suddenly subjected to a dead load-constant Stress is called
    a) Visco-elasticity b) Creep
    c) Retardation time d) None

62. The rate at which the retarded elastic deformation takes place in a material creeping under dead load is
   a) Visco-elasticity b) Creep
   c) Retardation time d) None
63. The ratio of shearing stress to shearing rate in Newtonian fluids is
   a) Viscosity coefficient b) Stress coefficient
   c) Strain coefficient d) None
64. Coefficient of viscosity divided by the mass density is called
   a) Viscosity b) Kinematic viscosity
   c) Surface tension d) All
65. The units for Kinematic viscosity are
   a) Cm/sec b) Kg/cm
   c) M/h d) $ft^2/sec$
66. Retardation time is the time required for the Kelvin model, representing creep behavior, to deform about ______ percent of its total deformation.
   a) 63% b) 73%
   c) 89% d) 93%
67. An ideal fluid in which the relationship between shear stress and shear rate is a straight line passing through the origin is a
   a) Newtonian liquid b) Non-Newtonian liquid
   c) Dilatant d) Pseudoplastic
68. A liquid in which the relationship between shear stress and shear rate is non-linear is a
   a) Newtonian liquid b) Non-Newtonian liquid
   c) Dilatant d) Pseudoplastic
69. A liquid in which the relationship between shear stress and shear rate is concave to the shear stress axis, the flow is called
   a) Newtonian liquid b) Non-Newtonian liquid
   c) Dilatant d) Pseudoplastic
70. A liquid in which the relationship between shear stress and shear rate is convex to the shear stress axis, the flow is called
   a) Newtonian liquid b) Non-Newtonian liquid
   c) Dilatant d) Pseudoplastic

71. Those fluids which show a decrease in shear stress with time of shear at a given shear rate are
    a) Thixotropic fluids
    b) Rheopectic fluids
    c) Dilatant
    d) Pseudoplastic
72. Those fluids which show an increase in shear stress with time of shear at a given shear rate are
    a) Thixotropic fluids
    b) Rheopectic fluids
    c) Dilatant
    d) Pseudoplastic
73. The types of deformation are
    a) Elastic deformation
    b) Flow
    c) a & b
    d) fluid
74. The classical ideal body representing elasticity, plasticity and viscosity properties is are
    a) Hookean body
    b) St. Venant body
    c) Newtonian liquid
    d) All
75. A typical example of the evaporation process is in the manufacture of
    a) Tomato puree
    b) Tomato ketchup
    c) Tomato paste
    d) All
76. Initial concentration of tomato juice is_______ % total solids.
    a) 1-2 %
    b) 3-4 %
    c) 5-6 %
    d) 10-15 %
77. The concentration of tomato paste is _____% total solids.
    a) 5-7%
    b) 25-27%
    c) 35-37%
    d) 45-47%
78. In non contact type heat exchanger, transfer of heat from ____ -pressure steam to the product occurs.
    a) High
    b) Low
    c) Atmosphere
    d) All
79. The evaporator in which produced vapors are discarded without further utilizing their inherent heat is called ____________evaporator.
    a) Single-effect
    b) Double -effect
    c) Triple-effect
    d) Multiple-effect

80. The advantage of keeping the product inside the evaporation chamber under vacuum is

   a) Vacuum causes the temperature difference between steam and product
   b) Product boils at relatively low temperatures
   c) To minimize heat damage
   d) All

81. The evaporator in which produced vapors are further utilized by reusing of vapors as the heating medium for another evaporator chamber is called ____________evaporator.

   a) Single-effect b) Double-effect
   c) Triple-effect d) Multiple-effect

82. The linear relationship exists between the boiling-point temperature of the solution and the boiling-point temperature of water at the same pressure is stated by

   a) Duhring's b) Hookean's
   c) Venant's d) All

83. The increase in boiling point of a solution over that of pure water, at a given pressure is called boiling-point __________.

   a) Falling b) Descend
   c) Elevation d) None

84. A simple method to estimate boiling-point elevation is the use of ___________ rule.

   a) Duhring's b) Hookean's
   c) Venant's d) All

85. In a rising-film evaporator, a______ -viscosity liquid food is allowed to boil inside vertical tubes.

   a) Low b) High
   c) Medium d) All

86. In a rising-film evaporator, a low viscosity liquid food is allowed to boil inside ______ m longvertical tubes.

   a) 1-5 m b) 10-15 m
   c) 100-150 m d) 150-200 m

87. In a rising-film evaporator, the upward movement of vapors causes a thin liquid film to move ________upward.
    a) Slowly
    b) Rapidly
    c) Gradually
    d) None
88. A temperature differential of at least _________°C between the product and the heating medium is necessary to obtain a well developed film in rising-film evaporator.
    a) 11°C
    b) 12°C
    c) 13C
    d) 14°C
89. Spray nozzles are used in _________ evaporator.
    a) Rising-film
    b) Falling-film
    c) A & b
    d) None
90. Function of spray nozzles in falling-film evaporator is
    a) Distribution of liquid in a uniform film
    b) Distribution of liquid in a uneven film
    c) a & b
    d) None
91. Which of the following evaporator allows a greater number of effects
    a) Falling-film
    b) Rising-film
    c) Forced-circulation
    d) None
92. Which of the following evaporator can handle more viscous liquids
    a) Falling-film
    b) Rising-film
    c) Forced-circulation
    d) None
93. Which of the following evaporator is best suited for highly heat-sensitive products
    a) Falling-film
    b) Rising-film
    c) Forced-circulation
    d) None
94. Typical residence time in a falling-film evaporator is
    a) 10-20 sec
    b) 20-30 sec
    c) 3-4 min
    d) 30-40 min
95. Typical residence time in a rising-film evaporator is
    a) 10-20 sec
    b) 20-30 sec
    c) 3-4 min
    d) 30-40 min

96. In which of the following evaporator a non-contact heat exchanger is used and liquid food is circulated at high rates.

a) Falling-film b) Rising-film

c) Forced-circulation d) None

97. A temperature differential of at least ________°C between the product and the heating medium is necessary in forced-circulation evaporator.

a) 1-2 °C b) 2-3 °C

c) 3-5 °C d) 14 °C

98. Pumps used in forced-circulation evaporators to maintain high circulation rates are

a) Axial flow pumps b) Radial flow pumps

c) Horizontal flow pumps d) All

99. Linear velocities of food in forced-circulation evaporators is

a) 0.3-1 m/s b) 2- 6 m/s

c) 9-10 m/s d) 10-15 m/s

100. Linear velocities of food in natural-circulation evaporators is

a) 0.3-1 m/s b) 2- 6 m/s

c) 9-10 m/s d) 10-15 m/s

## Answers

| | | | | | | | | | |
|---|---|---|---|---|---|---|---|---|---|
| 1. a | 2. b | 3. d | 4. d | 5. a | 6. c | 7. d | 8. c | 9. a | 10. b |
| 11. d | 12. b | 13. b | 14. c | 15. b | 16. c | 17. b | 18. a | 19. c | 20. a |
| 21. b | 22. d | 23. d | 24. a | 25. b | 26. c | 27. c | 28. a | 29. a | 30. b |
| 31. a | 32. c | 33. d | 34. a | 35. b | 36. b | 37. a | 38. d | 39. b | 40. a |
| 41. d | 42. a | 43. b | 44. c | 45. d | 46. a | 47. b | 48. c | 49. d | 50. d |
| 51. b | 52. a | 53. b | 54. c | 55. d | 56. d | 57. d | 58. a | 59. b | 60. a |
| 61. b | 62. c | 63. a | 64. b | 65. d | 66. a | 67. a | 68. b | 69. c | 70. d |
| 71. a | 72. b | 73. c | 74. d | 75. d | 76. c | 77. c | 78. b | 79. a | 80. d |
| 81. d | 82. a | 83. c | 84. a | 85. a | 86. b | 87. b | 88. d | 89. b | 90. a |
| 91. a | 92. a | 93. a | 94. b | 95. c | 96. c | 97. c | 98. a | 99. b | 100. a |

1. The type of evaporators that use the principles of rising falling-film, falling-film, wiped-film, and forced circulation are ______.

   a) Plate　　b) Tubular

   c) Circular　　d) Globular

2. The type of evaporator's configuration often provides features that make it more acceptable is

   a) Plate　　b) Tubular

   c) Circular　　d) Globular

3. A rising/falling-film plate evaporator requiring _____ floor area than atubular unit.

   a) Equal　　b) More

   c) Less　　d) More or less

4. The difference between the boiling point of the solution and that of pure water, at the same pressure is called boiling point _____________.

   a) Rise　　b) Fall

   c) Constant　　d) None

5. The drying process refers to the grain drying process in which all grains are fully exposed to the drying air under constant drying conditions is _____.

   a) Thin layer　　b) Bed layer

   c) Thick layer　　d) a & b

6. Generally up to _____ cm thickness of grain bed (with a recommended air-grain ratio) is taken as thin layer.

   a) 10 cm　　b) 20 cm

   c) 30 cm　　d) 40 cm

7. All commercial flow dryers are designed on _____ layer drying principles.

   a) Thin layer　　b) Bed layer

   c) Thick layer　　d) a & b

8. Grain isotherms are generally ___ shaped and attributed to multi-molecular adsorption.

   a) S-shaped　　b) U-shaped

   c) V-shaped　　d) Z-shaped

9. Many solid materials (foods) exhibit different equilibrium moisture characteristics depending upon whether the equilibrium is reached by adsorption / sorption or desorption of the moisture. This phenomenon is known as ___________.

   a) Hysteresis  b) Psychrometry

   c) Rheology  d) All

10. The moisture-contained by a substance which exerts equilibrium vapour pressure, less than that of the pure liquid at the same temperature is called_____________ moisture.

    a) Bound  b) Unbound

    c) Critical  d) Imbibed

11. The correct statement about bound moisture

    a) Inside the cell walls of the plant structure

    b) Moisture in loose chemical combination with the cellulosic material

    c) Moisture bound in small capillaries

    d) All

12. In the constant-rate period the rate of evaporation under any given set of air conditions is ______of the solid.

    a) Dependent  b) Independent

    c) No effect  d) All

13. The moisture contained by a substance which exerts equilibrium vapour pressure equal to that of the pure liquid at the same temperature is

    a) Bound  b) Unbound

    c) Critical  d) Imbibed

14. In the constant-rate period the rate of the solid is the same as the rate of evaporation from a free liquid surface under the

    a) Same condition  b) Different condition

    c) a & b  d) None

15. Double bonds in Vitamin A are susceptible to oxidation by

    a) Hydrogen  b) Free radicals

    c) Peroxides  d) a & b

16. The rate of drying during constant-rate period is dependent upon
    a) Difference between the temperature of air and temperature of the wetted surface at constant air velocity and relative humidity
    b) Difference in humidity between air stream and wet surface at constant air velocity and temperature
    c) Air velocity at constant air temperature and humidity
    d) All
17. The drying takes place by surface evaporation and moisture moves by vapour pressure difference is
    a) Constant-rate period    b) Falling-rate period
    c) Raising-rate period    d) None
18. The moisture content of the solid at which the drying rate ceases to be constant is known as
    a) Critical moisture    b) Constant moisture
    c) Unbound    d) Bound
19. Cereal grains are usually dried entirely under
    a) Falling-rate period    b) Raising-rate period
    c) Constant-rate period    d) None
20. In the drying process the period characterized by increasing temperatures both at the surface and within the solid is
    a) Falling-rate period    b) Raising-rate period
    c) Constant-rate period    d) None
21. Vitamin A present in animal foods as
    a) Retinol    b) *Tran* $\beta$-carotene
    c) *Cis* $\beta$-carotene    d) Carotinic acid
22. Vitamin A present in plant foods as
    a) Retinol    b) *Tran* $\beta$-carotene
    c) *Cis* $\beta$-carotene    d) Carotinic acid
23. Losses of Vitamin-A due to oxidation is promoted by traces of
    a) Copper    b) Iron
    c) Both Cu & Fe    d) None

24. Heat converts part of the *trans* isomer to _____which has lower potency is
    a) Neo-β-carotene-U
    b) Neo-β-carotene-V
    c) Neo-β-carotene-W
    d) All
25. Thiamin (Vitamin B1) present in animal and plant tissues as
    a) Free thiamin
    b) Bound thiamin
    c) Pyrophosphate
    d) All
26. Thiamin is destroyed by
    a) Sulphur dioxide
    b) Potassium bromate
    c) Thiaminase
    d) All
27. The enzyme catalyses thiamin destruction by phenols in plant tissues is
    a) Phynolase
    b) Polyphenoloxidase
    c) Tyrosinase
    d) All
28. Riboflavin occurs as free form in milk but in other foods it is bound with
    a) Lipids
    b) Statch
    c) Phosphate
    d) Proteins
29. The conditions that destroy Vitamin $B_2$ are
    a) Alkaline
    b) Light
    c) Excessive heat
    d) All
30. Vitamin $B_2$ is stable to
    a) Air
    b) Acids
    c) Alkalines
    d) a & b
31. Niacin occurs as (as nicotinamide adenine dinucleotide andnicotinamide adenine dinucleotide phosphate) and as.
    a) Nicotinamide
    b) Nicotinic acid
    c) NADP
    d) All
32. Niacin is bound to polysaccharides and peptides and therefore not available inmany cereals unless liberated by conditions
    a) Heat
    b) Alkaline
    c) Acids
    d) a & b
33. In the body, the amino acid tryptophan is converted to
    a) Valine
    b) Lysine
    c) Niacin
    d) Leucine

34. Niacin equivalent is free niacin plus _____ of the tryptophan
    a) One sixth b) One sixteenth
    c) One sixtieth d) None
35. The vitamin expressed as pteroylglutamic equivalents is
    a) Niacin b) Folic acid
    c) Vitamin $B_6$ d) Riboflavin
36. Richest sources of folic acid are
    a) Dark green leaves b) Liver
    c) Kidney d) All
37. The vitamin for which it is more difficult to assay than other vitamin is
    a) Folic acid b) Vitamin $B_6$
    c) Niacin d) a & b
38. Pyridoxine (Vitamin B6) is occurs in the form of
    a) Pyridoxal b) Pyridoxine
    c) Pyridoxamine d) All
39. Form of Vitamin $B_6$ found in plants are
    a) Pyridoxal b) Pyridoxine
    c) Pyridoxamine d) a & b only
40. Form of Vitamin $B_6$ found in animal tissues is/are
    a) Pyridoxal b) Pyridoxine
    c) Pyridoxamine d) All
41. The functional group of proteins and amino acids react with Vitamin $B_6$ and causes its loss when it is heated or during storage is
    a) Hydrogen b) Sulphydryl
    c) Carboxylic d) Carboxylic
42. In milk the loss of Vitamin $B_{12}$ is due to interaction with vitamin C and sulphydrylcompounds in the presence of
    a) Hydrogen b) Carboxylic
    c) Carboxylic d) Oxygen
43. Vitamin C occurs as
    a) Ascorbic acid b) Dehydroascorbic acid
    c) Hydroascorbic acid d) both a & b

44. Dehydroascorbic acid is very heat labile in
    a) Presence of oxygen b) Absence of oxygen
    c) Both a & b d) None
45. The plant enzyme that cause Vitamin C destruction is
    a) Cytochrome oxidase b) Peroxidase
    c) Phenolase d) All
46. The elements that catalyze oxidation of Vitamin C in air is
    a) Copper b) Iron
    c) Zinc d) Both a & b
47. Sulphur dioxide perfectly protects against oxidation of
    a) Vitamin A b) Vitamin B
    c) Vitamin C d) Vitamin D
48. The vitamin retention that sometimes used as an indicator of the severity of processing is
    a) Vitamin A b) Vitamin B
    c) Vitamin C d) Vitamin D
49. Vitamin C is used as
    a) Antioxidant b) Stabiliser
    c) Flour improver d) All
50. Vitamin D occurs in foods as
    a) Ergocalciferol (D2) b) Cholecalciferol (D3)
    c) Both a & b d) None
51. The form of Vitamin D (synthetic type) added to some milk products, baby foods and margarine is
    a) Ergocalciferol (D2) b) Cholecalciferol (D3)
    c) Both a & b d) None
52. Vitamin E occurs as eight compounds ___tocopherols and ____tocotrienols.
    a) Six & two b) Five & three
    c) Four & four d) Seven & one
53. Vitamin E activity usually expressed as ________equivalents.
    a) T*rans* $\beta$-tocopherol b) C*is* $\beta$-tocopherol
    c) $\alpha$-tocopherol d) All

54. Vitamin E generally stable during processing, except frying in which it is destroyed by

a) Nascent oxygen | b) Singlet oxygen
c) Superoxides | d) Peroxides

55. Function/s of Aluminium is/are

a) Texture modifier | b) Leavening agent
c) Dough improver | d) a & b

56. Function of Bromine is

a) Texture modifier | b) Leavening agent
c) Dough improver | d) Forms gels

57. Function of Calcium is

a) Texture modifier | b) Leavening agent
c) Dough improver | d) Forms gels

58. Sources of Calcium are

a) Dairy products | b) Green leafy vegetables
c) Tofu | d) All

59. Sources of copper

a) Organ meats | b) Seafood
c) Nuts & seeds | d) All

60. Sources of Iodine are

a) Organ meats | b) Seafood
c) Nuts & seeds | d) All

61. Sources of Iron are

a) Iodised salt | b) Seafood
c) Plants and meat | d) All

62. Sources of Iron are

a) Cereals | b) Legumes
c) Meat | d) All

63. Sources of Magnesium are

a) Whole grains | b) Nuts
c) a & b | d) All

64. Sources of Manganese are
    a) Whole grains b) Fruits
    c) Vegetables d) All
65. Sources of Nickel are
    a) Cereals b) Legumes
    c) Animal foods d) a & b
66. Sources of Magnesium are
    a) Legumes b) Green leafy
    c) Vegetables d) All
67. Sources of Potassium are
    a) meats b) Vegetables
    c) Fruits d) All
68. Sources of Selenium are
    a) Organ meats b) Seafood
    c) Cereals d) All
69. Sources of Sodium are
    a) Milk b) Food additives
    c) Raw foods d) a & b
70. Sources of Zinc are
    a) Meats b) Cereals
    c) a & b d) None
71. Sodium aluminum sulfate is a
    a) Leavening agent b) Dough improver
    c) Firming agent d) All
72. Dough improver is
    a) KBrO3 b) KIO3
    c) Aluminum sulfate d) Both a & b
73. The chemical compound that firms canned vegetables when added to canning brine is
    a) Aluminum b) Calcium
    c) Sodium d) All

74. The Catalyst present in lipid peroxidation, ascorbic acid oxidation and nonenzymaticoxidative browning is

a) Copper b) Sodium

c) Aluminum d) Calcium

75. Enzyme cofactor for polyphenoloxidase is

a) Copper b) Sodium

c) Aluminum d) Calcium

76. Catalyst that catalyses lipid peroxidation in foods is

a) Copper b) Fe2+

c) Fe3+ d) All

77. Colour of fresh meat depends on valence of ___________ in myoglobin and hemoglobin.

a) Iron b) Sodium

c) Aluminum d) Calcium

78. The valence of Fe in red coloured myoglobin and brown coloured hemoglobin is

a) 1+ b) 2+

c) 3+ d) 4+

79. The valence of Fe in brown coloured myoglobin and brown coloured hemoglobin is

a) 1+ b) 2+

c) 3+ d) 4+

80. The valence of sulpher that reacts with iron to form black FeS in canned foods is

a) -1 b) -2

c) -3 d) -4

81. Enzyme cofactor for lipoxygenase, cytochromes, ribonucleotide reductase is

a) Iron b) Sodium

c) Aluminum d) Calcium

82. Removal of Mg from chlorophyll changes colour from green to________ colour.

a) Red b) Olive-brown

c) Yellow d) Green

83. Cofactor for pyruvate carboxylase and superoxide dismutase enzymes is
    a) Manganese b) Sodium
    c) Aluminum d) Calcium
84. The catalyst used in hydrogenation of vegetable Oils is
    a) Na b) Ba
    c) Ni d) Bo
85. Which of the following is a inorganic acidulent used in food preservation?
    a) $CH_3CO_3$ b) $CH_3COOH$
    c) $H_3PO_4$ d) $H_3NO_3$
86. Sodium tripolyphosphate improves ________retention incured meats.
    a) Gas b) Moisture
    c) Mineral d) Salts
87. Phosphates are used as _______ in comminuted meats and in processed cheeses.
    a) Emulsification b) Emulsification aid
    c) Hydrogenation d) Flavour modifier
88. Which of the following may be used as a table salt substitute.
    a) PCl b) KCl
    c) BCl d) NaCl
89. Keshan disease (endemic cardiomyopathy in China) is associated with ________deficiency.
    a) Potassium b) Nickel
    c) Selenium d) Sodium
90. Enzyme cofactor for glutathione peroxidase activity is
    a) Selenium b) Nickel
    c) Potassium d) Sodium
91. Leavening agents is/are
    a) SodiumBicarbonate b) Sodium Aluminium Sulphate
    c) Sodiumacid Pyrophosphate d) All
92. Sulpher containing amino acid is/are
    a) Methionine b) Cystine
    c) Glycine d) a & b

93. Sulphur dioxide and sulphites inhibit __________ browning.

a) Enzymatic b) Nonenzymatic

c) a & b d) None

94. The compound used in the lining of cans for protein rich foods to lessen formation of black FeS during heating is

a) ZnO b) CuO

c) NO d) $H_2O_2$

95. The compound can be added to green beans to help stabilise the colour during canning.

a) Ni b) Zn

c) Cu d) Fe

96. The p*K*a expresses, on a logarithmic scale, the relative strength of a __________ acid or base.

a) Weak b) Strong

c) Neutral d) a & b

97. p*K*a =

a) log 1/*K*a b) - log *K*a

c) Log *K*a d) a & b

98. The extent of acception or regection of a hydrogen ion which affects the pH is expressed as a ____ constant (*K*a).

a) Dissociation b) Association

c) Equilibrium d) None

99. Approved organic acids used as antimicrobial agents have maximum activity in ____ foods.

a) Basic b) Low acid

c) Medium acid d) high acid

100. All regulatory approved organic acids used as antimicrobials have pKa values ________.

a) =5 b) >5

c) < 5 d) None

## Answers

| | | | | | | | | | |
|---|---|---|---|---|---|---|---|---|---|
| 1. a | 2. a | 3. c | 4. a | 5. a | 6. b | 7. a | 8. a | 9. a | 10. a |
| 11. d | 12. b | 13. a | 14. a | 15. d | 16. d | 17. a | 18. a | 19. a | 20. a |
| 21. a | 22. b | 23. c | 24. a | 25. d | 26. d | 27. b | 28. c | 29. d | 30. d |
| 31. d | 32. d | 33. c | 34. c | 35. b | 36. d | 37. d | 38. d | 39. d | 40. c |
| 41. b | 42. d | 43. d | 44. c | 45. d | 46. d | 47. c | 48. c | 49. d | 50. b |
| 51. a | 52. c | 53. c | 54. d | 55. d | 56. c | 57. d | 58. d | 59. d | 60. d |
| 61. d | 62. d | 63. c | 64. d | 65. d | 66. d | 67. d | 68. d | 69. d | 70. c |
| 71. a | 72. d | 73. b | 74. a | 75. a | 76. d | 77. a | 78. b | 79. c | 80. b |
| 81. a | 82. b | 83. a | 84. c | 85. c | 86. b | 87. b | 88. b | 89. c | 90. a |
| 91. d | 92. d | 93. c | 94. a | 95. b | 96. a | 97. d | 98. a | 99. d | 100. c |

1. Acid present in Cranberries
   a) Acetic acid  b) Benzoic acid
   c) Citric acid  d) Malic acid
2. Acid present in apple
   a) Citric acid  b) Benzoic acid
   c) Acetic acid  d) Malic acid
3. Acid present in plums
   a) Benzoic acid  b) Acetic acid
   c) Citric acid  d) Malic acid
4. Acid present in citrus fruits
   a) Acetic acid  b) Benzoic acid
   c) Citric acid  d) Malic acid
5. Acid present in prunes
   a) Acetic acid  b) Benzoic acid
   c) Citric acid  d) Malic acid
6. Acid present in cinnamon
   a) Citric acid  b) Benzoic acid
   c) Acetic acid  d) Malic acid
7. Acid present in swiss cheese
   a) Propionic acid  b) Benzoic acid
   c) Acetic acid  d) Malic acid
8. Acid present in cloves
   a) Acetic acid  b) Benzoic acid
   c) Citric acid  d) Malic acid
9. Acid present in brown berries
   a) Sorbic acid  b) Benzoic acid
   c) Acetic acid  d) Malic acid
10. The book "Art of Conserving all kinds of Animal and Vegetable Matter for several Years" is written by
    a) Charles Elton  b) Benzman
    c) Devidson  d) Nicolas Appert

11. Preservation mechanism present in Cooling, chill distribution andstorage is
    a) Low temperature to retard growth
    b) Low temperature and reduction of water activity to prevent growth
    c) Reduction in water activity sufficient to delay or prevent growth
    d) Low oxygen tension to inhibit strict aerobes and delay growth of facultative anaerobes
12. Preservation mechanism present in Freezing, frozen distribution andstorage is
    a) Reduction in water activity sufficient to delay or prevent growth
    b) Low temperature and reduction of water activity to prevent growth
    c) Low temperature to retard growth
    d) Low oxygen tension to inhibit strict aerobes and delay growth of facultative anaerobes
13. Preservation mechanism present in Drying, is
    a) Reduction in $a_w$
    b) Reduction in size
    c) Reduction in shape
    d) Osmosis
14. Preservation mechanism present in curing is
    a) Reduction in water activity sufficient to delay or prevent growth
    b) Low temperature and reduction of water activity to prevent growth
    c) Low temperature to retard growth
    d) Low oxygen tension to inhibit strict aerobes and delay growth of facultative anaerobes
15. Preservation mechanism present in conserving is
    a) Low temperature to retard microbial growth
    b) Low temperature and reduction of water activity to prevent growth
    c) Reduction in water activity sufficient to delay or prevent growth
    d) Low oxygen tension to inhibit strict aerobes and delay growth of facultative anaerobes
16. Preservation mechanism present in modified atmosphere packaging is
    a) Low temperature to retard growth
    b) Low temperature and reduction of water activity to prevent growth

c) Reduction in water activity sufficient to delay or prevent growth

d) Low oxygen tension to inhibit strict aerobes and delay growth of facultative anaerobes

17. Preservation mechanism present in addition of acids to foods is

a) Reduction in $a_w$ b) Reduction in $p^H$

c) Reduction in shape d) Osmosis

18. Preservation mechanism present in carbon dioxide-enriched modified atmosphere packaging

a) Specific inhibition of some micro-organisms by carbon dioxide

b) Low temperature and reduction of water activity to prevent growth

c) Reduction in water activity sufficient to delay or prevent growth

d) Low oxygen tension to inhibit strict aerobes and delay growth of facultative anaerobes

19. Preservation mechanism present in lactic fermentation process is

a) Reduction of pH value in situ by microbial action

b) Additional inhibition of microbes by the lactic and acetic acids formed

c) Reduction in water activity sufficient to delay or prevent growth

d) a & b

20. Inactivation of enzymes is the main preservation mechanism present in

a) Blanching b) Drying

c) Concentration d) Curing

21. Preservation mechanism present in Pascalization is

a) Low temperature and reduction of water activity

b) Additional inhibition of microbes by the lactic and acetic acids formed

c) Pressure-inactivation of vegetative bacteria, yeasts and moulds

d) Reduction of pH value in situ by microbial action

22. Preservation mechanism by compartmentalization and nutrient limitation within the aqueous droplets in water-in-oil type foods is

a) Blanching b) Drying

c) Emulsification d) Concentration

23 The main objective of grilling

a) Improvement of digestibility

b) Improvement of flavour

c) Destruction of pathogenic micro-organisms

d) All

24. Preservation method in which delivery of heat sufficient to inactivate target microorganismsto the desired extent is

a) Pasteurization b) Appertization

c) Conserving d) a & b

25. The main objective of baking

a) Improvement of digestibility

b) Improvement of flavour

c) Destruction of pathogenic micro-organisms

d) All

26. Correct statement about Appertization

a) Organisms that survive appertization process are nonpathogenic

b) Incapable of microbes developing within the product under normal conditions of storage

c) Appertized products have a long shelf-life even when stored at ambient temperatures

d) All

27. Preservation mechanism present in blanching is

a) Inactivation of enzymes

b) Expulsion of oxygen from tissues

c) Delivery of ionizing radiation

d) a & b

28. Preservation method in which delivery of ionizing radiation at a dose sufficient to inactivate target microorganisms to the desired extent is

a) Radurization b) Radicidation

c) Radappertization d) All

29. Elimination of micro-organisms to achieve 'commercial sterility' is termed as
    - a) Grilling
    - b) Pasteurization
    - c) Appertization
    - d) Baking
30. Improvement of digestibility, flavor of food and destruction of pathogenic micro-organisms are the main objectives of
    - a) Cooking
    - b) Boiling
    - c) Frying
    - d) All
31. Cooking method example for combinations of cooking methods
    - a) Baking
    - b) Braising
    - c) Brewing
    - d) Blanching
32. The temperatures applied in deep fat frying
    - a) 180-220 °C
    - b) 220-250°C
    - c) 250-270°C
    - d) 270-300°C
33. The temperatures applied in baking
    - a) 120-260 °C
    - b) 240-250 °C
    - c) 250-270°C
    - d) 270-300°C
34. The temperatures applied in pasteurization
    - a) ≤100°C
    - b) =100°C
    - c) ≥100°C
    - d) <100°C
35. Most vegetative cells are killed almost instantaneously at
    - a) ≤100°C
    - b) =100°C
    - c) ≥100°C
    - d) None
36. The temperatures applied in sterilization
    - a) ≤100°C
    - b) =100°C
    - c) ≥100°C
    - d) None
37. The temperatures applied in appertization
    - a) ≤100°C
    - b) =100°C
    - c) ≥100°C
    - d) None
38. Generally d"100°C temperatures are applied in
    - a) Blanching
    - b) Concentration
    - c) Drying
    - d) All

39. The type of steaming method used for aluminium foil cooking is
    a) Dry steaming b) Wet steaming
    c) Waterless steaming d) All
40. Microbial thermal death is generally fallows a ______order process.
    a) Zero b) First
    c) Second d) Third
41. The time at a given temperature for the surviving population to be reduced by 1 log cycle is
    a) D value b) Z value
    c) F value d) None
42. Which of the following statement is correct ?
    a) Gram-positives are more heat resistant than Gram-negatives.
    b) The temperature at which a D value applies is indicated by a subscript, e.g. $D_{65}$.
    c) Most vegetative cells are killed almost instantaneously at 100°C.
    d) All
43. Which of the following statement is correct?
    a) Bacterial spores are usually far more heat resistant than vegetative cells.
    b) D values are measured and expressed at temperatures appropriate to pasteurization.
    c) Thermophiles produce the most heat resistant spores while those of psychrotrophs.
    d) All
44. A parameter which expresses the integrated lethal effect of a heat process in terms of minutes at a given temperature is
    a) Z value b) F value
    c) D value d) All
45. The F value will depend on the ______value of the organism of concern;
    a) Z value b) T value
    c) D value d) None
46. If z value is 10°C then 1 minute at 111°C has an F121 value is
    a) 1 b) 0.1
    c) 0.01 d) 100

47. if z value is 5°C then 1 minute at 111°C has an F121 value is

a) 1
b) 0.1
c) 0.01
d) 100

48. When stating F value it is necessary to specify

a) Z value
b) Temperature
c) R.H
d) a & b

49. Minimum heat process lethality applied to low-acid canned foods, that it should produce 12 decimal reductions in the number of surviving *C. botulinum* spores is called

a) Botulinum cook
b) 12D
c) A or b
d) 12F

50. The defect in which both ends of the can are permanently and firmly bulged and do not yieldreadily to thumb pressure is called

a) Flat
b) Soft swell
c) Hard swell
d) Springer

51. The defect in which both ends of can are bulged but not tightly, they yield to thumb pressure

a) Flat
b) Soft swell
c) Hard swell
d) Springer

52. The can defect in which no evidence of swelling occurs is

a) Flat
b) Soft swell
c) Hard swell
d) Springer

53. The can defect in which one end flat, the other bulged. When the bulged end is pressed in then the flat one springs out istz

a) Flat
b) Soft swell
c) Hard swell
d) Springer

54. A can with a normal appearance which when brought down sharply on a flat surface causes a flat end to flip out and bulged end can be forced back by very slight pressure is

a) Flipper
b) Soft swell
c) Hard swell
d) Springer

55. Mode of action of UV, in causing death to micro-organisms is
   a) Under UV excite electrons from their ground state into higher energy orbitals making the molecules less reactive
   b) Under UV excite electrons from their ground state into higher energy orbitals making the molecules more reactive
   c) Under UV excite electrons from their higher energy into ground state orbitals making the molecules more reactive.
   d) None

56. Generally z value for chemical reactions such as vitamin loss, browning reactions and enzyme inactivation is typically ____°C.
   a) 5–25°C b) 25–40°C
   c) 45–50°C d) 55–60°C

57. The microwave region of the electromagnetic spectrum occupies frequencies between
   a) $10^9$ Hz up to $10^{12}$ Hz b) $10^{12}$ Hz up to $10^{16}$ Hz
   c) $10^{16}$ Hz up to $10^{20}$ Hz d) None

58. Domestic microwave ovens use _______MHz.
   a) 2450 MHz b) 4520 MHz
   c) 5420 MHz d) 24500 MHz

59. Wavelength of UV radiation acts as germicidal at
   a) 230 nm b) 240 nm
   c) 250 nm d) 260 nm

60. Nucleic acid bases having strong absorption of UV radiation at the wavelength of
   a) 230 nm b) 240 nm
   c) 250 nm d) 260 nm

61. The most commonly used cobalt isotope for production of Gamma ã-rays is
   a) $^{57}$Co b) $^{58}$Co
   c) $^{59}$Co d) $^{60}$Co

62. $^{60}$Co emits high-energy g-rays (1.1 MeV) which can penetrate food up to a depth of
   a) 10 cm b) 20 cm
   c) 30 cm d) 40 cm

63. The processes where the objective is the elimination of a pathogen is
   a) Radurization   b) Pascalization
   c) Radicidation   d) None
64. The processes aiming to prolong shelf-life is
   a) Radurization   b) Pascalization
   c) Radicidation   d) None
65. For removal of Salmonella from meat and poultry is done in
   a) Radurization   b) Pascalization
   c) Radicidation   d) None
66. Application of high pressures in food processing is called
   a) Pascalization   b) Radurization
   c) Radicidation   d) None
67. Vegetative bacteria and fungi can be reduced by at least one log cycle by400MPa applied for ____ min.
   a) 5 min.   b) 15 min.
   c) 25 min.   d) 35 min.
68. Which of the following statement is correct
   a) Chilled foods are those foods stored at temperatures near 0–5°C
   b) Freezing point of foods is being lower than that of pure water due to the solutes present
   c) Frozen foods are storage are generally less than -18 °C
   d) All
69. Which of the following statement is correct about freezing
   a) No microbial growth is possible at freezing temperature
   b) Residual microbial or endogenous enzyme activity such as lipases can persist
   c) Blanching is done before freezing
   d) All
70. Reason for blanching of fruits and vegetables before freezing
   a) Inactivate endogenous polyphenol oxidases
   b) To prevent discoloration of product during storage.
   c) To reduce the lipase activity
   d) All

71. A non-microbiological quality defect that may arise in frozen foods, where surface discolouration occurs due to sublimation of water is

a) Freezer burn
b) Botulism
c) Patulin
d) Black ear

72. In frozen foods Freezer burn is prevented by

a) Wrapping in water impermeable material
b) Glazing with a layer of ice
c) Wrapping in water permeable material
d) a & b

73. Benzoic acid occurs naturally present in

a) Cherry bark
b) Cranberries
c) Greengage plums
d) All

74. pKa value of Benzoic acid

a) 4.19
b) 5.19
c) 6.19
d) 7.19

75. pKa value of nitrous acid (HNO2) is

a) 3.4
b) 5.19
c) 6.19
d) 7.19

76. Bacterial inhibition by nitrite____________ with decreasing pH.

a) Increases
b) Decreases
c) No effect
d) All

77. Sulfur dioxide is active against

a) Bacteria
b) Yeasts
c) Moulds
d) All

78. Sulfurous acid ($H_2SO_3$) is a dibasic acid with pKa values of

a) 1.86 and 6.91
b) 2.86 and 4.91
c) 3.86 and 6.91
d) 1.86 and 5.91

79. An unsaturated fatty acid, 2,4-hexadienoic acid, found naturally in the berries of the mountain ash is

a) Propionic acid
b) Malic acid
c) Benzoic acid
d) Sorbic acid

80. pKa value of Sorbic acid
    a) 4.8
    b) 4.9
    c) 5.0
    d) 5.1
81. pKa value of Propionic acid
    a) 4.8
    b) 4.9
    c) 5.0
    d) 5.1
82. Sorbic acid is active against
    a) Yeasts
    b) Moulds
    c) Catalase-positive bacteria
    d) All
83. Natamycin, formerly know as
    a) Nisin
    b) Pimaricin
    c) Penicillin
    d) Streptomycin
84. The bacterium *Streptomyces natalensis* produces
    a) Nisin
    b) Penicillin
    c) Natamycin
    d) Streptomycin
85. The processing in which food is vacuum packed before being given a pasteurizationtreatment which gives it a longer shelf-life under chill storage is
    a) Hot fill
    b) Cuisine sous-vide
    c) Hydrothermal
    d) Curing
86. Intermediate moisture foods (IMF) are commonly defined as those foods with an $a_w$ between
    a) 0.85 and 0.3
    b) 0.95 and 0.6
    c) 0.95 and 0.5
    d) 0.85 and 0.6
87. Moisture content of Intermediate moisture foods (IMF)
    a) 5–20%
    b) 20–30%
    c) 35–70%
    d) 15–50%
88. Which of the following is a Intermediate Moisture Food
    a) Cakes
    b) Fish sauce
    c) Jams
    d) All
89. Which of the following is not a Intermediate Moisture Food
    a) Fresh fruits
    b) Sweetened condensed milk
    c) a & b
    d) Dried fruits

90. stability of sweetened condensed milk is a result of

a) High sugar content b) Low $a_w$

c) Appertization d) a & b

91. IMFs are generally made by a process of

a) Adsorption b) Desorption

c) Acidification d) a & b

92. Novel humectans is

a) Glycerol b) Sorbitol

c) Propylene glycol d) All

93. In hygienic plant design, the area Level 1 means

a) Production areas b) Non-production areas

c) Packaging areas d) None

94. In hygienic plant design, the area Level-2 means

a) Non-production areas

b) Food processing operations dealing with undecontaminated product or finished goods

c) Food processing operations dealing with decontamination processes

d) None

95. In hygienic plant design, the area Level-3 means

a) Non-production areas

b) Food processing operations dealing with undecontaminated product or finished goods

c) Food processing operations dealing with decontamination processes

d) None

96. Which of the following area is referred to as 'low-risk or good manufacturing practice (GMP) area in designing of hygienic plant

a) Level-1 b) Level-2

c) Level-3 d) All

97. In food industry, the recommended artificial lighting is

a) 500-600 lux b) 700-800 lux

c) 5800-900 lux d) 300-500 lux

98. Freezer burn is minimized by packaging in ____ materials.
    a) Moisture-proof    b) Hygroscopic
    c) Fatty    d) All

99. Height of the barrier to divide low- and high-risk floors in food industry construction
    a) 40 cm    b) 20 cm
    c) 50 cm    d) 60 cm

100. Dust control in the food industry serves main purpose like
    a) To prevent environmental pollution
    b) To protect operators from inhaling fine particles, e.g. from milling operations
    c) To prevent the spread of dust in processing areas which may lead to cross-contamination
    d) All

## Answers

| | | | | | | | | | |
|---|---|---|---|---|---|---|---|---|---|
| 1. b | 2. d | 3. a | 4. c | 5. b | 6. b | 7. a | 8. b | 9. a | 10. d |
| 11. a | 12. b | 13. a | 14. a | 15. c | 16. d | 17. b | 18. a | 19. d | 20. a |
| 21. c | 22. c | 23. d | 24. d | 25. d | 26. d | 27. d | 28. c | 29. c | 30. d |
| 31. b | 32. a | 33. a | 34. d | 35. b | 36. c | 37. c | 38. d | 39. c | 40. b |
| 41. a | 42. d | 43. d | 44. b | 45. a | 46. b | 47. c | 48. d | 49. c | 50. c |
| 51. b | 52. a | 53. d | 54. a | 55. c | 56. b | 57. a | 58. a | 59. d | 60. d |
| 61. d | 62. b | 63. c | 64. c | 65. c | 66. a | 67. a | 68. d | 69. d | 70. d |
| 71. a | 72. d | 73. d | 74. a | 75. a | 76. d | 77. d | 78. a | 79. d | 80. a |
| 81. b | 82. d | 83. b | 84. c | 85. b | 86. d | 87. d | 88. d | 89. a | 90. d |
| 91. d | 92. d | 93. b | 94. b | 95. c | 96. b | 97. a | 98. a | 99. d | 100. d |

1. Particle diameters less than _____ µm are of concern for respiratory damage to humans.
   a) 10 µm b) 20 µm
   c) 30 µm d) 40 µm
2. The type of filter used for pre-collector of large quantities of dust from conveying air
   a) Dry cyclones b) Bag filters)
   c) Rigid element filters d)
3. The type of filter used for heavy continuous dust burden at constant pressure drop
   a) Mechanically shaken b) Reverse jet
   c) Static d) Cartridge
4. Discharge burden achieved by reverse jet filter is
   a) 1-2 $mg/m^3$ b) 0-1 $mg/m^3$
   c) 1-3 $mg/m^3$ d) 7-15 $mg/m^3$
5. The type of filter used for pollution control of fine particles in large exhaust systems (e.g.boilers)
   a) Electrostatic precipitators b) Reverse jet
   c) Static d) Cartridge
6. Humidity levels maintained for Cheese maturation is
   a) less than 95% b) 95%
   c) More than 95% d) 0 %
7. The most common method of dehumidifying air is to pass it through
   a) Cooling coil b) Heating coil
   c) Freezing coil d) All
8. The purpose of fogging a production area is to reduce the numbers of____ micro-organisms.
   a) Airborne b) Water borne
   c) Food borne d) All
9. Fumigation is used in
   a) Ripening rooms b) Production areas
   c) Process lines d) All

10. Widely used type (grade) of stainless steel in food industries
    a) AISI-311 Type
    b) AISI-316 Type
    c) AISI-317 Type
    d) AISI-318 Type
11. Improved corrosion resistance of Type 316 stainless steel is due to presence of
    a) Carbon
    b) Molybdenum
    c) Ni
    d) Na
12. In food processing equipments the material gasket is made up of
    a) Polytetrafluoroethylene
    b) Polytrifluoroethylene
    c) Polydifluoroethylene
    d) All
13. Which of the following steel grade is good resistance against salt and alkalis at moderatetemperatures
    a) AISI-304 Type
    b) AISI-316 Type
    c) AISI-317 Type
    d) AISI-318 Type
14. Which of the following steel grade is commonly used in environments where product fluids do not contain any chlorides?
    a) AISI-304 Type
    b) AISI-316 Type
    c) AISI-317 Type
    d) AISI-318 Type
15. Which of the following steel grade is recommended for components of equipment such as valves, pump castings, rotors and shafts?
    a) AISI-304 Type
    b) AISI-316 Type
    c) AISI-317 Type
    d) AISI-318 Type
16. Which of the following steel grade is recommended for pipe work and vessels due to its enhanced weldability ?
    a) AISI-304 Type
    b) AISI-316L Type
    c) AISI-316*l* Type
    d) AISI-318 Type
17. The effectiveness of CIP systems thus depends on
    a) Time
    b) Temperature
    c) Chemical concentration
    d) All
18. In CIP system relatively high volume of solution has to be applied to soiled surfacesfor at least
    a) 5 min to 1 hr
    b) 50 min to 1 hr
    c) 50 min to 2 hr
    d) 65 min to 3 hr

19. Recommended Cleaning procedure for cleaning of sugars, organic acids, salt is

    a) Mildly alkaline detergent b) Strong alkali detergent
    c) Chlorinated alkaline d) All

20. Recommended Cleaning procedure for cleaning of high-protein foods (meat)

    a) Mildly alkaline detergent b) Strong alkali detergent
    c) Chlorinated alkaline d) All

21. Recommended Cleaning procedure for cleaning of starchy foods, tomatoes,

    a) Mildly alkaline detergent b) Strong alkali detergent
    c) Chlorinated alkaline d) All

22. Recommended Cleaning procedure for cleaning of fatty foods (fat, butter, margarine, oils)

    a) Mildly alkaline detergent b) Strong alkali detergent
    c) Chlorinated alkaline d) a & b

23. Recommended Cleaning procedure for cleaning of heat-precipitated water hardness, milkstone

    a) Acid cleaner b) Strong alkali detergent
    c) Chlorinated alkaline d) a & b

24. Recommended disinfectant for CIP (clean in place)

    a) Chlorine b) Iodophors
    c) Acid sanitiser d) All

25. Concentration of Quaternary ammonium compounds for cleaning and disinfecting of food contacting surfaces

    a) 5-10% b) 10-20%
    c) 0.5-1% d) 0.05-2%

26. Concentration of Iodophor for cleaning and disinfecting of food contacting surfaces

    a) 0.005-0.01% b) 0.05-0.1%
    c) 0.5-1% d) 5-10%

27. Concentration of Active chlorine for cleaning and disinfecting of food contacting surfaces

    a) 0.15-0.3%  b) 0.015-0.03%

    c) 0.3-0.5%  d) 0.5-3%

28. Concentration of Active oxygen compound for cleaning and disinfecting of food contacting surfaces

    a) 0.01-0.2%  b) 0.1-0.5%

    c) 0.03-0.5%  d) 0.3-5%

29. The most commonly used grade of stainless steel for piping

    a) AISI 403  b) AISI 304

    c) AISI316  d) b & c

30. Cr/Ni ratio present in AISI 304 grade of stainless steel is

    a) 18/10  b) 18/12

    c) 18/12/3  d) 33/11

31. Cr/Ni/Mo proportion in AISI316 grade of stainless steel is

    a) 18/10  b) 18/12

    c) 18/12/3  d) 11/33

32. Stainless steel grade used for chloride-containing products especially when the products are hot.

    a) AISI 403  b) AISI 304

    c) AISI 316  d) b & c

33. The most widely used plastics material for rigid pipes

    a) PVC  b) ABS

    c) PP  d) All

34. Expand the ABS in polymer science

    a) Acrylonitrile Butadiene System  b) Anti Breaking System

    c) a & b  d) None

35. Time range required for effectiveness of CIP systems

    a) 1 min - 5 min  b) 5 min - 1 hr

    c) 50 min - 10 hr  d) 10 min – 50min

36. Required linear velocity of CIP fluid in pipelines for effectiveness of CIP systems is at least
    a) 0.15 m/s b) 1.5 m/s
    c) 15 m/s d) 150 m/s
37. Required flow rate of CIP fluid in tank for effectiveness of CIP systems is at least
    a) 1 $m^3$/h per tank b) 10 $m^3$/h per tank
    c) 100 $m^3$/h per tank d) 10 $m^3$/sec per tank
38. At what pH, chlorine is predominantly present as HOCl ?
    a) PH 3-7.5 b) PH 1-3
    c) PH 5-6 d) pH 7-10
39. The matching of two colors, such as when judging the color of red hot metals is a
    a) Primary measurement b) Secondary measurement
    c) Tertiary measurement d) All
40. An instrument which translates pressure changes into length changes is comes under
    a) Primary measurement b) Secondary measurement
    c) Tertiary measurement d) All
41. Measurement of temperature of an object by thermocouple is a
    a) Primary measurement b) Secondary measurement
    c) Tertiary measurement d) All
42. A device/system is calibrated against primary standards, the procedure is termed
    a) Primary calibration b) Secondary calibration
    c) Tertiary calibration d) All
43. A device is used for further calibrating another device of lesser accuracy, then the procedure is termed
    a) Primary calibration b) Secondary calibration
    c) Tertiary calibration d) All
44. The degree of correctness of the measured value with respect to the true value is called
    a) Accuracy b) Precision
    c) Degree d) Value

45. The degree of repeatability of several independent measurements of the desired input at the same reference conditions is called

a) Precision b) Accuracy

c) Degree d) Value

46. Law of thermodynamics which states that if two bodies are in thermal equilibrium with a third body, then they are all in thermal equilibrium with each other is

a) Zeroth b) First

c) Second d) Third

47. Law of thermodynamics which states energy cannot be created or destroyed in an isolated system is

a) Zeroth b) First

c) Second d) Third

48. Law of thermodynamics which states that entropy of any isolated system always increases.

a) Zeroth b) First

c) Second d) Third

49. Law of thermodynamics which states the entropy of the crystal is equal to absolute zero.

a) Zeroth b) First

c) Second d) Third

50. Triple point of water (equilibrium between solid, liquid and vapor phases of water) is

a) 273.16 °K b) 0.01°C

c) 1°C d) a & b

51. Bimetallic thermometers are usually employed in the range of

a) 30 to 50°C b) -30 to 550°C

c) -70 to 750°C d) 700 to 750°C

52. Bellows type of mechanical pressure gauge is used for measurement of

a) High pressures b) Low pressures

c) Medium pressures d) All

53. Base-metal thermocouples use the combination of pure metals and alloys of iron, copper andnickel and are used for temperature up to _______°K.

a) 0 °K  b) 50 °K

c) 450 °K  d) 1450 °K

54. Rare-metal thermocouples use a combination of pure metals and alloys of platinum fortemperatures up to ______°C

a) L6°C  b) L60°C

c) L600°C  d) 16000°C

55. Rare-metal thermocouples use a combination of tungsten, rhodium and molybdenum for temperatures up to______°C

a) 3°C  b) 30°C

c) 300°C  d) 3000°C

56. The atmospheric pressure at sea level is 1.013 X105 N/m2 or of mercury.

a) 760 mm Hg  b) 1.01 bar

c) 101325 pascals  d) All

57. The fluid pressure above the reference value of a perfect vacuum or the absolute zero pressure is called

a) Absolute pressure  b) Gauge pressure

c) Atmospheric pressure  d) All

58. The difference between the absolute pressure and the local atmospheric pressure is

a) Absolute pressure  b) Gauge pressure

c) Atmospheric pressure  d) All

59. The pressure exerted by the weight of air in the atmosphere of earth is

a) Absolute pressure  b) Gauge pressure

c) Atmospheric pressure  d) All

60. Fluid flow in which the behavior of the fluid in motion, do not change with time is called

a) Steady Flow  b) Unsteady Flow

c) Uniform Flow  d) Non-uniform Flow

61. Fluid flow in which the behavior of the fluid in motion, change with time is called

a) Steady Flow  b) Unsteady Flow

c) Uniform Flow  d) Non-uniform Flow

62. Fluid flow in which the velocity of flow of fluid does not change, both in magnitude anddirection, from point to point in the flowing fluid, for any given instant of time is called

a) Steady Flow b) Unsteady Flow
c) Uniform Flow d) Non-uniform Flow

63. Fluid flow in which the velocity of flow of fluid changes, both in magnitude anddirection, from point to point in the flowing fluid, for any given instant of time is called

a) Steady Flow b) Unsteady Flow
c) Uniform Flow d) Non-uniform Flow

64. A venturi meter is a device which is used for measuring the ________ of fluid.

a) Velocity b) Rate of flow
c) Friction d) Direction

65. The basic principle on which a venturi meter works is

a) Temperature difference b) Pressure difference
c) Mass difference d) All

66. An orifice meter is a device which is used for measuring the ________ of fluid.

a) Velocity b) Rate of flow
c) Friction d) Direction

67. Generally the orifice diameter is kept as ________ times the pipe diameter.

a) 0.5 b) 1.0
c) 1.5 d) 2.0

68. A commonly used variable area flow meter

a) Orifice b) Venturi meter
c) Notch d) Rotameter

69. A simple device used for measuring the velocity of fluid flow is

a) Orifice b) Venturi meter
c) Notch d) Pitot tube

70. 1 Rankine degree (°R) is equals to ; such that

a) °R = °F+100 b) °R = °F+273
c) °R = °F+460 d) °R = °F+560

71. Methane ($CH_4$) percentage in gobar gas is

a) 55 -65% b) 30 - 40%

c) 20 - 30% d) 10 - 20%

72. Carbon dioxide ($CO_2$) percentage in gobar gas is

a) 55 -65% b) 30 - 40%

c) 20 - 30% d) 10 - 20%

73. The calorific value of biogas is ______kcal/$m^3$ of gas.

a) 50 to 55 kcal/$m^3$ b) 500 to 550 kcal/$m^3$

c) 5,000 to 5,500 kcal/$m^3$ d) None

74. The gobar gas production may be ______ $m^3$/kg of wet dung.

a) 0.036 $m^3$/kg b) 0.36 $m^3$/kg

c) 3.6 $m^3$/kg d) 36 $m^3$/kg

75. The steam containing moisture or particles of water in suspension is called

a) Wet steam b) Dry saturated steam

c) Superheated steam d) All

76. When the wet steam is further heated, and it does not contain any suspended particles of water, it is known as

a) Wet steam b) Dry saturated steam

c) Superheated steam d) All

77. When the dry steam is further heated at a constant pressure, thus raising its temperature, it is said to be

a) Wet steam b) Dry saturated steam

c) Superheated steam d) All

78. The amount of heat absorbed by 1 kg of water, when heated at aconstant pressure, from the freezing point (0°C) To the temperature of formation of steam

a) Sensible heat b) Latent heat

c) Enthalpy d) All

79. The specific heat of water at constant pressure is usually taken as

a) 2.4 kJ/kg K b) 3.2 kJ/kg K

c) 4.2 kJ/kg K d) None

80. The amount of heat absorbed to evaporate 1 kg of water, at its boiling point or saturation temperature without change of temperature is

a) Sensible heat b) Latent heat
c) Enthalpy d) All

81. Amount of heat absorbed by water from freezing point to saturation temperature plus the heat absorbed during evaporation is

a) Sensible heat b) Latent heat
c) Enthalpy d) All

82. The Indian Boilers Act is begin in

a) 1923 b) 1933
c) 1943 d) 1953

83. The ratio of the mass of water vapor to the mass of dry air contained in the moist air is called

a) Humidity Ratio b) Moiture Ratio
c) Volume Ratio d) Density Ratio

84. The temperature measured by a (dry) thermometer immersed in a vapor-gas mixture.

a) Dry bulb temperature b) Wet-Bulb Temperature
c) Dew-Point Temperature d) All

85. Temperature measured by placing a thermometer having a water-moistened wick covered bulb into a fast moving stream of ambient air is

a) Dry bulb temperature b) Wet-Bulb Temperature
c) Dew-Point Temperature d) All

86. The temperature at which a given unsaturated air vapor mixture becomes saturated is called

a) Dry bulb temperature b) Wet-Bulb Temperature
c) Dew-Point Temperature d) All

87. The partial pressure exerted by the water vapor molecules in moist air is

a) Vapor Pressure b) Vacuum Pressure
c) Normal Pressure d) Dew Pressure

88. Sorption Isotherm represents the relationship between

a) EMC & $a_w$ b) Temperature & time
c) Colour & time d) All

89. A process in which tiny particles or droplets are surrounded by a coating or embedded in a homogeneous or heterogeneous matrix is

a) Microencapsulation b) Vaporization

c) Polymerization d) Splitting

90. Which of the following is a encapsulating wall material

a) Sodium alginate b) Pre-biotic compounds

c) Marine protein isolates d) All

91. The material for encapsulation is homogenized with the carrier material usually at a ratio of

a) 1:1 b) 1:2

c) 1:3 d) 1:4

92. The outer material used in spray cooling

a) Vegetable oil b) Animal oil

c) Fossils oil d) All

93. The type of microencapsulation used for the encapsulation of volatile and unstable flavors in glassy carbohydrate matrices is

a) Cocrystallization b) Spray drying

c) Extrusion d) Fluidized-bed coating

94. Preservation method used for the dehydration of almost all heat sensitive materials and aromas is

a) Sun drying b) Freeze-drying/Lyophilization

c) Spray drying d) Drum drying

95. Separation of a liquid phase of coating material from a polymeric solution followed by the coating of that phase as a uniform layer around suspended core particles is called

a) Enrobing b) Encapsulation

c) Evaporation d) Coacervation

96. Which of the following encapsulation process utilizes sucrose as a matrix for the incorporation of core materials

a) Spray drying b) Spray drying

c) Cocrystallization d) Fluidized-bed coating

97. Which of the following encapsulation process is used for Vitamin C encapsulation?

a) Extrusion  b) Spray-cooling/spray chilling

c) Fluidized-bed coating  d) Both b & c

98. Single or multilayered vesicles of phospholipids containing either aqueous-based or lipophilic compounds are called

a) Lysosomes  b) Liposomes

c) Microsomes  d) Ribosomes

99. Placing of food beneath a red-hot source of heat, usually a glowing metal grid is called

a) Brazing  b) Broiling

c) Scorching  d) Boiling

100. Nanomaterial used for light activated sterilization of surfaces is

a) Zinc oxide  b) Silver

c) $NaHCO_3$  d) KCl

## Answers

1. a 2. a 3. b 4. d 5. b 6. c 7. a 8. a 9. d 10. b
11. b 12. a 13. a 14. a 15. b 16. b 17. d 18. a 19. a 20. c
21. a 22. d 23. a 24. d 25. d 26. a 27. b 28. c 29. d 30. a
31. c 32. c 33. d 34. a 35. b 36. b 37. b 38. a 39. a 40. b
41. c 42. a 43. b 44. a 45. a 46. a 47. b 48. c 49. d 50. d
51. b 52. b 53. d 54. c 55. d 56. d 57. a 58. b 59. c 60. a
61. c 62. d 63. b 64. b 65. b 66. b 67. a 68. d 69. d 70. c
71. a 72. b 73. c 74. a 75. a 76. b 77. c 78. a 79. c 80. b
81. c 82. a 83. a 84. a 85. b 86. c 87. a 88. a 89. a 90. d
91. d 92. a 93. c 94. b 95. d 96. c 97. d 98. b 99. b 100. a

1. Which of the following is used for sterilization?
   a) $H_2O_2$
   b) Co-extrusion heat process
   c) Ethylene oxide
   d) All
2. Which of the followings is not used for sterilization?
   a) Gamma-radiation
   b) UV light
   c) a & b
   d) Vis light
3. Which of the following sterilization method has residual problem?
   a) $H_2O_2$
   b) Co-extrusion heat process
   c) Ethylene oxide
   d) Gamma-radiation
4. Which of the following sterilization method is chemically safe?
   a) Hydrogen peroxide
   b) Co-extrusion heat process
   d) Gamma-radiation
5. Packaging material used in ethylene oxide sterilization method is ?
   a) Glass
   b) Plastic containers
   c) Pre-formed plastic
   d) a & b
6. Which of the following combination of laminates for retortable pouch is correct ?
   a) Nylon 15μm/ Cast polypropylene 60 μm
   b) Nylon 15μm/ Cast polypropylene 70 μm
   c) Nylon 25μm/ Cast polypropylene 70 μm
   d) All
7. Requirements of sterilization temperatures for retort pouch is
   a) 100 – 111°C
   b) 111 – 1116°C
   c) 116 – 145°C
   d) 151 – 165°C
8. Oxygen permeability in retort pouch is ______ cc/$m^2$/24 h/at atmosphere
   a) 0
   b) 10
   c) 100
   d) 1000
9. Moisture vapour transmission rate in retort pouch is ______ g/ M2/24 h
   a) 0
   b) 0.1
   c) 1
   d) 10

10. Seal strength in retort pouch is ______ Kg/100 mm.
    a) 0.02-0.035 b) 0.2-0.35
    c) 2-3.5 d) 20-35
11. Bond strength in retort pouch is ________g/10 ml.
    a) 0.15-0.50 b) 1.5-5.0
    c) 15-50 d) 150-500
12. Which of the following is can coating?
    a) Fruit enamel b) C-enamel
    c) Citrus enamel d) All
13. Fruit enamel type of can coating is used for _____.
    a) Berries
    b) Cherries
    c) Fruits require protection from metallic salts
    d) All
14. C-enamel type of can coating is used for _____.
    a) Corn b) Peas
    c) Sulphur bearing food products d) All
15. Citrus enamel type of can coating is used for _____.
    a) Citrus scices b) Citrus juice
    c) Citrus concentrate d) All
16. Beverage can enamel type of can coating is used for _____.
    a) Vegetable juices b) Red fruit juices
    c) Non carbonated beverages d) All
17. In fruit enamel the coating material is
    a) Oleoresinous b) Oleoresinous with zinc oxide
    c) Modified Oleoresinous d) None
18. In Beverage can enamel the coating material is
    a) Oleoresinous
    b) Zinc oxide
    c) Two coated with resinous base coat and Vinyl top coat
    d) None

19. In C-enamel the coating material is
    a) Oleoresinous
    b) Oleoresinous with zinc oxide
    c) Modified Oleoresinous
    d) None
20. In citrus enamel the coating material is
    a) Oleoresinous
    b) Oleoresinous with zinc oxide
    c) Modified oleoresinous
    d) None
21. The type of steel base used for packaging food products is __________.
    a) Type - MR
    b) Type - MS
    c) Type - L
    d) All
22. The type of steel base used for strongly corrosive food products is __________.
    a) Type - MR
    b) Type - MS
    c) Type - L
    d) Type - MC
23. Example for strongly corrosive food product/s is are__.
    a) Berries
    b) Sauerkraut
    c) Pickles
    d) All
24. The type of steel base used for moderately corrosive food products is __________.
    a) Type - MR
    b) Type - MS
    c) Type - L
    d) Type - MC
25. Example for moderately corrosive food product/s is are__.
    a) Apricots
    b) Grape fruit
    c) Peaches
    d) All
26. The type of steel base used for non-corrosive food products is __________.
    a) Type - MR
    b) Type - MS
    c) Type - L
    d) Type - MC or MR
27. Example for non-corrosive food product/s is are__.
    a) Dehydrated frozen foods
    b) Nuts
    c) Both a & b
    d) None

28. National Standards on Packaging code for foodstuffs and perishables is according to the _____code.
    a) FPO b) PFA
    c) BIS d) AGMARK
29. Pasteurized flavoured milk is packed in
    a) Glass bottles b) LDPE lined cartons
    c) Aseptic cartons d) All
30. Sterilized flavoured milk is packed in
    a) Poly papers b) Poly-laminated paper packs
    c) Glass bottles d) All
31. Sterilized milk is packed in
    a) Poly papers b) Poly-laminated paper packs
    c) Glass bottles d) All
32. The fermented milk products should be cooled before dispatch and maintained at a temperature below ______°C, to develop good flavour and texture.
    a) 1°C b) 10°C
    c) 100°C d) 121°C
33. Milk powder is packed in
    a) Hermetically sealed containers
    b) Hermetically sealed tinplate containers
    c) Hermetically sealed tinplate containers in N2 or a mixture of N2 and $CO_2$ gas
    d) All
34. Cereal weaning foods are packed in
    a) Hermetically sealed containers
    b) Hermetically sealed tinplate containers
    c) Hermetically sealed tinplate containers in N2 or a mixture of N2and $CO_2$ gas
    d) All
35. Raw vegetables and fruits are packed in
    a) Gunny bags b) Plastic crates
    c) Corrugated fiberboard boxes. d) All

36. Onion and garlic are packed in

a) Gunny bags b) Wooden crates

c) Corrugated fiberboard boxes d) All

37. Tomatoes are packed in

a) Baskets b) Wooden boxes

c) Corrugated fiberboard boxes. d) All

38. Chillies are packed in

a) Gunny bags b) Bamboo crates

c) Corrugated fiberboard boxes d) All

39. Juices, jams, jellies and marmalades are packed in

a) Glass bottles b) Open top cans

c) Tinplate containers d) All

40. Meat is wrapped in ____sheets or bags and delivered in clean, rust free and closed containers.

a) Polypropylene b) Polyethylene

c) Polymethylene d) None

41. whole carcasses and cooked meat products may are vacuum packed using suitable wrapping materials such as ____________.

a) PET b) LDPE

c) HDPE d) a & b

42. Egg powder is gas packed in nitrogen in suitable tinplate containers or flexible packaging materials such as ______.

a) PET b) LDPE

c) HDPE d) a & b

43. The fresh fish is packed in ___ lined insulated plywood.

a) Polypropylene b) Polyethylene

c) Polymethylene d) None

44. The thickness of insulation in plywood intended for fish packaging is ____________mm.

a) 15-30 mm b) 30-55 mm

c) 55-75 mm d) 75-100 mm

45. The packaging material used for bread is____.

a) LDPE coated poster paper b) Waxed paper

c) Grease –proof paper d) All

46. The packaging material used for biscuits is____.

a) Cello/LDPE b) BOPP/LDPE

c) PET/LDPE d) All

47. The packaging material used for Cakes is____.

a) Waxed paper b) Glassine/LDPE

c) Grease-proofpolyethylene d) All

48. The packaging material used for protein-rich extruded foods is____.

a) HDPE bags b) LDPE liner bag

c) BOPP/LDPE pouch d) All

49. Peanut butter is packed in

a) Glass jars b) Polystyrene tubs

c) a & b d) None

50. Flours and starches are packed in either or LDPE coated raffia bags.

a) LDPE coated jute bags b) LDPE coated raffia bags

c) a & b d) None

51. Edible spray dried potato flour is packed in

a) HDPE b) Metallized polyester bags

c) BOPP/LDPE d) All

52. Oils and fats are packed in

a) HDPE b) Food grade PVC

c) PET d) All

53. Cereal grains are packed in

a) PE coated jute bags b) HDPE coated jute bags

c) Raffia bags d) PET

54. Cereal flours are packed in

a) LDPE coated jute bag b) Single sound A-twill

c) B-twill jute bags d) All

55. Sugar is packed in
    a) PE coated Hessian bags b) PET coated raffia bags
    c) B-twill jute bags d) All
56. Cube sugar of net weight of_____ kg is wrapped in butter paper or kraft paper and packed in cartons.
    a) 0.5 kg b) 1.0 kg
    c) 1.5 kg d) 2.0 kg
57. Honey is packed in
    a) Acid resistant lacquered tin container
    b) Polyethylene
    c) Glass container
    d) All
58. Tea is packed in
    a) LDPE b) PET
    c) BOPP d) All
59. Roasted and ground coffee are packed in
    a) Paper/LDPE b) PET/LDPE
    c) BOPP/LDPE d) All
60. Chocolates is packed in
    a) Grease-proof paper
    b) Aluminium foil
    c) Laminates made of paper/LDPE
    d) All
61. Carbonated beverages should be filled in glass containers conforming to IS no. is
    a) IS 1107:1986 b) IS 1007:1986
    c) IS 1010:1996 d) IS 1107:1996
62. After filling of carbonated beverages, the containers should be hermetically sealed with clean, newcrown corks conforming to IS no. is
    a) IS 1964:1987 b) IS 1994:1987
    c) IS 1954:1967 d) IS 1954:1977

63. Beer is filled in _____ml glass bottles or in PET bottles.

a) 250 ml
b) 550 ml
c) 650 ml
d) 1000 ml

64. Beer is packed in bottle or cans. These bottles or cans should be packed in

a) Wooden cases
b) Corrugated fiber board boxes
c) Plastic bag
d) a & b

65. The process in which surface fibers are reorients in the base sheet of paper through the use of pressure & which serves to smooth the surface, control surface texture and develop a glossy finish is called___.

a) Calendering
b) Casting
c) Forging
d) Boring

66. Papers produced by calendering process are known as ________finished (MF).

a) Manual
b) Machine
c) Motor
d) Mortise

67. The containers made from sheets of paperboard which have been cut and scored for bending into desired shapes are called ________ cartons

a) Corrugated
b) Folding
c) Cart
d) straight

68. Thickness of paper boards used for preparation of folding cartons is ______ μm.

a) 100 - 100 μm
b) 300 - 1000 μm
c) 300 - 1100 μm
d) 700 - 1700 μm

69. Folding Carton Boxes are grouped into____types.

a) 8
b) 6
c) 4
d) 2

70. Characteristics of Group-A type folding carton boxes are

a) Long seam glued
b) Rectangular surfaces
c) a & b
d) None

71. Characteristics of Group-B type folding carton boxes are

a) Non-long seam glued
b) Rectangular surfaces
c) a & b
d) None of above

72. Characteristics of Group-C type folding carton boxes are
    a) Long seam glued
    b) Non-rectangular surfaces
    c) a & b
    d)
73. Characteristics of Group-D type folding carton boxes are
    a) Non-long seam glued
    b) Non-rectangular surfaces
    c) a & b
    d)
74. Characteristics of Group-E type folding carton boxes are
    a) Long seam glued
    b) Rectangular surfaces
    c) Integrated folding
    d)
75. Characteristics of Group-F type folding carton boxes are
    a) Long seam glued
    b) Rectangular surfaces
    c) Integrated folding
    d) None
76. Glass is a____________.
    a) Solid
    b) Liquid
    c) Amorphous
    d) None
77. Glass is a____________.
    a) Organic
    b) Inorganic
    c) a & b
    d) None
78. The principal ingredient of glass is silica derived from
    a) Sand
    b) Flint
    c) Quartz
    d) All
79. In glass manufacturing, silica is melted to temperatures of ___°C to form fused silica glass.
    a) L273°C
    b) L372°C
    c) 7123°C
    d) 1723°C
80. Which of the following is a corrosion form of glass surface?
    a) Etching
    b) Leaching
    c) Weathering
    d) All
81. The process in which glass surface is subject to alkaline attack which slowly destroys the silica network, releasing other glass components is called
    a) Etching
    b) Leaching
    c) Weathering
    d) None

82. The process in which glass surface is subject to acid attack in which hydrogen ions exchange for alkali or other positively charged mobile ions is called

a) Etching b) Leaching

c) Weathering d) None

83. The process in which glass surface is subject to surface bloom is called

a) Etching b) Leaching

c) Weathering d) None

84. The basic parts of a glass container are

a) Finish b) Body

c) Bottom d) All

85. The part of the bottle that holds the cap or closure, the glass surrounding the opening in the container is

a) Finish b) Body

c) Bottom d) All

86. The largest part of the container and lies between the finish and the bottom is

a) Finish b) Body

c) Bottom d) All

87. The part of container that is made in the bottom plate part of the glass-container mold is

a) Finish b) Body

c) Bottom d) All

88. The closures of glass bottles or jars consist of

a) Cap b) Cork

c) a & b d) None

89. The closures of glass bottles or jars, function of cork or plug is to _______ the jar or bottle.

a) Break b) Seal

c) Open d) None

90. Main function of closure of glass bottle or jar is
    a) Provides an effective hermetic (air-tight) seal
    b) Provides easy opening and resealing of the container
    c) Provides evidence of inviolability
    d) All
91. In packaging science, the abbreviation ROPP means
    a) Round on pilfer proof b) Roll on pinch proof
    c) Roll on pilfer proof d) All
92. In Roll on pilfer proof (ROPP) closures, a tell tale ring around the bottom of the neck of bottles ____ opening.
    a) After b) Before
    c) While d) None
93. The flat center area in the top of the closure of bottle or jar is called
    a) Panel b) Radius
    c) Skirt d) Curl
94. The rounded area at the outer edge of the panel connecting he panel and the skirt of the closure of bottle or jar is called
    a) Panel b) Radius
    c) Skirt d) Curl
95. The flat, almost vertical portion on the side of the closure, which serves as the gripping surface and may be smooth, knurled or fluted of the closure of bottle or jar is called
    a) Panel b) Radius
    c) Skirt d) Curl
96. The round or rolled portion at the bottom of the skirt that adds rigidity to the cap and serves to protect the cut edge of the metal of the closure of bottle or jar is called
    a) Curl b) Lug
    c) Thread d) Face
97. horizontal inward protrusion from the curl that is seated under the thread on the glass finish and holds the cap in position of the closure of bottle or jar is called
    a) Curl b) Lug
    c) Thread d) Face

98. He spiral groove on the skirt of a continuous thread closure that meshes with the tread on the glass finish of the closure of bottle or jar is called

| a) Curl | b) Lug |
|---|---|
| c) Thread | d) Face |

99. He part, outside of the cap of the closure of bottle or jar is called

| a) Curl | b) Lug |
|---|---|
| c) Thread | d) Face |

100. The part, inside of the cap of the closure of bottle or jar is called

| a) Reverse | b) Lug |
|---|---|
| c) Thread | d) Face |

## Answers

| | | | | | | | | | |
|---|---|---|---|---|---|---|---|---|---|
| 1. d | 2. d | 3. c | 4. c | 5. b | 6. d | 7. c | 8. a | 9. a | 10. c |
| 11. d | 12. d | 13. d | 14. d | 15. d | 16. d | 17. a | 18. c | 19. b | 20. c |
| 21. d | 22. c | 23. d | 24. a | 25. d | 26. d | 27. c | 28. c | 29. d | 30. d |
| 31. d | 32. c | 33. d | 34. d | 35. d | 36. d | 37. d | 38. d | 39. d | 40. b |
| 41. d | 42. d | 43. b | 44. a | 45. d | 46. d | 47. d | 48. d | 49. c | 50. c |
| 51. d | 52. d | 53. c | 54. d | 55. a | 56. d | 57. d | 58. d | 59. d | 60. d |
| 61. a | 62. b | 63. c | 64. d | 65. a | 66. b | 67. b | 68. c | 69. b | 70. c |
| 71. c | 72. c | 73. c | 74. c | 75. d | 76. c | 77. b | 78. d | 79. d | 80. d |
| 81. a | 82. b | 83. c | 84. d | 85. a | 86. c | 87. c | 88. c | 89. b | 90. d |
| 91. c | 92. a | 93. a | 94. b | 95. c | 96. a | 97. b | 98. c | 99. d | 100. a |

1. The actual sealing component of the closure, which makes an intimate contact with the glass finish to form an effective seal is
   a) Gasket b) Liner
   c) a & b d) Thread
2. Gaskets are made from
   a) Rubber b) Plastisols
   c) Iron d) a & b
3. Resistance to _____ of glass is determined by transferring glass containers from hot water bath to cold water bath.
   a) Observance b) Breakage
   c) Adherence d) All
4. Time and temperature combinations for hot water bath, for dipping of glass containers to determine the resistance to breaking of glass are
   a) 121°C/15 min b) 100°C/5 min
   c) 63°C/5 min d) 53°C/15 min
5. Glass is optically
   a) Isotropic b) Anisotropic
   c) Polytropic d) All
6. The optical properties of glass relate to the degree of penetration of light and the subsequent effect of that transmission which is a function of
   a) Wavelength b) Frequency
   c) Energy d) None
7. In packaging science, the abbreviation ECCS means
   a) Electrolytically chloride-coated strengthened
   b) Electrolytically chloride -coated steel
   c) Electrolytically chromium-coated strengthened
   d) Electrolytically chromium-coated steel
8. In packaging science, the abbreviation TFS means
   a) Titaniumn -full steel b) Tin -full steel
   c) Tin-free steel d) None
9. The metal commonly used for the packaging of foods along with steel is
   a) Aluminium b) Tin
   c) Chromium d) All

10. Tinplate refers to _____ carbon mild steel sheet.

    a) Low b) High

    c) Free d) None

11. Tinplate refers to low carbon mild steel sheet, varying in thickness from around ________ mm.

    a) 5 to 15 b) 1 to 1.5

    c) 0.15 to 0.5 d) 0.015 to 0.05

12. Tinplate refers to low carbon mild steel sheet, with a coating of tin between _____gsm.

    a) 17 to 28 b) 2.8 to 17

    c) 0.28 to 0.17 d) 0.028 to 0.17

13. The principal method of electroplating is

    a) Acid stannous sulfate process b) Halogen process

    c) Both a & b d) None

14. The electroplating method, acid stannous sulfate process is also called as

    a) Ferro process b) Halogen process

    c) Ferrostan process d) None

15. The temperature range maintained for cathodic deposition in a dilute chromium plating electrolyte in the production of ECCS is

    a) 5 to 7°C b) 50 to 70°C

    c) 500 to 700°C d) 70 to 150°C

16. The concentration of CrO3 for cathodic deposition in a dilute chromium plating electrolyte in the production of ECCS is

    a) 0.05 g $L^{-1}$ b) 0.5 g $L^{-1}$

    c) 5 g $L^{-1}$ d) 50 g $L^{-1}$

17. The concentration of H2SO4 for cathodic deposition in a dilute chromium plating electrolyte in the production of ECCS is

    a) 0.5 g L-1 b) 0.5 g L-1

    c) 0.5 g L-1 d) 0.5 g L-1

18. Aluminum foil is a thin-rolled sheet of alloyed aluminum varying in thickness from about __ µm.

    a) 0.4 to 15 µm b) 4 to 150 µm

    c) 40 to 1500 µm d) 400 to 15000 µm

19. Aluminum foil is essentially impermeable to gases and water vapor when it is thicker than__µm.

a) 0.254 µm  b) 2.54 µm

c) 25.4 µm  d) 254.0 µm

20. The aluminium tube is formed by the _____ impact extrusion of an aluminum slug using a plunger.

a) Cold  b) Hot

c) Lukewarm  d) None

21. Compared to the metal can, the retort pouch enabling retorting times to be reduced by up to ____%

a) 30%  b) 40%

c) 50%  d) 60%

22. An outer layer of three-layer pouch structure is

a) PET  b) HDPE

c) LDPE  d) BOP

23. Thickness of PET layer in three-layer pouch structure is ____ µm.

a) 12 µm  b) 13 µm

c) 14 µm  d) 15 µm

24. The function of PET layer in three-layer pouch structure is

a) Strength  b) Toughness

c) A & b  d) None

25. A middle layer of three-layer pouch structure is

a) Paper  b) Aluminium foil

c) LDPE  d) HDPE

26. Thickness of aluminium foil in three-layer pouch structure is ____µm.

a) 7 to 10 µm  b) 0 to 100 µm

c) 70 to 100 µm  d) 70 to 100 µm

27. The function of aluminium foil in three-layer pouch structure is

a) Heat sealability  b) Strength

c) Compatibility  d) All

28. The term aseptic implies
    a) Absence or exclusion of any unwanted organisms from the product
    b) Absence or exclusion of any unwanted organisms from package
    c) Absence or exclusion of any unwanted organisms from specific areas
    d) All
29. The term hermetic implies term (strictly air tight) is used to indicate
    a) Mechanical properties to exclude the entrance of micro-organisms into a package
    b) Mechanical properties to exclude the entrance of gas into (or from) the package.
    c) Mechanical properties to exclude the entrance of water vapor into (or from) the package.
    d) All
30. The first aseptic packaging of food (specifically milk in metal cans) was carried out in Denmarkby _______ before 1913.
    a) Nicolas b) Nielsen
    c) Naaren d) Nimpher
31. The first food that aseptically packed in metal cans
    a) Egg b) Milk
    c) Wine d) water
32. Which of the following is a sterilization processes is
    a) HTST b) UHT
    c) LTLT d) both a & b
33. After sterilization the sterile low viscous food product is cooled to an appropriate temperature, typically ______ °C.
    a) 20°C b) 30°C
    c) 40°C d) 50°C
34. After sterilization the sterile high viscous food product is cooled to an appropriate temperature, typically ______ °C.
    a) 20°C b) 30°C
    c) 40°C d) 50°C

35. The sterilization processes for packaging material is are in common use, eitherindividually or in combination:, and treatments.
    a) Irradiation
    b) Heat
    c) Chemical
    d) Any one of these options or in combination
36. Ionizing radiation technique uses gamma rays from
    a) Cobalt-60 b) Cesium-137
    c) a & b d) Cesium-173
37. A radiation dose of ____ kGy or more is generally accepted to be sufficient to ensure sterility.
    a) 0.25 b) 2.50
    c) 25.0 d) 250
38. In Pulsed electric field the duration of the pulses ranges from ________sec., and the flashes are typically applied at a rate of 1 to 20 flashes per second.
    a) 1 µsec to 0.1 sec b) 10 µsec to 1 sec
    c) 100 µsec to 10 sec d) 1 sec to 100 sec
39. In Pulsed electric field the flashes are typically applied at a rate of ______flashes per second.
    a) 100 to 200 b) 10 to 20
    c) 1 to 2 d) 1 to 20
40. UV radiation has a wavelength of ______ nm.
    a) 150 to 300 b) 200 to 315
    c) 315 to 415 d) 415 to 515
41. Optimum wavelength of UV radiation for effectiveness destruction of microbes is ______nm.
    a) 253.7 nm b) 235.7 nm
    c) 225.7 nm d) 275.3 nm
42. Time temperature combinations for heat sterilization processes are
    a) 121°C for 20 min b) 170°C for 60 min
    c) Both a & b d) None

43. Time temperature combinations for steam heat sterilization processes are
    a) 121°C for 20 min  b) 170°C for 60 min
    c) Both a & b  d) None
44. Time temperature combinations for dry heat sterilization processes are
    a) 121°C for 20 min  b) 170°C for 60 min
    c) Both a & b  d) None
45. Example for chemical sterilizing agent is
    a) Hydrogen peroxide  b) Peracetic acid (PAA)
    c) Both a & b  d) None
46. A liquid sterilant, which is particularly effective against spores is
    a) Hydrogen peroxide  b) Peracetic acid (PAA)
    c) Both a & b  d) None
47. The criteria for an aseptic packaging system is
    a) Cleaned properly after use.
    b) Effectively sterilized before use
    c) Able to carry out the filling, sealing and critical transfer operations in a sterile environment.
    d) All
48. The most widespread consumer package for aseptic products is
    a) Paperboard laminate carton  b) Parchment paper
    c) Metal cans  d) Butter paper
49. In aseptic packaging equipment the concentration $H_2O_2$ for sterilization purpose is _____%
    a) 35%  b) 45%
    c) 55%  d) 65%
50. In aseptic packaging equipment the temperature of the $H_2O_2$ bath for sterilization purpose is _____ °C.
    a) 40°C  b) 50°C
    c) 60°C  d) 70°C

51. In aseptic packaging equipment the temperatures applied to evaporate $H2O2$ on material is _____°C.

a) 80°C
b) 100°C
c) 125°C
d) 150°C

52. The test measures the ability of a paper or paper board to withstand pneumatic or hydraulic pressure built up is

a) Burst Strength
b) Tear Resistance
c) Tensile strength
d) Grease resistance

53. The test measures the ability of a paper or paper board to withstand tearing is

a) Burst Strength
b) Tear Resistance
c) Tensile strength
d) Grease resistance

54. The force applied parallel to the plane of the specimen of specified width and length under specified condition of loading is

a) Burst Strength
b) Tear Resistance
c) Tensile strength
d) Grease resistance

55. The resistance of packaging material measured by exposing one of the test specimen creased to grease containing red dye is

a) Burst Strength
b) Tear Resistance
c) Tensile strength
d) Grease resistance

56. The transmission rate that is normally determined by measuring the change in pressure at constant volume is

a) GTR
b) WVTR
c) OTR
d) None

57. In WVTR experiment the quantity of water vapour in grams that will permeate from one side to other side of the film of an area of one square meter in ___ hours.

a) 1 hour
b) 12 hours
c) 24 hours.
d) 48 hours

58. In WVTR experiment the quantity of water vapour in grams that will permeate from one side to other side of the film of an area of one square meter when the relative humidity difference between the two sides is maintained at ___ % gradient.

a) 75 %
b) 80 %
c) 85 %
d) 90 %

59. In WVTR experiment the quantity of water vapour in grams that will permeate from one side to other side of the film of an area of one square meter at the surrounding temperatures of 37.8°C.

a) 3.8°C b) 38°C
c) 37.8°C d) 38.7°C

60. The test designed to measure the ability of packaging material of the films to withstand fracture by shock the test is

a) Impact Strength b) Tear Resistance
c) Tensile strength d) Tear Resistance

61. The test designed to measure the ability of packaging material to withstand surface wear and rubbing is

a) Abrasion Resistance b) Grease resistance
c) Tensile strength d) Grease resistance

62. The requirement for the fast degradation process of bioplastics is

a) Temperature b) Humidity
c) Type of microbes d) All

63. Acceptable bioplastic is

a) Cellulose b) Starch
c) Poly-beta-hydroxyalkanoates d) All

64. Cellulose is fusible and soluble in hydrogen bond-breaking solvents such as

a) N-methylmorpholine-N-oxide b) Water
c) Alcohol d) All

65. Starch normally consists of

a) Amylose b) Amylopectin
c) Hemicellulose d) a & b

66. Starch normally consists of __________ % amylopectin.

a) 20-30% b) 70-80%
c) 80-90% d) 90-95%

67. Starch normally consists of__________% amylase.

a) 20-30% b) 70-80%
c) 80-90% d) 90-95%

68. The property of Poly-beta-hydroxyalkanoate (PHB) is

a) 100% resistance to water b) 100% biodegradability

c) Thermoplastic process ability d) All

69. The property of Polylactide Acid (PLA) Plastic is

a) Excellent biodegradability b) Process ability

c) Biocompatibility d) All

70. Self-cooling beer using _____ technology makes it possible to drink cool beer anywhere.

a) Zeolite b) CaO

c) Li d) MgO

Self-heating coffee container based on the _____ exothermic reaction is becoming available.

a) Zeolite b) CaO

c) Li d) MgO

71. Packaging using electronics technology by using ________ battery power sources that enhances thebranding citation in a crowded consumer product category is an influencing feature.

a) Be b) B

c) Li d) Na

72. The common nanocomposites used in the food packaging industry are

a) Polymer clay nanoclay

b) Silica nanocomposites of nanosilver

c) a & b

d) None

73. The effect of nanoclay in polymers is

a) Nucleating agent in foams b) Higher cell density

c) Flame retardant d) All

74. Characteristic of nanoclay in polymers is

a) Stiffness b) Strength

c) Smaller cell size d) All

75. Nanosilver is composed of _______ water with silver in suspension which has excellent antibacterial properties.
  a) Ionized  b) De-ionized
  c) Mineral  d) Distilled

76. Durethanda transparent plastic film containing nanoparticles of clay that block__________from entering into the food particles and also make plastic more durable and heat resistant.
  a) Oxygen  b) Moisture
  c) Carbon dioxide  d) All

77. Embedded nano crystals which create a molecular barrier that helps to prevent the escape of ________.
  a) Oxygen  b) Moisture
  c) Carbon dioxide  d) All

78. Property of antimicrobial-kodak packaging system is also
  a) Absorbs oxygen  b) Anti-microbial
  c) Oxygen resistant  d) a & b

79. Nanoparticle films with embedded sensors in packaging in order to detect pathogens is called
  a) Pathogen detector  b) Electronic nose
  c) Electronic tongue  d) All

80. Absolute pressure devices measure approximately
  a) 10000 millibar or 1 bar  b) 1000 millibar or 1 bar
  c) 100 millibar or 1 bar  d) 1 millibar or 0.1 bar

81. Which of the following pressure devices measure a reading of 0 millibar and registered as negative values.
  a) Gauge  b) Bourdon tube
  c) Manometer  d) All

82. An instant infusion plant typically consists of
  a) Preheating section  b) Instant infusion section
  c) Final cooling section  d) All

83. Feeding rate of product for instant infusion process is
  a) 50–20,000 l/h  b) 50–2000 l/h
  c) 50–200 l/h  d) 5–20 l/h

84. Maximum dry matter content of product for instant infusion process is

a) 42% TS
b) 52% TS
c) 62% TS
d) 72% TS

85. In ohmic heating, ________ current is passed through liquid-particulate food where the heat is generated due to electrical resistance of liquid-particulate food.

a) AC
b) DC
c) Both AC & DC
d) None

86. In ohmic heating, the rate of heating is directly proportional to the ________of the electric field strength.

a) Fourth power
b) Cubic power
c) Square
d) Unit

87. In ohmic heating, the electric field strength can be varied by adjusting the ______.

a) Electrode gap
b) Applied voltage
c) Food composition
d) a & b

88. Factor effecting the rate of heating produced in ohmic heating is

a) Electrical conductivity of food
b) Temperature dependence
c) Electrode gap
d) All

89. In ohmic heating electroporation across cell walls occurs at frequency of ______ Hz.

a) 30–40 Hz
b) 40–50 Hz
c) 50–60 Hz
d) 60–70 Hz

90. Vegetative forms of yeasts and moulds are inactivated by pressures between ________MPa.

a) 50-100
b) 100-200
c) 200 - 300
d) 300-400

91. Gram negative bacteria can be inactivated by pressures of about ______MPa

a) 100
b) 200
c) 300
d) 400

92. Gram positive bacteria can be inactivated by pressures of about ______ MPa

a) 100 b) 200

c) 300 d) 400

93. Most pressure sensitive groups of bacteria is

a) *Vibrio paraheamolyticus* b) *E. coli*

c) *Lactobacillus* d) *Micrococcus*

94. Which of the following combination of enzyme & its role is correct

a) Polyphenoloxidase (PPO)-enzymatic browning

b) Lipoxygenase (LOX) - change in flavour, colour and nutritional value,

c) Pectinmethylesterase (PME) -cloud estabilization & consistency changes

d) All

95. *Vibrio paraheamolyticus* in clam juice is inactivated by pressure of ______MPa for 10 min.

a) 140 MPa b) 150 MPa

c) 160 MPa d) 170 MPa

96. *Vibrio paraheamolyticus* in buffer is completely inactivated by pressure of 170 MPa for ____minutes.

a) 10 min b) 20 min

c) 30 min d) 40 min

97. Which of the following combination of enzyme & its role is not correct

a) Papain – tenderization

b) Peroxidase (POD) - unfavourable flavours

c) Polyphenoloxidase (PPO)-nonenzymatic browning

d) Lipoxygenase (LOX) - change in flavour, colour and nutritional value,

98. Superchilling is a process by which the temperature of a food product is lowered to _____ °C below the initial freezing point of the product.

a) 0–1°C b) 1–2°C

c) 3–4°C d) 4–5°C

99. In cryogenic freezing by liquid nitrogen the temperatures obtained are

a) -192°C b) -194°C

c) -195°C d) -196°C

100. In cryogenic freezing by solid carbon dioxide the temperatures obtained are

a) -78°C b) -79°C

c) -80°C d) -81°C

101. Ozone is a powerful antimicrobial substance due to the its potential ________ capacity.

a) Reducing b) Neutralizing

c) Oxidizing d) All

102. Advantage of ozone in the food industry is

a) Surface hygiene b) Sanitation of equipment

c) Plant hygiene d) All

103. Application of ozone in the food industry is

a) Reuse of waste water

b) Lowering biological oxygen demand (BOD) food plant waste

c) Lowering chemical oxygen demand (COD) food plant waste

d) All

104. Ozone solubility in water is ______ times that of oxygen at 0–30°C.

a) 10 b) 13

c) 15 d) 20

Ozone solubility in ______water is progressively more.

a) Cold b) Hot

c) Lukewarm d) None

105. In extraction knowledge the abbreviation SFE means

a) Super food extraction b) Supercritical food extraction

c) Supercritical fluid extraction d) None

106. CO2 acts as supercritical fluid at is above ___°C and ______ MPa.

a) 31.1°C & 7.38 MPa b) 3.1°C & 73.8 MPa

c) 3°C & 7 MPa d) 3.1°C & 7.38 MPa

107. Ice cold water (0°C) Will dissolve ______ grams of $CO_2$ per liter of water at atmospheric pressure.

a) 2.4 g b) 2.8 g

c) 3.4 g d) 4.3 g

108. Best tool for preservation of agricultural and food products is

a) Drying b) Cryogenic freezing

c) Curing d) Radiation

109. The process that converts waste animal tissue into stable, value-added materials is

a) Extraction b) Leaching

c) Rendering d) Bleaching

110. Tallow a product derived from beef waste by rendering process is used for

a) Filing b) Lubrication

c) Painting d) Casting

## Answers

| | | | | | | | | | |
|---|---|---|---|---|---|---|---|---|---|
| 1. c | 2. d | 3. b | 4. c | 5. d | 6. a | 7. d | 8. c | 9. d | 10. a |
| 11. c | 12. b | 13. c | 14. c | 15. b | 16. d | 17. a | 18. b | 19. c | 20. a |
| 21. d | 22. a | 23. a | 24. c | 25. b | 26. c | 27. d | 28. d | 29. d | 30. b |
| 31. b | 32. d | 33. a | 34. c | 35. d | 36. c | 37. c | 38. a | 39. d | 40. b |
| 41. a | 42. c | 43. a | 44. b | 45. c | 46. b | 47. d | 48. a | 49. a | 50. d |
| 51. c | 52. a | 53. b | 54. c | 55. d | 56. a | 57. c | 58. c | 59. c | 60. a |
| 61. a | 62. d | 63. d | 64. a | 65. d | 66. b | 67. a | 68. d | 69. d | 70. a |
| 71. c | 72. c | 73. d | 74. d | 75. b | 76. d | 77. a | 78. d | 79. c | 80. b |
| 81. a | 82. d | 83. a | 84. b | 85. a | 86. c | 87. d | 88. d | 89. c | 90. c |
| 91. c | 92. d | 93. a | 94. d | 95. d | 96. c | 97. c | 98. b | 99. d | 100. c |
| 101. c | 102. d | 103. d | 104. b | 105. a | 106. c | 107. a | 108. b | 109. c | 110. b |

1. The study of inter-related properties of air–water vapour systems is called
   a) Rheology b) Photometry
   c) Psychrometry d) None
2. Factor/s that control the capacity of air to remove moisture from a food is/ are
   a) Amount of air that passes over the food
   b) Air temperature
   c) Amount of water vapour already carried by the air
   d) All of the above
3. Amount of water vapour in air is expressed as
   a) Absolute humidity b) Relative humidity
   c) Specific humidity d) All of the above
4. Temperature of the air, measured by a thermometer bulb, is
   a) Dry-bulb temperature b) Wet-bulb temperature
   c) Relative temperature d) None
5. Temperature of the air, measured by a thermometer bulb is
   a) Dry-bulb temperature b) Wet-bulb temperature
   c) Relative temperature d) None
6. Drying system in which liquid foods are formed into a stable foam by the addition of a stabiliser and aeration with nitrogen or air then dried is called
   a) Fluid mat drying b) Fly mat drying
   c) Fry mat drying d) Foam mat drying
7. In Fluidised-bed drier mesh trays contain a bed of particulate foods up to
   a) 12 cm deep b) 13 cm deep
   c) 14 cm deep d) 15 cm deep
8. In spray driers: A fine dispersion of food is pre-concentrated to
   a) 10–20% moisture b) 30–50% moisture
   c) 40–60% moisture d) 50–70% moisture
9. Which of the following is a two-stage atomiser
   a) Centrifugal atomiser b) Pressure nozzle atomiser
   c) Two-fluid nozzle atomiser d) Ultrasonic nozzle atomiser

10. Baking involves
    a) Mass transfer
    b) Heat transfer
    c) Both a & b
    d) None
11. Advantage/s of direct heating ovens
    a) Control over baking conditions
    b) Short baking times
    c) High thermal efficiencies
    d) All of the above
12. Which of the oven does not circulate the food through the oven on trays
    a) Rotary-hearth ovens
    b) Reel ovens and
    c) Multi-cycle tray ovens
    d) None
13. Frying involves
    a) Destruction of enzymes
    b) Reduction in water activity
    c) Preservative effect
    d) All of the above
14. Breakdown product of oil, produced at high temperatures, which forms a blue hazeabove the oil and is a source of atmospheric pollution is the
    a) Cyanide
    b) Acrelein
    c) Osazone
    d) None
15. Rate of free fatty acid development in frying oil depends on
    a) Rate of turnover of the oil
    b) Temperature of the frying oil
    c) Amount of water, from the frying food
    d) All of the above
16. Percentage of free fatty acid is encountered in a batch frying operation
    a) 0.05–0.08
    b) 0.5–0.8
    c) 1–2
    d) 2-5
17. Key elements of good frying oil are
    a) Good oxidative stability
    b) Good thermal stability
    c) a & b
    d) None
18. If the turnover rate of the frying oil is higher then, the production of free fatty acids is
    a) Equal
    b) Higher
    c) Slower
    d) None

19. frying oils or shortenings should have a minimum smoke point of
   a) 120ºC b) 220ºC
   c) 320ºC d) 420ºC

20. The safe level of lechithins in a frying medium is generally considered tobe less than
   a) 1mg/100g oil b) 100mg/100g oil
   c) 200mg/100g oil d) 10mg/100g oil

21. Monoglycerides content in good quality frying oil should be less than
   a) 0.1% b) 0.2%
   c) 0.3% d) 0.4%

22. Which of the oils prone to oxidation and off-flavour development due to presence of linolenic acid
   a) Soya-bean and b) Rapeseed
   c) a & b d) None

23. Dielectric and ohmic heating are used to preserve foods, whereas infrared radiation is used to change
   a) Colour b) Flavour
   c) Aroma d) All the above

24. Microwave radiation falls in electromagnetic spectrum between the Frequencies of
   a) 3 and 30 MHz b) 30 and 3,000 MHz
   c) 300 and 300,000 MHz d) None

25. Dielectric heating falls in electromagnetic spectrum between the Frequencies of
   a) 1 and 100 MHz. b) 100 and 1000 MHz.
   c) 1000 and 10000 MHz. d) None

26. The permitted Microwave wave length for heating / drying purposes is
   a) 915 MHz b) 2450 MHz
   c) 9150 MHz d) 245 MHz

27. The permitted Dielectric Heating wave length for heating / drying purposes is
   a) 915 MHz b) 2450 MHz
   c) 9150 MHz d) 245 MHz

28. Which of the following is a number relating the ability of a material to carry alternating current to the ability of vacuum to carry alternating current.

a) Dielectric constant (k) b) Thermal constant

c) Gas constant d) Mass constant

29. Relative dielectric constant of Water (0°c) Is

a) 8 b) 18

c) 58 d) 88

30. Relative dielectric constant of Water (+20°c) Is

a) 81 b) 71

c) 61 d) 51

31. Relative dielectric constant of Olive oil is

a) 3.1 b) 9.1

c) 15.2 d) 16.2

32. Relative dielectric constant of air is

a) 1.00059 b) 10.0059

c) 10.059 d) 1.59

33. The energy conversion in microwave and dielectric is mainly by

a) Ionic conduction b) Dipole rotation

c) a & b d) None

34. Which of the following is not a dielectric material

a) Water

b) Oils

c) Waste food materials containing moisture

d) All of the above

35. Which of the following material/s heat well under Microwave and D.E Heating

a) Proteins b) Alkali halides

c) Hydrocarbons d) None

36. Which of the following statement/s is/are correct

a) Higher the initial temperature of food the faster is heating

b) Microwave absorption rate is much higher for water than for ice

c) Microwaves penetrate deeper into ice than into water

d) all of the above

37. Heating method in which an alternating electric current is passed through a food, and the electrical resistance of the food causes the power to be translated directly into heat is called

a) Conduct heating b) Microwave heating

c) Ohmic heating d) None

38. Ohmic heating also called

a) Electro heating b) Resistance heating

c) a & b d) None

39. Conventional filtration process is used for separation of suspended particles having size

a) >10 μm b) = 10 μm

c) <10 μm d) None

40. Membrane filtration separates substances of molecular sizes

a) >10 μm b) = 10 μm

c) <10 μm d) None

41. First generation membranes were made up of

a) Cellulose acetate b) Polyanides,

c) PVC d) PAN

42. Cellulose acetate has the limitation of temperature

a) <35°C b) =35°C

c) >35°C d) 35-100°C

43. The membrane is made up of _____ sections

a) 1 b) 2

c) 3 d) 4

44. Thickness of separation membrane in manufacture of membrane is

a) 9.1 to 25 μm b) 90.1 to 0.25 μm

c) 900.1 to 0.025 μm d) None

45. In flocculation process the potential which arises from the presence of charges in particles and the medium at the zone of shear is called

a) Flocculative potential b) Zeta potential

c) Net potential d) All

46. In ultrasonication the whole process of gas bubble nucleation, growth and collapse due to the action of intense sound waves is called

a) Accoustication b) Bubbling

c) Cavitation d) Cleavage

47. Cabbage and other brassica vegetables (such as cauliflower and brussel sprouts) contain active compounds called

a) Hydrocyanates b) Thiocyanates

c) Isothiocyanates d) Triterpenes

48. Seaweed produces natural toxins called as

a) Halocystines b) Microcystins

c) Microcyanate d) Microtoxins

49. Shiitake mushroom that contains an immune-boosting ingredient called

a) Limacine b) Lentinan

c) Linalool d) Lysine

50. Sources of Omega-3 acid are

a) Salmon & flaxseed b) Sardines & tuna

c) Nuts and seeds d) All

## Answers

| | | | | | | | | | |
|---|---|---|---|---|---|---|---|---|---|
| 1. c | 2. d | 3. d | 4. a | 5. b | 6. d | 7. d | 8. c | 9. d | 10. c |
| 11. d | 12. d | 13. d | 14. b | 15. d | 16. b | 17. c | 18. c | 19. b | 20. d |
| 21. d | 22. c | 23. d | 24. c | 25. a | 26. a | 27. a | 28. a | 29. d | 30. a |
| 31. a | 32. a | 33. c | 34. d | 35. a | 36. d | 37. c | 38. c | 39. a | 40. c |
| 41. a | 42. a | 43. b | 44. b | 45. b | 46. c | 47. c | 48. b | 49. b | 50. d |

# 10

# Mathematics, Aptitude and Reasoning

1. When a number is increased by 24, it becomes 110% of itself. What is the number?

   a) 288 b) 360
   c) 216 d) 240

2. Reduce 3596 / 4292 to lowest terms.

   a) 29/37 b) 17/43
   c) 31/37 d) 19/23

3. A does 60% of a work in 15 Days. He then calls B, and they together ûnish the remaining work in 5 Days. How long B alone would take to do the whole work?

   a) 25 Days b) 20 Days
   c) 80 Days d) 24 Days

4. If $(8 - 10x) - (13x - 2) = -9$, then the value of x is

   a) -19/23 b) 19/23
   c) -1/23 d) 1/23

5. If $xy = -18$ and $x^2 + y^2 = 85$, then ûnd the value of $(x + y)$.

   a) 8 b) 10
   c) 9 d) 7

6. A number is greater than twice its reciprocal by 31/4. Find the number.

   a) 7 b) 8
   c) 9 d) 6

7. If $2x -3(x + 2) < 5 - 2x < -x + 2$, then ûnd the value of x.

   a) 2 b) 0
   c) 10 d) 12

8. If 3.352 - (9.759 - x) - 19.64 = 7.052, then what is the value of x?

   a) -6.181 b) 13.581

   c) 33.099 d) 39.803

9. Square of (7 - 4x) is

   a) $16x^2 - 28x + 49$ b) $49 - 28x - 16x^2$

   c) $49 - 56x - 16x^2$ d) $16x^2 - 56x + 49$

10. The third proportional of two numbers 9 and 15 is

    a) 21 b) 30

    c) 25 d) 45

11. What will be the roots of the quadratic equation $x^2 - 25x + 156 = 0$?

    a) 12, 13 b) 25, 1

    c) 9, 16 d) 31, 6

12. In a flight of 600 km, an aircraft was slowed down due to bad weather. Its average speed for the trip was reduced by 200 km/hr and the time of flight increased by 30 minutes. The duration of the flight is

    a) 1 hour b) 2 hours

    c) 3 hours d) 4 hours

13. A circle is inscribed in a square. If the length of the diagonal of the square is 14"2 cm, what is the area (in sq cm) of the circle?

    a) 308 b) 462

    c) 154 d) 616

14. A man completes a journey in 10 hours. He travels first half of the journey at the rate of 21 km/hr and second half at the rate of 24 km/hr. Find the total journey in km.

    a) 220 km b) 224 km

    c) 230 km d) 234 km

15. The third proportional of two numbers 9 and 24 is

    a) 39 b) 48

    c) 72 d) 64

16. The compound interest accrued on a certain sum at the rate of 20% per annum compounded annually for 2 years is Rs. 3344. Find the sum.

    a) Rs. 7600 b) Rs. 7500

    c) Rs. 7560 d) Rs. 7660

17. On the occasion of Independence Day, some chocolates were to be distributed equally among 180 children. Due to some reasons, 40 children could not be present and hence each of the remaining children got 4 more extra chocolates. What was the original number of chocolates?

a) 2540 b) 2520

c) 2420 d) 2460

18. Vikas is older than Vimal by 14 years. Four years ago, the ratio of the ages of Vikas and Vimal was 17: 10 respectively. What will Kunal's age 9 years hence, If Kunal's present age is half of that of Vimal?

a) 25 Years b) 24 Years

c) 21 Years d) 27 Years

19. The present population of a village is 25600. After two years it will be 30976. What is the rate of annual growth of population?

a) 12.5 % b) 8%

c) 12% d) 10%

20. A man on tour travels first 160 km at 64 km/hr and the next 160 km at 80 km/hr. The average speed for the first 320 km of the tour is

a) 35.55 km/hr b) 36 km/hr

c) 71.11 km/hr d) 71 km/hr

21. The product of two numbers is 2028 and their H.C.F. is 13. The number of such pairs is

a) 1 b) 2

c) 3 d) 4

22. If each side of a square is increased by 25%, find the percentage change in its area?

a) 65.25 b) 56.25

c) 65 d) 56

23. 7 8 20 69 292?

a) 1285 b) 1458

c) 1485 d) 1385

24. 11 16 23 32 43?

a) 66 b) 48

c) 52 d) 56

25. 12 21 39 66 102?

a) 142 b) 145

c) 247 d) 147

26. The total price of 6 shirts and 7 trousers is Rs.4130 while that of 4 shirts and 9 trousers is Rs.4270. Find out the price of 3 shirts.

a) Rs.850 b) Rs.840

c) Rs.860 d) Rs. 810

27. A starts a business with an investment of Rs.13600. After four months B and C join the partnership withinvestment of Rs.17600 and Rs.15200 respectively. The total profit at the end of one year is Rs.46550.Find C's share in the profit

a) Rs. 16300 b) Rs.13200

c) Rs.13300 d) Rs.13800

28. A boat rows downstream and covers 12.6 km in 18 minutes and 8.25 km upstream in 15 minutes. Find the speed of current

a) 5.4 kmph b) 3.2 kmph

c) 4 kmph d) 4.5 kmph

29. Two varieties of rice are available the first variety at the rate of Rs.30 per kg and the second at the rate of Rs.36 per kg. 10 kg of first variety and 15 kg of second variety are mixed together. At what rate per kg should the mixture be sold so that the seller gets a profit of 20% on the whole transaction?

a) Rs.40.32 b) Rs. 42.30

c) Rs.45.50 d) Rs. 40.68

30. A man took loan from a bank at the rate of 12% p.a. simple interest. After 3 years he had to pay Rs. 5400 interest only for the period. The principal amount borrowed by him was

a) Rs. 2000 b) Rs.10000

c) Rs. 15000 d) Rs.20000

31. The square number is 48 more than 22 times of itself. Find the number

a) 25 b) 22

c) 24 d) 28

32. The area of a square is twice to that of a rectangle. The length of rectangle is 4 cm less than 24 cm long side of square. What will be the perimeter of the rectangle?

a) 68.8 cm b) 86.8cm

c) 88.8 cm d) 87.8 cm

33. Two pipes A and B are fitted with a tank. Pipe A can fill 3?5 of the tank in 27 hours while pipe B can fill the empty tank completely in 30 hours. In what time will both the pipes together fill the empty tank completely?

a) 16 hours b) 18 hours

c) 27 hours d) 20 hours

34. It takes eight hours for a 600 km journey, if 120 km is done by train and the rest by car. It takes 20 minutes more, if 200 km is done by train and the rest by car. The ratio of the speed of the train to that of the cars is

a) 2:3 b) 3:2

c) 3:4 d) 4:3

35. In a class the average weight of 32 students is 48.5 kg. 10 new students join the class and hence the average weight of all the students increases by 7.5 kg. Find the average weight of 10 new students.

a) 82 kg b) 75 kg

c) 80 kg d) 70 kg

36. A 300 metre long train crosses a person running at the rate of 7 kmph in the direction opposite to that of train in 36 seconds. Find the speed of train.

a) 27 m/sec b) 25 m/sec

c) 20 m/sec d) 28 m/sec

37. A boat sails 15 km of a river towards upstream in 5 hours. How long will it take to cover the same distance downstream, if the speed of current is one-fourth the speed of the boat in still water

a) 1.8h b) 3h

c) 4h d) 5h

38. A is twice as good a work man as B. A and B together complete a piece of work in 28 Days. In how many Days will A alone do the same piece of work?

a) 40 Days b) 42 Days

c) 35 Days d) 35 Days

39. A and B together have a sum of Rs.1680. Half of the total amount is equal to the 7? 10 th part of total amount with B. Find the total sum with B.

a) Rs. 1260 b) Rs. 600

c) Rs. 1200 d) Rs. 1160

40. I prefer coffee_______ tea.

a) Than b) To

c) For d) So

41. I am studying in _______ University.

a) The b) A

c) An d) None

42. This is _______ H.C.L prouct

a) The b) A

c) An d) None

43. Don't sit _______ the fire.

a) By b) To

c) In d) For

44. The boy jumped ______ the canal.

a) To b) Into

c) In d) So

45. I am teaching Chemistry ________ English.

a) Than b) Beside

c) Besides d) So

46. Don't run _______ the road.

a) On b) Through

c) Along d) Across

47. Ramesh is _____ Tendulkar of our college

a) The b) A

c) For d) An

48. In the following question, a sentence has been given in Active/Passive voice. Out of four alternatives suggested, select the one, which best expresses the same sentence in Passive/Active voice. **The candidates are writing an essay.**

a) An essay was wrote by the candidates.

b) An essay is being written by the candidates.

c) An essay has had been written by the candidates.

d) An essay had been written by the candidates.

49. In the following question, a sentence has been given in Direct/Indirect speech. Out of the four alternatives suggested, select the one, which best express the same sentence in Indirect/Direct speech. **Bunty said, "I have done my class work".**

a) Bunty said, he had done his class work.

b) Bunty said that he had done his class work.

c) Bunty was saying, he has done his class work.

d) Bunty said that he did his class work.

50. Improve the bracketed part of the sentence. If I **(just had)** the money I would have bought the car.

a) Have had  b) Was having

c) Had  d) No improvement

51. Select the antonym of "**diligence**"

a) Lethargy  b) Exertion

c) Vigour  d) Assiduity

52. In the following question, the sentence given with blank to be filled in with an appropriate word. Select the correct alternative out of the four and indicate it by selecting the appropriate option." **It does not suit you to cast ____________ on the character of such a pure woman".**

a) Praise  b) Abuses

c) Vituperations  d) Aspersions

53. Select the word with the correct spelling.

a) Marothon  b) Seagulls

c) Pungancy  d) Palpatted

54. Sclcct thc word with thc corrcct spelling.

a) Garotte  b) Dillution

c) Variably  d) Exhumming

55. Select the antonym of " **homogeneous** "

a) Akin  b) Ditto

c) Disparate  d) Cognate

56. Improve the bracketed part of the sentenceBefore I met your brother I **(have had)** a poor opinion of him.

a) Had had  b) Had

c) Did have  d) No improvement

57. The government is planning to set ......... family welfare centers for slums incities.

    a) Up
    b) With
    c) For
    d) In

58. Economic independence and education have .......... women more assertive.

    a) Prepared
    b) Made
    c) Marked
    d) Resulted

59. In the modern world, the .......... of change and scientific innovation is unusuallyrapid.

    a) Supplies
    b) Context
    c) Pace
    d) Fantasy

60. The unprecedented economic growth of China has ......... worldwide attention.

    a) Perceived
    b) Proposed
    c) Attracted
    d) Astonished

61. Each business activity .......... employment to people who would otherwise beunemployed.

    a) Taking
    b) Finds
    c) Creates
    d) Provides

62. The economic disparity

    a) Has grown rapid in
    b) Era of globalization
    c) And free market forces.
    d) No error

63. Research shows that people

    a) Is more sensitive to perceiving
    b) Messages that are consistent
    c) With their opinions and attitudes.
    d) No error

64. Many poverty alleviation schemes

    a) Are not applicable of
    b) Dwellers in metro cities
    c) As they are above the poverty line.
    d) No error

65. Rather than considering its
    a) Human capital as a drain on
    b) Resources, India needs to
    c) Resource develop its into a huge opportunity.
    d) No error
66. Which word does NOT belong with the others
    a) Parsley
    b) Basil
    c) Dill
    d) Mayonnaise
67. If a credit card bill
    a) Is paid in full and
    b) On time, none finance
    c) Charges are levied
    d) No error
68. Training have a
    a) Ignore the errors of punctuation, if any
    b) Development of various
    c) Skills and abilities
    d) No error
69. Lack of ability to
    a) Read or write is just one of
    b) The all barriers that keep
    c) The poor people under developed
    d) No error
70. Science and technology have
    a) Bccomc dominant factors
    b) Affecting our economic, cultural
    c) And spiritual development.
    d) No error
71. A social business sclls products
    a) At a prices that make it self-sustaining, pays
    b) No dividends and reinvestmen
    c) The profits in the business
    d) No error
72. If all the numbers are deleted from the above arrangement then which of the following will be seventh to the left of sixth from the right?
    a) H
    b) J
    c) M
    d) $

73. How many such numbers are there in the above arrangement each of which is immediately preceded by a symbol?

a) One
b) Two
c) Three
d) Four

74. '9W' is to 'GH#' and '$7' is to '2BM' in the same way as '4J' is to .......... in the arrangement.

a) E@8
b) 58#
c) B2L
d) 58Z

75. How many such symbols are there in the above arrangement each of which is immediately followed by a letter?

a) None
b) One
c) Two
d) Three

76. If all the symbols are deleted from the above arrangement then which of the following will be the fourth to the left of twelfth from the right?

a) 9
b) 3
c) W
d) M

77. **Statements:**

All chillies are garlics. Some garlics are onions.

All onions are potatoes. No potato is ginger.

**Conclusions:**

I. No onion is ginger. II. Some garlics are potatoes.

III. Some chillies are potatoes.

a) Only I follows
b) Only II follows
c) Only I & II follow
d) Only I & III follow

78. **Statements:**

Some keys are locks. Some locks are numbers.

All numbers are letters. All letters are words.

**Conclusions:**

I. Some locks are letters. II. Some words are numbers.

III. All numbers are words.

a) Only I & II follow
b) Only II & III follow
c) Only I & III follow
d) None of these

79. **Statements:**

Some windows are doors. All doors are walls.

No wall is roof. All roofs are shelters.

**Conclusions:**

I. Some windows are walls. II. No wall is shelter.

III. No door is shelter.

a) None follows
b) Only II & III follow
c) Only I & III follow
d) Only I follows

80. **Statements:**

All bottles are jars. Some jars are pots.

All pots are taps. No tap is tank.

**Conclusions:**

I. No pot is tank. II. Some jars are tanks.

III. Some bottles are pots.

a) Only I & III follow
b) Only I & II follow
c) Only II & III follow
d) None of these

81. **Statements:**

Some fish are crocodiles. Some crocodiles are snakes.

No snake is tortoise. All tortoises are frogs.

**Conclusions:**

I. No snake is frog. II. Some snakes are fish.

III. Some fish are frogs.

a) None follows
b) Only I & II follow
c) Only II & III follow
d) Only I & III follow

82. Which of the following is the correct position of L?

a) To the immediate right of N
b) To the immediate right of Q
c) To the immediate left of M
d) To the immediate left of Q

83. Which of the following pair of persons represents O's neighbours?

a) L&N
b) P&K
c) M&P
d) N&P

84. Which of the following groups has the first person sitting between the other two persons?

a) PKJ
b) JQL
c) QNL
d) None of these

85. Who is to the immediate right of K?

a) J
b) P
c) Q
d) Cannot be determined

86. Who is to the immediate left of O?

a) P
b) L
c) Q
d) M

87. Ram told Ramu the person I met yesterday in the market was the only son of the brother in law of my friend's sister's mother. How that boy is related to Ram's friends?

a) Cousin
b) Nephew
c) Brother
d) Brother in law

88. Abhishek travels 12 km. towards North, then he turns to his right and walks 6 km. He again turns to his right and moves 8 km forward. Now in which direction is he from his starting point?

a) North
b) South
c) North/East
d) West

89. Amit start from point A and walks 1km towards south, turn left and walks 1km, than he turn left again and walks 1 km. Now in which direction he is facing?

a) South
b) East
c) West
d) North

90. Bag A contains 3red, 4white balls and Bag B contains 6 red, 5 white balls select a bag at random and draw two balls from it then what is the probability to get 1 red and 1 white.

a) 0.5585
b) 0.5445
c) 0.5335
d) 0.8585

91. A bag contains 10 white and 10 black balls A ball is drawn from it. The probability that it will be white is_____________

a) $\frac{1}{10}$
b) 1
c) 0.8
d) 0.5

92. The chance that a leap year contains 53mondays is________

a) $\frac{2}{7}$ b) $\frac{1}{7}$

c) $\frac{3}{7}$ d) $\frac{6}{7}$

93. A bag contains 6 white and 4 black balls. 2 balls are drawn at random. Find the probability that they are of same colour

a) 1/2 b) 7/15

c) 8/15 d) 9/15

94. A bag contains 2 red, 3 green and 2 blue balls. Two balls are drawn at random. What is the probability that none of the balls drawn is blue?

a) $\frac{10}{21}$ b) $\frac{11}{21}$

c) $\frac{2}{7}$ d) $\frac{5}{7}$

95. If on an average 9 ships out of 10 return safely to a port, the probability of one ship returns safely is___________

a) $\frac{1}{10}$ b) $\frac{8}{10}$

c) $\frac{9}{10}$ d) $\frac{7}{10}$

96. A man can kill a bird once in three shots. The probabilities that a bird is not killed is__________

a) $\frac{1}{3}$ b) $\frac{2}{3}$

c) 1 d) 0

97. Thc valuc of P(s) where S is the sample space is

a) -1 b) 0

c) 1 d) 0.5

98. If $P(A) = \frac{7}{8}$ then $P(A^c)$ is equal to _______

a) 1 b) 0

c) $\frac{7}{8}$ d) $\frac{1}{8}$

99. If the probability of a horse A winning a race is $\frac{1}{6}$ and the probability of a horse B winning the same race is $\frac{1}{4}$, What is the probability that none of them will win.

a) $\frac{5}{12}$ b) $\frac{7}{12}$

c) $\frac{5}{8}$ d) $\frac{4}{12}$

100. Two Cards are drawn from a well shuffled pack of 52 cards. Find the probability that they are both aces if the first card is replaced.

a) $\frac{1}{221}$ b) $\frac{1}{171}$

c) $\frac{1}{169}$ d) $\frac{1}{160}$

## Answers

| | | | | | | | | | |
|---|---|---|---|---|---|---|---|---|---|
| 1. d | 2. c | 3. a | 4. b | 5. d | 6. b | 7. c | 8. c | 9. d | 10. c |
| 11. a | 12. a | 13. c | 14. b | 15. d | 16. a | 17. b | 18. c | 19. d | 20. c |
| 21. b | 22. b | 23. c | 24. d | 25. d | 26. b | 27. c | 28. d | 29. a | 30. c |
| 31. c | 32. a | 33. b | 34. c | 35. c | 36. a | 37. b | 38. b | 39. c | 40. b |
| 41. b | 42. c | 43. a | 44. b | 45. c | 46. d | 47. a | 48. b | 49. b | 50. c |
| 51. a | 52. d | 53. b | 54. b | 55. c | 56. b | 57. a | 58. b | 59. c | 60. c |
| 61. d | 62. b | 63. b | 64. b | 65. d | 66. d | 67. 3 | 68. a | 69. c | 70. 5 |
| 71. c | 72. a | 73. c | 74. 5 | 75. 4 | 76. b | 77. c | 78. d | 79. d | 80. d |
| 81. a | 82. b | 83. c | 84. d | 85. a | 86. d | 87. a | 88. c | 89. d | 90. a |
| 91. d | 92. a | 93. b | 94. a | 95. c | 96. b | 97. c | 98. d | 99. c | 100. c |

1. Which of the following matrix does not have an inverse?

a) $\begin{bmatrix} 1 & 2 \\ 2 & 4 \end{bmatrix}$ b) $\begin{bmatrix} 2 & 1 \\ 2 & 4 \end{bmatrix}$

c) $\begin{bmatrix} 1 & 2 \\ 4 & 2 \end{bmatrix}$ d) $\begin{bmatrix} 1 & 4 \\ 2 & 2 \end{bmatrix}$

2. If A and B are Hermitian matrices, then AB + BA is

a) Symmetric b) Skew symmetric

c) Hermitian d) Skew Hermitian

3. If $A = \begin{bmatrix} 3 & 2 \\ 5 & 6 \end{bmatrix}$ and $B = \begin{bmatrix} 1 & 3 \\ 2 & 5 \end{bmatrix}$ such that $2A + 3B + X = 0$, then X = - ______________________

a) $\begin{bmatrix} -9 & -13 \\ -16 & -27 \end{bmatrix}$ b) $\begin{bmatrix} -9 & 13 \\ -16 & -27 \end{bmatrix}$

c) $\begin{bmatrix} -9 & -13 \\ -16 & 27 \end{bmatrix}$ d) $\begin{bmatrix} 9 & -13 \\ 16 & -27 \end{bmatrix}$

4. The rank of a Unit (identity) matrix of order 4 is_________

a) 2 b) 4

c) 3 d) 1

5. If the matrix $\begin{bmatrix} 1 & 0 & -1 & 2 \\ 0 & 0 & 2 & 4 \\ 0 & 0 & \lambda-3 & \mu-10 \end{bmatrix}$ is Echelon form of the given system of equations, then the system of equations has infinite number of solutions if

a) $\lambda = 3, \mu = 10$ b) $\lambda = 8, \mu = 10$

c) $\lambda \neq 8, \mu = 10$ d) $\lambda = 3, \mu \neq 10$

6. The characteristic equation of the square matrix $A = \begin{bmatrix} 1 & 2 \\ 2 & -1 \end{bmatrix}$ is $\lambda^2 - 5 = 0$ then $A^{-1}$ is

a) $\frac{1}{5}\begin{bmatrix}1 & 2\\2 & -1\end{bmatrix}$ b) $\frac{1}{5}\begin{bmatrix}-1 & 2\\2 & 1\end{bmatrix}$

c) $\frac{1}{5}\begin{bmatrix}1 & -2\\2 & 1\end{bmatrix}$ d) $\frac{1}{5}\begin{bmatrix}1 & 2\\2 & 1\end{bmatrix}$

7. The matrix $\begin{bmatrix}0 & 6 & -6\\-6 & 0 & 6\\6 & -6 & 0\end{bmatrix}$ is a _______ matrix.

a) Symmetric b) Skew-symmetric

c) Orthogonal d) Nilpotent

8. If $A = \begin{bmatrix}2 & 0 & 0\\0 & 2 & 0\\0 & 0 & 2\end{bmatrix}$ and $B = \begin{bmatrix}1 & 2 & 3\\0 & 1 & 3\\0 & 0 & 2\end{bmatrix}$ then det (A B) has the value.

a) 4 b) 8

c) 16 d) 32

9. If the matrix $A = \begin{bmatrix}2 & 3\\5 & -1\end{bmatrix}$ is expressed as the sum of a symmetric and a skew symmetric. Then the skew-symmetric matrix is

a) $\begin{bmatrix}2 & 1\\2 & 4\end{bmatrix}$ b) $\begin{bmatrix}2 & 4\\4 & 1\end{bmatrix}$

c) $\begin{bmatrix}0 & -1\\1 & 0\end{bmatrix}$ d) $\begin{bmatrix}4 & 2\\2 & -1\end{bmatrix}$

10. The rank of a matrix in Echelon form is equal to

a) Number of non-zero columns b) Number of non-zero rows

c) Number of diagonal elements d) Number of non-zero elements

11. The necessary and sufficient condition that the system of equations AX = B is Consistent if_______

a) $\rho(A/B) = \rho(A)$ b) $\rho(A/B) \neq \rho(A)$

c) $\rho(A/B) < \rho(A)$ d) $\rho(A/B) > \rho(A)$

12. If Square matrix A is skew symmetric if

a) $AA^T = I$ b) $A = -A^T$

c) $A = A^T$ d) $AA^{-1} = I$

13. If $A = \begin{bmatrix} 1 & 2 \\ 2 & -1 \end{bmatrix}$ find A+ is ___________________

a) 125 I b) 625 I

c) 5 I d) 25 I

14. If $\lambda$ is an eigen value of a non-singular matrix A , then $\frac{|A|}{\lambda}$ is an Eigen value of

a) Inverse of A b) Adj A

c) A d) $\lambda A$

15. If $A = \begin{bmatrix} 4 & 2 \\ -1 & -1 \end{bmatrix}$ then $(A-2I)(A-3I)$ is equal to

a) $\begin{bmatrix} 0 & 4 \\ 2 & 10 \end{bmatrix}$ b) $\begin{bmatrix} 0 & -4 \\ 2 & 10 \end{bmatrix}$

c) $\begin{bmatrix} 0 & 4 \\ 2 & -10 \end{bmatrix}$ d) None of these

16. If A is square matrix = $(A^T)^T$ =

a) $AA^T$ b) $A^T$

c) A d) I

17. If $A^T A = AA^T = I$ then A is called

a) Symmetric matrix b) Skew symmetric

c) Orthogonal d) Unitary

18. The rank of 3x3 matrix whose elements are all 2 is

a) 2 b) 3

c) 1 d) 0

19. If A is non-singular and if B is any matrix such that AB is defined, then the rank of AB is

a) Rank of A b) Rank of B

c) Can't define d) None

20. Function f(x, y) has Minimum value if

a) $\ln - m^2 > 0, l > 0$ b) $\ln - m^2 > 0, l < 0$

c) $\ln - m^2 < 0, l < 0$ d) $\ln - m^2 = 0$

21. If the function f(z) satisfies Laplacian equations then it is

a) Homogenous function b) Analytic function

c) Quadratic function d) Harmonic function

22. The value '*c*' of CMV theorem for $f(x) = x^2$, $g(x) = x^3$ in the interval [1,2] is

a) 13/9 b) 12/9

c) 14/9 d) 15/9

23. The value '*c*' of CMV theorem for function f(xZ) and f'(x) in (l, e) where f (x) = log x is

a) 1 - e/e b) e - 1/e

c) e/e - 1 d) e/1 - e

24. If $u = \frac{x+y}{1-xy}, v = \tan^{-1} x + \tan^{-1} y$, then the function relation between u,v is

a) u + v = tan v b) u = tan v

c) u-v = tan v d) v = tan u

25. If u = $\tan^{-1}(y/x)$ then $\frac{\partial u}{\partial x}$ at (0,1)

a) 1 b) -1

c) 2 d) -2

26. If u = x(1-y), v = xy then $J.J' =$

a) 0 b) 2

c) 1 d) u

27. The saddle point to the surface Z= xy is

a) (0,-1) b) (1,0)

c) (0,0) d) (-1,0)

28. The functions relation between u = $e^x \sin y$ and v = x + log sin y

a) u = log v b) v = log v

c) u = 2 log u d) v = 2 log u

29. If u,v,w are functions of x,y,w and x,y,w are functions of r, $\theta, \varnothing$ then

$$\frac{\partial(u,v,w)}{\partial(x,y,z)} \times \frac{\partial(x,y,z)}{\partial(r,\theta,\varnothing)} =$$

a) $\frac{\partial(u,v,w)}{\partial(r,\theta,\varnothing)}$
b) $\frac{\partial(\mathrm{r},\theta,\varnothing)}{\partial(\mathrm{u,v,w})}$
c) $\frac{\partial(x,y,z)}{\partial(u,v,w)}$
d) 1

30. Value of $\int_1^a \int_1^b \frac{1}{xy} dxdy$ is

a) log(ab)
b) $\log\left(\frac{a}{b}\right)$
c) (log a)(log b)
d) $\frac{\log a}{\log b}$

31. If $u = x(1-y), v = xy$ then $J =$

a) y
b) x+y
c) x-y
d) x

32. $\int_1^e \int_0^{\log y} \frac{1}{\log y} dxdy =$

a) $e$
b) $e^{-1}$
c) $e-1$
d) $e+1$

33. $\int_0^\infty \int_x^\infty \frac{e^{-x}}{x} dxdy =$

a) 1
b) -1
c) 2
d) -2

34. If $f(0)=1, f'(0)=0, f''(0)=-1, f'''(0)=0, f^{iv}(0)=1$........ then $f(x)=$

a) $1-\frac{x^2}{\lfloor 2}+\frac{x^4}{\lfloor 4}-\frac{x^6}{\lfloor 6}+$.........
b) $1+\frac{x}{\lfloor 1}+\frac{x^2}{\lfloor 2}+\frac{x^3}{\lfloor 3}+$..........

c) $1-\frac{x^2}{\lfloor 2}+\frac{x^4}{\lfloor 4}+\frac{x^6}{\lfloor 6}+.........$ d) $x-\frac{x^3}{\lfloor 3}+\frac{x^5}{\lfloor 5}-\frac{x^7}{\lfloor 7}+.........$

35. The maclaurin's serious expansion for log (1-x) =

a) $x-\frac{x^2}{2}+\frac{x^3}{3}-\frac{x^4}{4}+.........$ b) $x+\frac{x^2}{2}+\frac{x^3}{3}+\frac{x^4}{4}+.........$

c) $-x-\frac{x^2}{2}-\frac{x^3}{3}-\frac{x^4}{4}-.........$ d) $-x+\frac{x^2}{2}-\frac{x^3}{3}+\frac{x^4}{4}-.........$

36. The stationary point of the function

$f(x,y)=x^2+y^2+xy+x-4y+5$ is

a) (2,-3) b) (2, 3)

c) (-2, 3) d) (-2, -3)

37. If the period of $f(x)$ is ***T***, then for a positive integer n, then $\boldsymbol{nT}$ is period of

a) $f(x)$ b) $\frac{1}{f(x)}$

c) $nf(x)$ d) $-f(x)$

38. Evaluate $I=\int_0^1\int_1^2\int_2^3 xyz\ dxdydz$

a) $\frac{15}{8}$ b) $\frac{8}{15}$

c) 0 d) Can't possible

39. If $f(x)$ is expanded as a Fourier series in $(-\pi,\pi)$ ,then $a_0 =$ ...........

a) $\frac{1}{\pi}\int_{-\pi}^{\pi} f(x)dx$ b) $\frac{1}{\pi}\int_{-\pi}^{\pi} f(x)\sin nxdx$

c) $\frac{1}{\pi}\int_{-\pi}^{\pi} f(x)\cos nxdx$ d) $\frac{1}{2\pi}\int_{0}^{2\pi} f(x)dx$

40. If $\bar{f} = f_1(y,z)\bar{i} + f_2(z,x)\bar{j} + f_3(x,y)\bar{k}$ then $\bar{f}$ is

a) Irrotational b) Solenoidal

c) Both irrotational & Solenoidal d) None

41. If $\bar{n}$ is the unit out ward drawn normal to any closed surface then $\int_v div\,\bar{n}\, dv =$

a) S
b) 2S
c) 3S
d) None

42. If S is any closed surface enclosing a volume v and $\bar{F} = xi + 2yj + 3zk$ then $\iint \bar{F}.\bar{n}ds =$

a) V
b) 3V
c) 6V
d) None

43. If $\bar{r} = x\bar{i} + y\bar{j} + z\bar{k}$ then $\nabla^2\left(\frac{1}{r}\right) =$

a) 0
b) 3X
c) $-4r^3$
d) 3 (x+y+z)

44. If $\bar{a}$ and $\bar{b}$ are irrotational vectors then $\bar{a} \times \bar{b}$ is

a) Solenoidal vector
b) Irrotational vector
c) Free vector
d) None

45. If $curl\,\bar{f}=0$ then $\bar{f}$ is

a) Solenoidal vector
b) Irrotational vector
c) Constant vector
d) Can't say

46. If $\vec{r} = x\vec{i} + y\vec{j} + z\vec{k}$ then $div\,\vec{r}$

a) 3
b) 0
c) $\vec{r}/r$
d) $\vec{r}/r^2$

47. If $\phi - xyz$ then the maximum directional derivative of $\phi$ at $(1,1,1)$ is

a) 0
b) $\sqrt{3}$
c) $\nabla_{\varnothing}$
d) 1

48. The value of $\oint_C (xy^2dx + x^2ydy)$ around $C: x^2 + y^2 = 4$ is

a) $4_\pi$
b) $2_\pi$
c) 0
d) $\pi$

49. The value of $\oint_C \left[(2xy - x^2)dx + (x^2 + y^2)dy\right]$ where $C$ is the closed region bounded by $y = x^2$ and $y^2 = x$

a) $4_\pi$ b) $2_\pi$

c) 0 d) $\pi$

50. If $r = \sqrt{x^2 + y^2 + z^2}$ then $\nabla(\log r) =$

a) $\vec{\gamma}/\gamma^2$ b) $\vec{\gamma}/\gamma$

c) $1/\gamma^2$ d) $-\vec{\gamma}/\gamma^2$

51. If $C$ is any simple closed curve, then $\int_C (z^2+8)dz =$ ______

a) 0 b) 1

c) 2 d) 3

52. The function $f(z)$ is analytic if it satisfies

a) Laplacian equations b) Linear equations

c) C-R equations d) None

53. Conjugate of the complex number 2+3i is

a) 2-3i b) -3i

c) 2+3i d) 2

54. C-R equations of a function f(z) are

a) $u_x = v_y$ and $u_y = v_x$ b) $u_x = v_x$ and $u_y = -v_y$

c) $u_x = v_y$ and $u_y = -v_x$ d) $u_x = u_y$ and $v_y = -v_x$

55. If $f(x+T) = f(x) \forall x \in R$, where $T$ is a constant, then $f(x)$ is .........

a) Periodic function b) Even function

c) Odd function d) Both b & c

56. The graph of an odd function is always symmetrical about……

a) Origin
b) X-axis
c) Y-axis
d) Z-axis

57. The Half Range Fourier Sine series expansion of the function

$$f(x)=\begin{cases} x & 0<x<\frac{\pi}{2} \\ \pi-x & \frac{\pi}{2}<x<\pi \end{cases}$$ is defined in $0<x<\pi$ then $a_0 =$ …………

a) $\frac{4}{\pi}$
b) $\frac{\pi}{2}$
c) $\frac{2\pi}{4}$
d) $\frac{2\pi}{4}$

58. Value of $\int_0^{10} {}^1/_x \, dx$ by Trapezoidal rule (talking n= 4) is

a) 2.623
b) 1.624
c) 2.378
d) 0.693

59. In Rungue Kutta method the value of $y_1$ is

a) $y_0 + h$
c) $y_0 + k$
c) $y_0$
d) 0

60. The P.I of $e^{2x}, e^{-2x}$ is

a) $e^{2x}, e^{-x}$
b) $e^{x}, e^{-x}$
c) $\left(D^3 - D\right)y = 2e^x$
d) $C_1 + C_2e^x + C_3e^{-x}$

61. The differential equation $\left(\frac{d^2y}{dx^2}\right)^2 - \frac{dy}{dx} = y^3$ has degree

a) ½
b) 2
c) 3
d) 4

62. The differential equation of the form $\frac{dy}{dx}$+ P(x).y = Q(X).$y^n$is called

a) Auxiliary equation
b) Linear equation
c) Euler equation
d) Bernoulli's equation

63. The differential equation $x^3\frac{d^3y}{dx^3}+5x\frac{d^2y}{dx^2}-2\frac{dy}{dx}=y^3$ has order

a) ½
b) 2
c) 3
d) 4

64. C.F of the differential equation $(D^2+1)y=e^x$ is

a) $c_1\cos x+c_2\sin x$
b) $c_1\cos x-c_2\sin x$
c) $c_1\cosh x+c_2\sinh x$
d) $c_1\cosh x-c_2\sinh x$

65. The General form of differential equation $P_p+q^Q=R$ is

a) Exact D.E
b) Linear P.D.E
c) Homogenous P.D.E
d) Bernoulli D.E

66. The general solution of $\frac{dy}{dx}=e^{x-y}$

a) $e^x+e^y=c$
b) $e^{-x}+e^{-y}=c$
c) $e^x-e^y=c$
d) $e^{-x}-e^{-y}=c$

67. An integrating factor of $x^2ydx+(y^3-x^3)dy=0$ is

a) $1/y^4$
b) $x^4$
c) $y^4$
d) $1/x^4$

68. An Integrating Factor of $(1+y^2)dx=(\tan^{-1}y-x)dy$ is

a) $\tan^{-1}y$
b) $_e-\tan^{-1}y$
c) $_e\tan^{-1}y$
d) $e^{\tan y}$

69. The differential equation satisfying the relation $y=ce^{\sin x}$ is

a) $y^1-y\cos x=0$
b) $y^1+y\cos x=0$
c) $y^1-y=0$
d) $y^1-2y\cos x=0$

70. An I.F. of $(x-y)dx-dy=0$ is

a) $e^x$ b) -1

c) 1 d) $e^y$

71. The complete solution of $p^2+q^2=1$ is

a) $z=ax-(1-a^2)y+b$ b) $z=ax+\sqrt{(1-a^2)}y+b$

c) $z=a^2+b^2$ d) $z=ax-\sqrt{(1-a^2)}y+b$

72. The partial differential equation obtained by eliminating arbitrary constants a and b from $z=a(x+y)+b$ is

a) p = q b) p = -q

c) pq = 0 d) p = 2q

73. The partial differential equation obtained by eliminating arbitrary constants a and b from $z=(x+a)(y+b)$ is

a) p =qz b) p =-qz

b) P =-qz d) p =-qz

74. The partial differential equation obtained by eliminating arbitrary constants a and b from $z=axy+b$ is

a) xp = yq b) xp = yq

c) py = xq d) py =-xq

75. The partial differential equation obtained by eliminating arbitrary constants a and b from $z=(x-a)^2+(y-b)^2$ is

a) $(p^2+q^2)=z$ b) $(p^2+q^2)=2z$

c) $(p^2+q^2)=4z$ d) $(p^2+q^2)-\,-4z$

76. The partial differential equation obtained by eliminating arbitrary function f from $z=f(x^2+y^2)$ is

a) xp = yq b) xp = -yq

c) py = xq d) py = xq

77. $z^2=p^2+pq$ is ________ Partial differential equation

a) Linear b) Non Linear

c) Bernoulli d) Legendre

78. The partial differential equation obtained by eliminating arbitrary function f from $z = xy + f(x^2 + y^2)$ is

a) xp-yq = z    b) xp + yq = z

c) $py - xq = y^2 - x^2$    d) $py + xq = y^2 - x^2$

79. The partial differential equation obtained by eliminating arbitrary function f from $z = f(x^2 - y^2)$ is

a) $xp = yq$    b) $xp = -yq$

c) $xp = -yq$    d) $py = xq$

80. The partial differential equation obtained by eliminating arbitrary function f from $z = e^y f(x+y)$ is

a) $P + q = z$    b) $q - P = z$

c) $P + z = q$    d) $Pq = z$

81. In a certain college 25% of boys and 10% of girls are studying mathematics the girls constitute 60% of the student body. What is the probability that the mathematics is being studied.

a) 4/25    b) 1/10

c) 3/10    d) 4/21

82. In tossing a fair die, the probability of getting an odd number or a number less than 4 is

a) 2    b) 1/2

c) 2/3    d) 3/4

83. An unbiased coin is tossed three times, then what is the probability to get at least 2heads

a) 1/2    b) 3/8

c) 1/8    d) 3/4

84. An unbiased coin is tossed four times, then what is the probability to get exactly two heads

a) 3/8    b) 5/8

c) 7/16    d) 1/16

85. 50 tickets serially numbered 1 to 50. One ticket is drawn from these at random. The probability of it being multiple of 3 or 4 is

a)12/25    b) 14/25

c) 2/5    d) 2/5

86. The distribution in which mean, median and mode are equal is________

a) Binomial Distribution b) Poisson Distribution

c) Negative Binomial distribution d) Normal Distribution

87. The value of the $\int_0^1 \frac{1}{1+x} dx$ = by Simpsons1/3 rule, taking n = 4

a) 0.96315 b) 0.63915

c) 0.69325 d) 0.7084

88. If dy/dx = $-y$, y(0) =1 and h=0.01then by the Euler's method the value of $y^1 is =$

a) 0.952 b) 0.99

c) 1.1 d) 1.905

89. If $y_0 = 0, y_1 = 10, y_2 = 18, y_3 = 25$ and h = 2 then the value of $\int_0^6 ydx$ by Simpson's 3/8 rule is

a) 81.75 b) 65.25

c) 18.34 d) 75.81

90. Δ f(x) is

a) f(x) b) f(x) – f(x+h)

c) f(x+h) – f(x) d) f(x-h)

91. Evaluate $\int_0^1 \sqrt{1+x^3 d} dx$ taking h = 0.1 using Trapezoidal rule

a) 1.1122 b) 2.1122

c) 1.2211 d) 2.2211

92. If dy/dx = $(x-y)$, y(0)=1 by Picard's method the value of series method $y'(1) =$

a) 0.905 b) 1.9

c) 2.1 d) 0.95

93. Consider below table to find "y" value at x=0.8, by using below methods

| x | 0 | 1 | 2 |
|---|---|---|---|
| y | 1 | 1.8 | 3.3 |

a) Newton's forward formulae

b) Newton'sbackward formulae

c) Gauss formulae
d) Lagrange's formula

94. If dy/dx = $x+y^2$ y(0)=1 and h=0.1 then the value of $K_2$ in fourth order Runge-kutta method is ___
a) 0.11
b) 1.25
c) 0.99
d) 0.55

95. The value of $\nabla$ =
a) 1-E
b) 1-
c) 1+E
d) 1+

96. The following of $E^{-n}$ f(x) is
a) f(x)
b) f(x-nh)
c) f(x+ h)
d) f(x+ nh)

97. The relationship between the operators E and D is
a) E = $e^{hD}$
b) D = $e^{hE}$
c) E = D
d) None

98. Evaluate $\int_0^1 \frac{1}{1+x^2}dx$ using Simpson's 3/8 rule taking h =1/6 is
a) 0.5487
b) 0.83915
c) 0.6487
d) 0.7854

99. If f(x) = $x^2+2x+1$ and h=3 then Δ f(x) =
a) 4x
b) 4x+8
c) 4x-8
d) 6x+15

100. Range of t-distribution is
a) $(-\infty,\infty)$
b) $(-\infty,0)$
c) $(0,\infty)$
d) None

## Answers

| | | | | | | | | | |
|---|---|---|---|---|---|---|---|---|---|
| 1. a | 2. c | 3. a | 4. b | 5. a | 6. a | 7. b | 8. c | 9. c | 10. b |
| 11. a | 12. b | 13. d | 14. b | 15. b | 16. c | 17. d | 18. c | 19. b | 20. a |
| 21. d | 22. c | 23. c | 24. d | 25. b | 26. c | 27. c | 28. b | 29. a | 30. c |
| 31. d | 32. c | 33. a | 34. c | 35. c | 36. c | 37. a | 38. a | 39. d | 40. a |
| 41. a | 42. c | 43. c | 44. b | 45. b | 46. a | 47. b | 48. d | 49. c | 50. a |

| | | | | | | | | | |
|---|---|---|---|---|---|---|---|---|---|
| 51. a | 52. c | 53. a | 54. c | 55. a | 56. a | 57. b | 58. c | 59. b | 60. d |
| 61. b | 62. d | 63. c | 64. a | 65. b | 66. c | 67. a | 68. c | 69. a | 70. a |
| 71. b | 72. a | 73. c | 74. a | 75. c | 76. c | 77. b | 78. c | 79. d | 80. b |
| 81. a | 82. c | 83. a | 84. a | 85. a | 86. d | 87. c | 88. b | 89. a | 90. c |
| 91. a | 92. a | 93. a | 94. a | 95. b | 96. b | 97. a | 98. d | 99. d | 100. a |

1. If A = $\begin{bmatrix} 1 & 2 \\ -4 & 3 \end{bmatrix}$ and B= $\begin{bmatrix} 1 & 0 \\ 0 & 3 \end{bmatrix}$ then $A^2 + B =$

a) $\begin{bmatrix} -6 & 8 \\ -16 & 4 \end{bmatrix}$  b) $\begin{bmatrix} 6 & 8 \\ 16 & 4 \end{bmatrix}$

c) $\begin{bmatrix} 6 & -8 \\ -16 & -4 \end{bmatrix}$  d) None

2. If f(-x) = f(x) then the function f(x) is said to

a) Odd  b) Even

c) Identity  d) None

3. The rank of a singular matrix of order 3 is

a) 3  b) $\leq 3$

c) $\leq 2$  d) 2

4. Trace and determinant of a Matrix A= $\begin{bmatrix} 2 & 0 & 1 \\ 0 & 3 & 0 \\ 1 & 0 & 2 \end{bmatrix}$ is

a) 6, 8  b) 7, 9

c) 7, 6  d) 9, 7

5. The diagonal elements of Hermitian matrix are

a) Real  b) Purely imaginary

c) Zero  d) None

6. If matrix A satisfies $A^T A = AA^T = 1$ then A is

a) Orthogonal  b) Symmetric

c) Skew- Symmetric  d) Unitary

7. Which of the following matrix Symmetric Matrix

a) $\begin{bmatrix} 1 & -4 \\ 4 & 3 \end{bmatrix}$  b) $\begin{bmatrix} 1 & -4 \\ -4 & 3 \end{bmatrix}$

c) $\begin{bmatrix} 1 & 4 \\ -4 & 3 \end{bmatrix}$  d) None

8. If r (A) = r(AB) = n then it has

a) No solution
b) Unique solution
c) Many solution
d) None

9. The rank of the matrix $\begin{bmatrix} 1 & 0 & 0 \\ 0 & 1 & 0 \\ 0 & 0 & 1 \end{bmatrix}$ is

a) 1
b) 2
c) 3
d) 0

10. The rank of the product of two non -singular matrices each of order 3 is

a) 3
b) $\leq 2$
c) $\leq 3$
d) 2

11. Find the trace (A-B) where A= $\begin{bmatrix} 2 & 0 & 1 \\ 0 & 3 & 0 \\ 1 & 0 & 2 \end{bmatrix}$ and B = $\begin{bmatrix} -2 & 0 & 14 \\ 0 & 3 & 0 \\ 5 & 0 & -2 \end{bmatrix}$

a) 6
b) 7
c) 8
d) None

12. The Eigen values of the matrix A= $\begin{bmatrix} 0 & 1 \\ -2 & -3 \end{bmatrix}$

a) -1, -2
b) 0,1
c) 0,2
d) 1,-1

13. The rank of the matrix is $\begin{bmatrix} 1 & 1 & 0 \\ 0 & 0 & 1 \\ 2 & 2 & 2 \end{bmatrix}$

a) 1
b) 2
c) 3
d) 0

14. If A = $\begin{bmatrix} 1 & 0 \\ 2 & -1 \end{bmatrix}$ then find is

a) $\begin{bmatrix} 1 & 0 \\ 2 & -1 \end{bmatrix}$
b) $\begin{bmatrix} -1 & 0 \\ 2 & -1 \end{bmatrix}$
c) $\begin{bmatrix} 1 & 0 \\ -2 & -1 \end{bmatrix}$
d) $\begin{bmatrix} 1 & 0 \\ 2 & 1 \end{bmatrix}$

15. If r (A) = r(AB) <n then it has

a) No solution b) Unique solution

c) Infinite solutions d) None

16. If A= $\begin{bmatrix} 3 & 2 \\ 1 & 4 \end{bmatrix}$ then $A(adjA)$ equals

a) $\begin{bmatrix} 1 & 1 \\ 1 & 1 \end{bmatrix}$ b) $\begin{bmatrix} 0 & 10 \\ 10 & 0 \end{bmatrix}$

c) $\begin{bmatrix} 1 & 0 \\ 0 & 1 \end{bmatrix}$ d) $\begin{bmatrix} 10 & 0 \\ 0 & 10 \end{bmatrix}$

17. Which of the following is false?

a) The non homogeneous system of equations is always Consistent

b) The homogeneous system of equations is always Consistent

c) The system $AX = B$ is said to be homogeneous if $B = 0$

d) The system $AX = B$ is said to be non homogeneous if $B \neq 0$

18. The rank of a Singular matrix of order 3 is

a) 3 b) $\leq 3$

c) $\leq 2$ d) 2

19. If $I$ is unit matrix of order n then $\det I$ is

a) N b) 0

c) 2 d) 1

20. Function f( x) has maximum at (a, b) if

a) $ln-m^2 > 0, l > 0$ b) $ln-m^2 > 0, l < 0$

c) $ln-m^2 = 0$ d) None

21. The nature of the quadratic form $x^2 + y^2 + z^2 - 2xy$ is

a) Positive definite b) Positive semi definite

c) Negative definite d) Indefinite

22. The index and signature of the quadratic form $x^2 + 3y^2 + 3z^2 - 2xy$ are

a) 2, 1 b) 3, 1

c) 3, 2 d) 3, 3

23. If the canonical form of a quadratic form $X^T AX$ is $3y_1^2 + 2y_2^2 + 7y_3^2$ then the signature of the Quadratic form is

a) 1
b) 2
c) 3
d) 0

24. If the eigen values of A are 0,1,2 then the Rank of the quadratic form is

a) 0
b) 1
c) 2
d) 3

25. $\int_0^1 x \sin^2\theta \cos^4 \theta d\theta =$ ______________

a) $\frac{\pi}{32}$
b) $\frac{\pi}{64}$
c) $\frac{\pi}{16}$
d) $\frac{\pi}{4}$

26. $\int_0^1 x(\log x)^3 dx =$ ______________

a) 3/8
b) -3/8
c) 3/4
d) -3/4

27. The primitive period of the function sin3x is

a) $2\pi/3$
b) $\pi/2$
c) $\pi/3$
d) $3\pi/2$

28. If f(x) = x in (0, 2) then the Fourier coefficient $a_a =$

a) $\pi$
b) $\frac{\pi}{4}$
c) $\frac{\pi}{2}$
d) None

29. If the function $f(x)$ is even then the Fourier expansion is $f(x)$

a) 0
b) $\frac{3\pi}{2}$
c) $\frac{\pi}{2}$
d) None

30. The Fourier transform of the function $f(x)$ is

a) $\int_{-\infty}^{\infty} f(x)e^{ix}dx$

b) $\int_{0}^{\infty} f(x)e^{ix}dx$

c) $\int_{-\infty}^{\infty} f(x)e^{ipx}dx$

d) $\int_{0}^{\infty} f(x)e^{ipx}dx$

31. If $A=\begin{bmatrix}0 & 1 & 1\\ 1 & 0 & 1\\ 1 & 1 & 0\end{bmatrix}$ then the quadratic form $X^T AX$ is

a) Positive definite

b) Positive semi definite

c) Negative definite

d) Indefinite

32. The quadratic form corresponding to the symmetric matrix

$$\begin{bmatrix}1 & -1 & 2\\ -1 & 2 & -2\\ 2 & -2 & 3\end{bmatrix}$$

a) $x^2+2y^2+3z^2-2xy+4xz-4yz$

b) $x^2-2y^2+3z^2+2xy+4xz+4yz$

c) $x^2+2y^2+3z^2-2xy-4xz-4yz$

d) $x^2+2y^2+3z^2-2xy+4xz$

33. $\int_{0}^{\infty} e^{-y^2}dy=$

a) $\pi/2$

b) $\sqrt{\pi}/2$

c) $\pi/4$

d) $\pi/\sqrt{2}$

34. The Maclaurin's series expansion for $\log(1+x)=$

a) $x-\frac{x^2}{2}+\frac{x^3}{3}-\frac{x^4}{4}+$..............

b) $x+\frac{x^2}{2}+\frac{x^3}{3}+\frac{x^4}{4}+$..............

c) $-x-\frac{x^2}{2}-\frac{x^3}{3}-\frac{x^4}{4}-$..............

d) $-x+\frac{x^2}{2}-\frac{x^3}{3}+\frac{x^4}{4}-$..............

35. If $u=2y-z, v=x+3z, w=x+4y$ then $\frac{\partial(u,v,w)}{\partial(x,y,z)}=$

a) -2

b) 6

c) 2

d) -6

36. Which of the following statement is TRUE?

a) Rolle's theorem is not applicable for $f(x) = x^2$ in $[-1,1]$

b) $f(x) = f(0) + xf(0) + ....... + \frac{x^n}{\lfloor n} f^n(0)$ is called Maclaurin's series expansion

c) $\frac{\partial(u,v)}{\partial(r,s)} \times \frac{\partial(r,s)}{\partial(x,y)} = \frac{\partial(r,s)}{\partial(x,y)}$

D) Cauchy's mean value theorem reduces to Rolle's Theorem when $g(x) = x$

37. $\int_0^\infty \int_0^\infty e^{-(x^2+y^2)} dxdy =$

a) $\pi/4$

b) $\pi/2$

c) $\pi/2$

d) $\pi^2/2$

38. If $f(0) = 1, f^i(0) = 0, f^{ii}(0) = -1, f^{iii}(0) = 0, f^{iv}(o) = 1........$ then $f(x) =$

a) $1 + \frac{x^2}{\lfloor 2} + \frac{x^4}{\lfloor 4} + \frac{x^6}{6} + ...........$

b) $1 + \frac{x}{\lfloor 1} + \frac{x^2}{\lfloor 2} + \frac{x^3}{3} + ...........$

c) $1 - \frac{x}{\lfloor 1} + \frac{x^4}{\lfloor 4} - \frac{x^6}{\lfloor 6} + ...........$

d) $x - \frac{x^3}{\lfloor 3} + \frac{x^5}{\lfloor 5} - \frac{x^7}{\lfloor 7} + ...........$

39. $\int_0^1 \int_1^2 \int_2^3 xyzdxdydz =$

a) 5/8

b) 3/8

c) 15/8

d) 9/8

40. $\nabla(r^n) =$

a) $n r^{n-1} \vec{r}$

b) $nr^{n-2}$

c) $n r^{n-2} \vec{r}$

d) $nr^{n-1}$

41. If $r^{\rightarrow} = xi^{\rightarrow} + yj^{\rightarrow} + zk^{\rightarrow}$ and $a^{\rightarrow}$ is a constant vector, then $\nabla(a^{\rightarrow}.r^{\rightarrow}) =$

a) 0

b) 3

c) $\vec{a}$

d) $\vec{r}$

42. If $\phi = x^2 + y^2 + z$ then the directional derivative of $\phi$ at $(1,0,1)$ in the direction of $\vec{a} = \vec{i} + \vec{j} + \vec{k}$ is

a) 2 b) 3

c) $1/\sqrt{3}$ d) $\sqrt{3}$

43. If $\phi = x^2 y + y^2 + z$ then unit normal vector of $\phi$ at $(2,1,1)$ is

a) $(2\hat{i}+2\hat{j}+\hat{k})/3$ b) $(4\hat{i}+6\hat{j}+\hat{k})/\sqrt{53}$

c) $(3\hat{i}+\hat{j}+\hat{k})/\sqrt{21}$ d) $(2\hat{i}+3\hat{j}+4\hat{k})/\sqrt{29}$

44. If $r = \sqrt{x^2 + y^2 + z^2}$, then $\nabla(\frac{1}{r}) =$

a) $\vec{r}/r^3$ b) $-\vec{r}/r^3$

c) $\vec{r}/r^2$ d) $-\vec{r}/r^2$

45. If $\vec{r} = x\vec{i} + y\vec{j} + z\vec{k}$, then $\nabla \times \vec{r} =$

a) $\vec{0}$ b) 3

c) r d) $\vec{r}$

46. If $\phi$ is a scalar point function then $curl(grad\phi) =$

a) 0 b) $\vec{0}$

c) $\nabla\phi$ d) 1

47. If $\vec{A} = (5x+y)\hat{i} + (cy-z)\hat{j} + (3x-7z)\hat{k}$ is Solenoidal, then $c =$

a) 3 b) 0

c) 1 d) 2

48. If $\vec{A}$ is vector point function then $\vec{A} =$

a) 2 b) 1

c) 0 d) $\vec{0}$

49. $S$ is any closed surface enclosing a volume $V$ and $\vec{F} = x\hat{i} + 2y\hat{j} + 3z\hat{k}$,

if $\iint_S \vec{F}.\vec{n}\, dS = kV$ then $k =$

a) 3
b) 4
c) 5
d) 6

50. Reciprocal of " i " is

a) 1/i
b) -i
c) 1
d) $i^2$

51. The function $f(z)$ is analytic if it satisfies

a) Laplacian equations
b) Linear equations
c) C-R equations
d) None

52. C-R equations of a function f(z) are

a) 2-3i
b) -3i
c) 2+3i
d) 2

53. The value of the derivative of $f(z) = 2z^2$ at $z = i$ is __________

a) 4i
b) 2i
c) -4i
d) -2i

54. If $\phi(x, y) = 2x - x^2 + ay^2$ is harmonic then a = __________

a) 1
b) -1
c) 2
d) -2

55. A function $u(x, y)$ is harmonic if __________

a) $u$ have continuous second derivative

b) $u$ have continuous second derivative and $\frac{\partial^2 u}{\partial x^2} + \frac{\partial^2 u}{\partial y^2} = 0$

c) $\frac{\partial^2 u}{\partial x^2} + \frac{\partial^2 u}{\partial y^2} = 0$

d) $\frac{\partial^2 u}{\partial x^2} - \frac{\partial^2 u}{\partial y^2} = 0$

56. If $f(z)$ is a regular function of $z$ then $(\frac{\partial^2}{\partial x^2}+\frac{\partial^2}{\partial y^2})|f(z)|^2 =$ ______

a) $|f^i(z)|^2 =$ b) $4|f^i(z)|^2 =$

c) $9|f^i(z)|^2 =$ d) $8|f^i(z)|^2 =$

57. Which of the following function is not analytic at $z=0$

a) $e^z$ b) $z^4$

c) $\cos^2 z$ d) $\log z$

58. If $f(z)=\frac{1}{2}\log(x2+y2)+i\tan-1(\frac{px}{y})$ is an analytic function, then p = ______

a) -1 b) 1

c) 2 d) -2

59. The order and degree of the differential equation $\left[1+(y')^2\right]^{1/5}=ky''$ are

a) 2, 3 b) 2, 5

c) 2, 4 d) 2, 1

60. $\frac{1}{D^2-1}$ sinx =

a) $-\frac{x\sin x}{2}$ b) $-\frac{x\cos x}{2}$

c) $\frac{\sin x}{3}$ d) $\frac{\cos x}{3}$

61. Integrating factor of Linear differential equation $\frac{dy}{dx}+\frac{y}{x}=e^x$ is

a) x b) $e^x$

c) 2x d) None

62. Complimentary function of $\frac{d^2y}{dx^2}-y=0$ is

a) $y=c_1e^x$ b) $y=e^x(c_1+xc_2)$

c) $y=c_1e^x+c_2e^{-x}$ d) None

63. The equation $\frac{dy}{dx}+\frac{ax+hy+g}{hx+by+f}=0$ is

a) Homogeneous
b) Variable Separable
c) Exact
d) Non Homogeneous

64. Order of Differential equation $[1+(\frac{dy}{dx})^2]=k\frac{d^3y}{dx^3}$

a) 1
b) 2
c) 3
d) Not exist

65. Degree of differential equation $\frac{d^3y}{dx^3}+2\frac{d^2y}{dx^2}+\frac{dy}{dx}+y=\sin x$

a) 1
b) 2
c) 3
d) Not exist

66. The differential equation of the form $\frac{dy}{dx}$ + P(x) y = Q(x) $y^n$ is called

a) Auxiliary equation
b) Linear equation
c) Euler equation
d) Bernoulli's equation

67. The Integrating factor of the differential equation $ydx-xdy=0$

a) x
b) $e^{-x}$
c) $1/x^2$
d) None

68. The general solution of $(d^2-m^2)y=0$ is

a) $(c_1+c_2x)e^{mx}=0$
b) $y=c_1\sin mx+c_2\cos mx$
c) $c_1e^{mx}+c_1xe^{\quad mx}$
d) None

69. If $f(D)=0$ be a linear differential equation with constant coefficient then its auxiliary equation is

a) $f(D-m)=0$
b) $f(m)=0$
c) $f(e^m)=0$
d) $f(x)=0$

70. If $Pp+qQ=R$ is the Lagrange's partial differential equation then $\frac{dx}{P}=\frac{dy}{Q}=\frac{dz}{R}$ is called

a) Auxiliary equation
b) Non linear equation
c) Linearequation
d) None

71. If the number of arbitrary constants to be eliminated is equal to the number of independent variables then the P.D.E is

a) First order b) Second order

c) Third order d) None

72. The equation of the form $z = px + qy + f(p,q)$ is of the form

a) Clairaut's form b) Exact form

c) Linear form d) Homogenous form

73. The partial differential equation $Pp + qQ = R$ is________equation

a) Quadratic equation b) Linear equation

c) Symmetric equation d) Non-Linear equation

74. $z = p + q + pq$ is

a) Linear P.D.E b) Non-Linear P.D.E

c) Bernoulli P.D.E d) Legendre D.E

75. The general solution of the equation $pe^y - qe^x = 0$ is

a) $z = a(e^x + e^y) + c$ b) $e^x + e^y = az + c$

c) $z = ae^x + ce^y$ d) None

76. The solution of the partial differential equation $z = px + qy - \sqrt{pq}$ is

a) $z = 3ax + ay + x^3 - y^3 + c$ b) $z = 3(ax + ay) + x^3 - y^3 + c$

c) $z = x^3 - y^3 + c$ d) $z = ax + by - \sqrt{ab}$

77. General solution of $p + q = 1$ is

a) $f(x-y, y-z) = 0$ b) $f(x+y, y+z) = 0$

c) $f(xy, yz) = 0$ d) None

78. The partial differential equation obtained by eliminating arbitrary function $f$ from $z = e^{my} f(x-y)$ is

a) $p + q = mz$ b) $p - q = mz$

c) $p + q = -mz$ d) $p - q = -mz$

79. One of the solutions of the partial differential equation $p \tan x + q \tan y = \tan z$ is

a) $\frac{\sin x}{\sin y} = c$ b) $\sin x - \sin y = c$

c) $\sin x - \sin y = c$ d) $\frac{\cos x}{\cos y} = c$

80. One of the solutions of the partial differential equation $y^2 p - xyp = x(z-2y)$ is

a) $x^2 + y^2 = c$ b) $x^2 - y^2 = c$

c) $xy = c$ d) $x + y = c$

81. The function $f(x) = kx$ in $0 \le x \le 1$ is valid probability density function, if k is_______

a) 2 b) 3

c) 4 d) 1

82. The function $f(x) = kx^3$ in $0 \le x \le 1$ is valid probability density function, if k is______

a) 5 b) 4

c) 3 d) 2

83. If $f(x) = \begin{cases} kx2, 0 < x < 3 \\ 0, elsewhere \end{cases}$ is a probability density function, then k is__________

a) 1/8 b) 2/9

c) 1/9 d) 8/9

84. If $f(x) = \begin{cases} kxe^{-x}, x > 0 \\ 0, elsewhere \end{cases}$ is the probability density function, then k is _________

a) 0 b) 2

c) 1 d) 3

85. Range of t-distribution is

a) $(-\infty, \infty)$ b) $(-\infty, 0)$

c) $(0, \infty)$ d) None

86. If the standard deviation of a chi-square distribution is 10, then its degree of freedom is

a) 25 b) 49

c) 50 d) 55

87. If $\bar{x}$ =47.5, μ=42.1, S =8.4, n=25 then t is

a) 3.21 b) 4.5

c) 3.12 d) 5.2

88. Which of the following is the highest range of " r " is

a) 0 and 1
b) -1 and 0
c) -1 and 1
d) 1 and 1

89. Regression coefficient is independent of

a) Origin
b) Scale
c) Origin & scale
d) None

90. When two regressions lines coincide then r is

a) 0
b) 1
c) 1.5
d) 0.5

91. If dy/dx = x+y, y(0)=1 by Picard's method, $y^{(1)}(0.1) =$

a) 1.1
b) 1.105
c) 0.905
d) 0.95

92. If dy/dx = $(x^2 - y)$, y(0)=1 by Taylor's series method y (0.1)=

a) 0.95
b) 1.05
c) 1.1
d) 0.905

93. If f(x) = $x^2 + 2x + 1$ and h=3 then " f(x) =

a) 4x
b) 4x+8
c) 4x-8
d) 6x+15

94. The relationship between the operators E and D is

a) $E = e^{hD}$
b) $D = e^{hE}$
c) E = D
d) $E = e^{D}$

95. Taking h=1 , $\Delta(e^x)$ is

a) $(e - 1)\, e^x$
b) $(e + 1)\, e^x$
c) $ec^x$
d) $e^x$

96. Given that

| x | 1 | 2 | 3 |
|---|---|---|---|
| f(x) | 3 | 8 | 15 |

$Than\Delta^2 f(1) =$

a) 1
b) 3
c) 4
d) 2

97. Value of $\int_0^{10} \frac{1}{x} dx$ by Trapezoidal rule (talking n=4) is

a) 0.26 b) 1.6

c) 2.6 d) 0.693

98. In Newton's backward interpolation formulae p =

a) $(x + x_0)/h$ b) $(x - x_0)/h$

c) $(x - x_n)/h$ d) $(x + x_n)/h$

99. The following formula is used for unequal interval of f(x) values

a) Newton's forward formula b) Lagrange's formula

c) Gauss formula d) Newton's Backward formula

100. The value of y(0.1) if $\frac{dy}{dx} = x^2 + y^2$, y (0) = 1 by Runge Kutta method

a) 1.1115 b) 0.9995

c) 1.5565 d) 2.5065

## Answers

| | | | | | | | | | |
|---|---|---|---|---|---|---|---|---|---|
| 1. a | 2. b | 3. c | 4. b | 5. a | 6. a | 7. b | 8. b | 9. c | 10. a |
| 11. c | 12. a | 13. b | 14. a | 15. c | 16. d | 17. a | 18. c | 19. d | 20. b |
| 21. b | 22. d | 23. c | 24 . c | 25. a | 26. b | 27. a | 28. d | 29. c | 30. c |
| 31. a | 32. a | 33. b | 34. a | 35. c | 36. b | 37. a | 38. c | 39. c | 40. c |
| 41. c | 42. d | 43. b | 44. b | 45. a | 46. b | 47. d | 48. c | 49. d | 50. b |
| 51. c | 52. c | 53. a | 54. a | 55. b | 56. b | 57. d | 58. a | 59. b | 60. b |
| 61. a | 62. c | 63. c | 64. c | 65. a | 66. a | 67. c | 68. c | 69. b | 70. a |
| 71. a | 72. a | 73. d | 74. b | 75. a | 76. d | 77. a | 78. a | 79. a | 80. a |
| 81. a | 82. b | 83. c | 84. c | 85. a | 86. c | 87. a | 88. c | 89. a | 90. b |
| 91. b | 92. d | 93. d | 94. a | 95. a | 96. d | 97. d | 98. c | 99. b | 100. a |

1. If A is involuntary matrix its possible Eigen values are
   a) 0, 0 b) 1, 0
   c) 1, -1 d) -1, 0
2. The diagonal elements of skew Hermit ion matrix are
   a) Real b) Purely imaginary
   c) Zero d) None
3. Sum of the characteristic roots of a matrix A is equal to
   a) Sum of the principal diagonal elements of A
   b) Trace of the matrix A
   c) Both (a) and (b)
   d) None of these
4. Eigen values of the unit matrix of order 3 are
   a) 0,0,1 b) 0,1,1
   c) 1,1,1 d) 0,1,-1
5. If 4, 6 are the Eigen values of a square matrix A, and then the Eigen values of $2A^2 - \frac{1}{2}A + 3l$ are
   a) 28, 72 b) 31, 62
   c) 33, 72 d) 33, 76
6. Eigen values of $\begin{bmatrix} 0 & i \\ i & 0 \end{bmatrix}$ are
   a) $i, i$ b) $i, -i$
   c) $1, -1$ d) $-1, -1$
7. $P^{-1}AP = D$ then the matrix P is called
   a) Square b) Model matrix
   c) Spectral matrix d) Diagonal matrix
8. If r (A) = r (AB) = n then it has
   a) No solution b) Unique solution
   c) Infinite solutions d) None
9. If $A = \begin{bmatrix} 1 & 2 & 3 \\ 3 & 4 & 4 \\ 7 & 10 & 12 \end{bmatrix}$ then Rank is
   a) 1 b) 2
   c) 3 d) Can't define

10. The determinant of an orthogonal matrix is
    a) <1 b) ± 1
    c) 0 d) >1
11. The rank of the product of two matrices each of order 3 is
    a) 3 b) $\leq 2$
    c) $\leq 3$ d) 2
12. If A is Skew symmetric matrix then A is
    a) $A = -A^0$ b) $A = A^0$
    c) $A = -A^T$ d) $A = A^T$
13. If $A^0 A = 1$ then A is called
    a) Hermition matrix b) Skew hermition matrix
    c) Orthogonal matrix d) Unitary matrix
14. Product of Eigen values of $A = \begin{bmatrix} 2 & 3 & 4 \\ 0 & 2 & 5 \\ 0 & 0 & 3 \end{bmatrix}$ is
    a) 7 b) 12
    c) 2 d) 0
15. Inverse of a singular matrix of order 3x3 is
    a) Square matrix b) Symmetric matrix
    c) Unit matrix d) Not defined
16. If matrix A satisfies $A^T A = AA^T = I$ then A is
    a) Orthogonal matrix b) Symmetric matrix
    c) Skew-Symmetric matrix d) Unitary Matrix
17. If $A^2 + A - I = 0$ then $A^{-1} =$ ________
    a) $A+I$ b) $A-I$
    c) $A^2 + A$ d) Can't define
18. Which of the following can be the Eigen values of a Hermition matrix
    a) 1,2,3 b) $\pm 3i, 0$
    c) $\pm 3i, 5$ d) $1, 1 \perp i$
19. State the conditions for the existence of $A+B$ If $A_{m\times n}$ and $B_{p\times q}$ are two matrices
    a) $m = p, n = q$ b) $m \neq p, n = q$
    c) $m = q, n = q$ d) $m = p, n \neq q$

20. If $ln-m^2 = 0$ at (a, b) of a function then

a) Has Maximum value at (a, b)  b) Has Minimum value at (a, b)

c) No conclusion at (a, b)  d) None of these

21. The determinant of an orthogonal matrix is

a) 1  b) 0

c) -1  d) $\neq 1$

22. Which one of the following is TRUE?

a) The inverse of a matrix is need not be unique.

b) If $A$ is a matrix of order $n$ and $A$ is non singular then $\rho(A) \neq n$

c) If $A$ is a matrix of order $n \times m$, then $\rho(A) \leq \min(m, n)$.

d) Rank of matrix is not unique.

23. Determine for what value of t $\lambda$ he system of equations $x+y+z=1, x+2y+4z=\lambda, x+4y+10z=\lambda^2$ have a solution.

a) $\lambda = 2, \lambda = 3$  b) $\lambda = 1, \lambda = 3$

c) $\lambda = 2, \lambda = 1$  d) $\lambda = 3, \lambda = 3$

24. If $A = \begin{bmatrix} 1 & 0 & 3 \\ 2 & 1 & -1 \\ 1 & -1 & 1 \end{bmatrix}$ then

a) $A^3 + 3A^2 + A - 9I = 0$  b) $A^3 - 3A^2 - A + 9I = 0$

c) $A^3 - 3A^2 + A + 9I = 0$  d) $A^3 - 3A^2 + A + 9I = 0$

25. If $A = \begin{bmatrix} 3 & 1 \\ -1 & 2 \end{bmatrix}$ then $A^2 =$

a) 2A-3I  b) 2A-3I

c) 2A-3I  d) 2A-3I

26. The Characteristic equation of $A = \begin{bmatrix} 1 & 1 \\ 2 & 5 \end{bmatrix}$ is

a) $\lambda^3 - 6\lambda + 3 = 0$  b) $\lambda^2 - 6\lambda + 3 = 0$

c) $\lambda^2 + 6\lambda + 3 = 0$  d) $\lambda^2 + 6\lambda - 3 = 0$

27. Determine for what values of $\lambda$ the reduced system $\begin{bmatrix} 1 & 1 & 1 \\ 0 & 1 & 3 \\ 0 & 0 & 0 \end{bmatrix}\begin{bmatrix} x \\ y \\ z \end{bmatrix} = \begin{bmatrix} 1 \\ \lambda - 1 \\ \lambda^2 - 3\lambda + 2 \end{bmatrix}$ will be Consistent?

a) $\lambda = 2, \lambda = 3$
b) $\lambda = 1, \lambda = 3$
c) $\lambda = 2, \lambda = 1$
d) $\lambda = 2, \lambda = 5$

28. The matrix $A = \left[\frac{\cos\theta \sin\theta}{-\sin\theta \cos\theta}\right]$ is

a) Symmetric
b) Skew symmetric
c) Orthogonal
d) Unitary

29. Find the rank of the matrix $\begin{bmatrix} 1 & 2 & 3 & 4 \\ -2 & 0 & 5 & 7 \end{bmatrix}$

a) 1
b) 2
c) 3
d) 4

30. If the eigen values of A are $1, 3+\sqrt{8}, 3-\sqrt{8}$ then the index and signature of the quadratic form $X^T AX$ is

a) 1, 2
b) 3, 1
c) 2, 1
d) 3, 3

31. The sum and product of the eigen values of $\begin{bmatrix} 2 & 2 & 1 \\ 0 & 3 & 1 \\ 0 & 0 & 1 \end{bmatrix}$ are

a) 6, 6
b) 6, 5
c) 9, 7
d) 7, 9

32. If 1,-5,3 are eigen values of a matrix $A_{3x3}$ then (A)=................

a) 0, 0, 1
b) 0, 1, 1
c) -15
d) 5

33. If $\lambda_1, \lambda_2, \lambda_3$ are eigen values of a non-singular matrix A, then the eigen values of $A^3$ are

a) $\lambda_1, \lambda_2, \lambda_3$
b) $\frac{1}{\lambda_1}, \frac{1}{\lambda_2}, \frac{1}{\lambda_3}$
c) $\frac{1}{\lambda_1^3}, \frac{1}{\lambda_2^3}, \frac{1}{\lambda_3^3}$
d) $\lambda_1^3, \lambda_2^3, \lambda_3^3$

34. The matrix of the quadratic form $x^2 + y^2 + z^2$ is

a) $\begin{bmatrix} 1 & 1 & 1 \\ 1 & 1 & 1 \\ 1 & 1 & 1 \end{bmatrix}$ b) $\begin{bmatrix} 1 & 0 & 1 \\ 0 & 1 & 0 \\ 0 & 0 & 1 \end{bmatrix}$

c) $\begin{bmatrix} 1 & 1 & 0 \\ 1 & 1 & 0 \\ 0 & 0 & 1 \end{bmatrix}$ d) $\begin{bmatrix} 1 & 0 & 0 \\ 1 & 1 & 0 \\ 0 & 0 & 1 \end{bmatrix}$

35. $\int_0^\infty \int_x^\infty \frac{e^{-x}}{x} dydx =$

a) 1 b) -1

c) 2 d) -2

36. If (4, 0), (6,0) & $(5, \neq 1)$ are stationary points of $f(x, y) = x^3 + 3xy^2 - 15x^2 - 15y^2 + 72x$ and $f(x, y) = x^3 + 3xy^2 - 15x^2 - 15y^2 + 72x$ then $f_{\min} =$

a) 108 b) 112

c) -7 d) 5

37. If u= $\tan^{-1} (y/x) than \frac{\partial u}{\partial x} at(0,1)$

a) 1 b) -1

c) 2 d) -2

38. If $u = 2x - y + 3z, v = 2x - y - z, w = 2x - y + z$, than $\frac{\partial(u, v, w)}{\partial(x, y, z)} =$

a) 0 b) 1

c) 2 d) -2

39. $\int_0^1 \int_0^1 \int_0^1 e^{x+y+z} dxdydz =$

a) $e^3$ b) $e^3 - 1$

c) $(e - 1)^3$ d) $(e - 1)^{-3}$

40. The stationary point of is

a) (0,3) b) (0,0)

c) (-3,0) d) (3,0)

41. The directional derivative of $f = x + y + z$ in the direction of $\vec{a} = \vec{i} + \vec{j} + \vec{k}$ at $(1,1,1)$ is

a) 2
b) 3
c) $\sqrt{1/5}$
d) $1/\sqrt{3}$

42. If $\phi = xyz$ then unit normal vector of $\phi$ at $(1,2,-1)$ is

a) $(-2\hat{i} - \hat{j} + 2\hat{k})/3$
b) $(4\hat{i} + 6\hat{j} + \hat{k})/\sqrt{53}$
c) $(3\hat{i} + \hat{j} + \hat{k})/\sqrt{11}$
d) $(-2\hat{i} + 3\hat{j} + 2\hat{k})/\sqrt{17}$

43. If $\vec{F} = (x+y)\vec{i} + \vec{j} - (x+y)\vec{k}$, then $\vec{F}.curl\,\vec{F} =$

a) 0
b) 1
c) 2
d) 3

44. $\nabla[f(r)] =$

a) $f'(r)\dfrac{\vec{r}}{r}$
b) $f'(r)$
c) $f'(r)\vec{r}$
d) $\vec{r}$

45. $grad(x^2 + y^2 + z^2)at(1,1,1)is$

a) $\hat{i} + \hat{j} + \hat{k}$
b) $2\hat{i} + 2\hat{j} + 2\hat{k}$
c) $\hat{i} + 2\hat{j} + \hat{k}$
d) $2\hat{i} + 3\hat{j} + 4\hat{k}$

46. The greatest rate of increase of $\phi = x^2yz^3$ at $(1,1,1)$ is

a) $\sqrt{11}$
b) $2\sqrt{11}$
c) $3\sqrt{11}$
d) $4\sqrt{11}$

47. If S is the surface of a unit sphere, then $\int_s (ax\,\vec{i} + by\,\vec{j} + cz\,\vec{k}).\vec{n}\,dS =$

a) $(a+b+c)\pi/3$
b) $(a+b+c)\pi/3$
c) $4\pi/3$
d) 0

48. The total work done by a force $\vec{F} = 2xy\,\vec{i} - 4z\,\vec{j} + 3x\,\vec{k}$ along $C: x = t^2, y = 2t + l, z = t^3$ from $t = 1$ to $t = 2$ is

a) 125/8 b) 247/12

c) 517/16 d) 638/5

49. The scalar potential of $\vec{F} = y^2\,\hat{i} + (2xy + z^2)\,\hat{j} + 2yz\,\hat{k}$ is

a) $x^2 + y^2 + z^2$ b) xyz

c) $xy^2 + yz^2$ d) xy + yz + zx

50. The cosine of angle between the surfaces $xy^2z = 3x + z^2$ and $3x^2 - y^2 + 2z = 1$ at $(1,-2,1)$

a) $-3/7\sqrt{6}$ b) $-1/2\sqrt{7}$

c) $-1/7\sqrt{5}$ d) $-1/5\sqrt{7}$

51. $\int_0^i ze^{z^2}\,dz =$ ____________

a) $e^{-1}$ b) $\frac{e^{-1} - 1}{2}$

c) $\frac{1}{2e}$ d) $\frac{1}{e-1}$

52. $\int_{(0,0)}^{(1,1)} \left[x^2 + y^2)dx - 2xydy\right]$ along $y = x$ is ______

a) 3 b) 2

c) 1 d) 0

53. $\int_0^{1+i} (x - y^2 + ix^3)dz$ along the real axis from $z = 0$ to $z = 1$ is ________

a) $2 + 4i$ b) $\frac{1}{2} + \frac{1}{4}i$

c) $\frac{1}{2} + 4i$ d) $-\frac{1}{2} - \frac{1}{4}i$

54. $\int\left[(2y+x^2)dx-(3x-y)dy\right]$ along the parabola $x=2t, y=t^2+3$ joining (0, 3) and (2, 4) is________

a) 22/3 b) 16.5

c) 44/3 d) 12.5

55. $\int_{(0,0)}^{(1,2)}\left[(x^2-y)dx-(y^2+x)dx\right]$ along the straight line from (0,1) to (1,2) is__________

a) 2/3 b) 4/9

c) 5/3 d) 5/9

56. If $f(z)=u+iv$ is an analytic function then $f'(z)=$

a) $\frac{\partial u}{\partial x}+i\frac{\partial v}{\partial y}$ b) $\frac{\partial u}{\partial x}-i\frac{\partial v}{\partial x}$

c) $\frac{\partial u}{\partial y}+i\frac{\partial v}{\partial x}$ d) $\frac{\partial u}{\partial x}-i\frac{\partial u}{\partial y}$

57. The imaginary part of $\log z$ is

a) $\log\sqrt{x^2+y^2}$ b) $\log(x^2+y^2)$

c) $\tan^{-1}(x/y)$ d) $\tan^{-1}(y/x)$

58. Which of the following is not an entire function

a) $e^z$ b) sin z

c) cos z d) log z

59. The C-R equations in polar form are

a) $\frac{\partial u}{\partial r}=\frac{1}{r}\frac{\partial v}{\partial \theta}, \frac{\partial v}{\partial r}=-\frac{1}{r}\frac{\partial u}{\partial \theta}$ b) $\frac{\partial u}{\partial r}=-\frac{1}{r}\frac{\partial v}{\partial \theta}, \frac{\partial v}{\partial r}=\frac{1}{r}\frac{\partial u}{\partial \theta}$

c) $\frac{\partial u}{\partial r}=\frac{\partial v}{\partial \theta}, \frac{\partial v}{\partial r}=-\frac{1}{r}\frac{\partial u}{\partial \theta}$ d) $\frac{\partial u}{\partial r}=\frac{\partial v}{\partial \theta}, \frac{\partial u}{\partial r}=-\frac{\partial v}{\partial r}$

60. The equation $\frac{dy}{dx}+\frac{ax+hy+g}{hx+by+f}=0$ is

a) Homogeneous b) Linear

c) Exact d) Auxiliary

61. Order of Differential equation $[1 + (\frac{dy}{dx})^2] = k \frac{d^3y}{dx^3}$

a) 1
b) 2
c) 3
d) Not exist

62. Degree of differential equation $\frac{d^3y}{dx^3} + 2(\frac{dy}{dx})^6 + y = 0$

a) 1
b) 2
c) 3
d) Not exist

63. P.I of $(D^2 - 1)y = \sin x \cos x$ is

a) $\frac{\sin 2x}{10}$
b) $\frac{\cos 2x}{10}$
c) $\frac{-\sin 2x}{10}$
d) $\frac{-\cos 2x}{10}$

64. If $u = 2x + 3y, v = 3x - 4y$ then $\frac{\partial(u,v)}{\partial(x,y)} =$

a) -14
b) 17
c) -12
d) 12

65. $\frac{1}{(D-1)^5} e^x =$

a) $\frac{x^4}{4!} e^x$
b) $\frac{e^x}{4!}$
c) $\frac{x^5}{5!} e^x$
d) $xe^x$

66. $\frac{1}{(D^2-1)} e^x =$

a) $\frac{1}{2} xe^x$
b) $-\frac{1}{2} xe^x$
c) $\frac{x^2}{2} e^x$
d) None

67. Particular Integral of $(1 - D)$ y = x is

a) x
b) 1 + x
c) $(x + 1)^2$
d) $(x + 2)^2$

68. $\frac{1}{D^2+D+1}$ sinx =

a) sinx
b) cosx
c) $\frac{\sin x}{3}$
d) - cosx

69. The solution of $(D^2+1)y=0$ satisfying the initial conditions $y(0)=1, y(\frac{\pi}{2})=2$ is

a) $y=2\cos x+\sin x$
b) $y=\cos x+\sin x$
c) $y=\cos x+2\sin x$
d) $y=2\cos x+2\sin x$

70. The general Solution of $z=px+qy-2p-3q$ is

a) $z=ax+by-2a-3b$
b) $z=ax+by+2a+3b$
c) $z=ax+by-a-b$
d) $z=ax+by+a+b$

71. The general Solution of $pq=xy$ is

a) $2z=ax^2+\frac{y^2}{a}+b$
b) $z=ax^2+\frac{y^2}{a}+b$
c) $z=ax^2+\frac{y^2}{a}+b$
d) $z=ax^2+y^2+b$

72. The general Solution of $px=qy$ is

a) $z=a\log xy+b$
b) $z=a\log x+\log y+b$
c) $z=ax+y+b$
d) $z=a(x+y)+b$

73. The partial differential equation of all planes passing through origin

a) $z=px+qy$
b) $z=px-qy$
c) $z=px-qy$
d) $z=p-q$

74. The partial differential equation of all planes having equal intercepts on x and y axis is

a) $p=q$
b) $p=-q$
c) $pq+1=0$
d) $pq=0$

75. The partial differential equation obtained by eliminating arbitrary constants a and b from $\log(az-1)=x+ay=b$

a) $p(q+1)=qz$
b) $p(q+1)=qz$
c) $q(p+1)=qz$
d) $q(p-1)=qz$

76. The partial differential equation obtained by eliminating arbitrary constants a and b from $z = axe^y + \frac{1}{2}a^2e^{2y} + b$

a) $p^2 + px = q$
b) $p^2 - px = q$
c) $q^2 + px = p$
d) $q^2 + qx = p$

77. The partial differential equation obtained by eliminating arbitrary constants a and b from $z = (x-a)^2 + (y-b)^2 + 1$ is

a) $p^2 + p^2 = 4z$
b) $p^2 - p^2 = 4z$
c) $p^2 + p^2 + 4 = 4z$
d) $p^2 + p^2 - 4 = 4z$

78. The partial differential equation obtained by eliminating arbitrary constants a and b from $2z = \sqrt{x+a} + \sqrt{y+b}$ is

a) $p+q = 8pqz$
b) $p-q = 8pqz$
c) $p+q = 4pqz$
d) $p-q = 4pqz$

79. The Solution of $p+q = pq$ is

a) $z = a+b$
b) $z = ax - by$
c) $z = ax + \frac{a}{a-1}y + b$
d) $z = ax + by$

80. The complete integral of $p+q = \sin x + \sin y$

a) $z = a + b - \sin a - \sin b$
b) $z = ax - ay - \cos x - \cos y + b$
c) $z = ax + ay + \cos x + \cos y + b$
d) $z = ax + ay$

81. Find the value of K for the following distribution.

| X | 0 | 1 | 2 | 3 | 4 |
|---|---|---|---|---|---|
| P(X= x) | K | 3K | 5K | 2K | K |

a) 1/4
b) 1/12
c) 3/10
d) 5/12

82. The mean of the probability distribution of the number of heads obtained in two flops of a balanced coin is

a) -1
b) 1
c) 2
d) 0

83. If X and Y are independent random variables, then $E(XY)$ is______

a) E(X) E(Y)
b) E(x) +E(Y)
c) E(X+Y)
d) E(X-Y)

84. A continuous random variable x that can assume any value between x=2 and x=5 has a probability density function f(x)=k(1+x) then find k

a) 4/27
b) 6/27
c) 2/27
d) 8/26

85. If in Binomial distribution np = 9 and npq= 2.25 then q is equal to ____

a) 0.25
b) 0.75
c) 1
d) 0.5

86. A renowned hospital usually admits 200 patients every day. One percent patients, on an average, require special room facilities. On one particular morning, it was found that only one special room is available. What is the probability that more than 3 patients would require special room facilities?

a) 0.1428
b) 0.1732
c) 0.2235
d) 0.3450

87. If the variance of a chi-square distribution is 60, then its degree of freedom is

a) 25
b) 49
c) 30
d) None

88. Coefficient of correlation =

a) $b_{xy}.b_{yx}$
b) Square root of $b_{xy}.\ b_{yx}$
c) Square root of $b_{xy}$
d) Square root of $b_{yx}$

89. If the mean and variance of binomial distribution are 6 and 2 respectively , then the Probability of two failures is

a) $4(\frac{2}{3})^7$
b) $4(\frac{2^2}{3^7})$
c) $17(\frac{2}{3})^7$
d) $17(\frac{2^2}{3^7})$

90. Range of F-distribution is

a) $(-\infty,\infty)$
b) $(-\infty,0)$
c) $(0,\infty)$
d) None

91. The $n^{th}$ order difference of a polynomial of $n^{th}$ degree is

a) Polynomial of$(n-1)^{th}$ degree b) Constant

c) Poly nominal of first degree d) Zero

92. Given that

| x | 1 | 2 | 3 | 4 |
|---|---|---|---|---|
| f(x) | 1 | 8 | 27 | 64 |

The value f(2.5) by Gauss forward

a) 16.625 b) 15.625

c) 17.625 d) 18.625

93. If $\Delta y = 1+2x+3x^2$ which of the following is not true?

a) $y= x^2 +x^3$ b) $\Delta^3 xy = 6$

c) $\Delta^4 y= 0$ d) $\Delta^2 = 6x+5$

94. If $f(x) = x^2+x+1$ then the value of $\Delta$ f(x) is (taking h=1)

a) 2x+3 b) 2

c) 2x+2 d) $x^2+1$

95. Given that

| x | 10 | 15 | 20 |
|---|---|---|---|
| f(x) | 19.97 | 21.51 | 22.47 |

a) -0.58 b) 0.96

c) 1.05 d) 1.54

96. Which of the following is not a step by step method

a) Taylor's series method b) Euler's method

c) R-K Method d) Modified Euler's method

97. The value of $\int_1^2 \frac{1}{x} dx$ by Simpson's $\frac{1}{3}$ rule (taking n = 4) is =

a) 0.593 b) 0.693

c) 0.393 d) 0.963

98. The value of $\int_0^{10} \frac{1}{x} dx$ by Trapezoidal rule (taking n=4) is

a) 2.3788 b) 3.3788

c) 4. 8837 d) 5. 8837

99. The value of $\int_0^1 \frac{1}{1+x^2} dx$ = by Simpson's $\frac{1}{3}$ rule (taking n=4) is

a) 0.7854 b) 0.8547

c) 0.5478 d) 0.4785

100. If dy/dx = x-y, y(0)=1 by Picard's method, $y^{(1)}(1) =$

a) 1.1 b) 1.05

c) 0.905 d) 0.95

## Answers

| | | | | | | | | | |
|---|---|---|---|---|---|---|---|---|---|
| 1. c | 2. c | 3. b | 4. c | 5. c | 6. b | 7. b | 8. b | 9. c | 10. b |
| 11. c | 12. c | 13. d | 14. b | 15. d | 16. a | 17. a | 18. a | 19. a | 20. c |
| 21. d | 22. c | 23. c | 24. b | 25. c | 26. a | 27. c | 28. c | 29. b | 30. d |
| 31. a | 32. c | 33. d | 34. d | 35. A | 36. a | 37. b | 38. a | 39. c | 40. d |
| 41. d | 42. a | 43. b | 44. a | 45. b | 46. d | 47. b | 48. d | 49. c | 50. a |
| 51. b | 52. d | 53. b | 54. b | 55. c | 56. d | 57. d | 58. d | 59. a | 60. c |
| 61. c | 62. a | 63. c | 64. a | 65. c | 66. a | 67. b | 68. d | 69. c | 70. a |
| 71. a | 72. a | 73. a | 74. a | 75. a | 76. a | 77. c | 78. c | 79. c | 80. b |
| 81. b | 82. b | 83. a | 84. c | 85. a | 86. a | 87. c | 88. b | 89. c | 90. c |
| 91. b | 92. b | 93. a | 94. b | 95. d | 96. c | 97. b | 98. a | 99. a | 100. c |